GAS WELL DELIQUIFICATION

GAS WELL DELIQUIFICATION

Second Edition

James F. Lea
Henry V. Nickens
Mike R. Wells

AMSTERDAM • BOSTON • HEIDELBERG
LONDON• NEW YORK • OXFORD
PARIS • SAN DIEGO • SAN FRANCISCO
SINGAPORE • SYDNEY • TOKYO
Gulf Professional Publishing is an imprint of Elsevier

ELSEVIER

Gulf Professional Publishing is an imprint of Elsevier
30 Corporate Drive, Suite 400, Burlington, MA 01803, USA
Linacre House, Jordan Hill, Oxford OX2 8DP, UK

 Recognizing the importance of preserving what has been written, Elsevier prints its books
on acid-free paper whenever possible.

Library of Congress Cataloging-in-Publication Data
Lea, James Fletcher.
  Gas well deliquification / James F. Lea, Henry V. Nickens, Mike R. Wells.
    p. cm.
  ISBN 978-0-7506-8280-0 (hardcover : alk. paper) 1. Gas wells. I. Nickens, Henry
Valma, 1947– II. Wells, Michael R. III. Title.
TN880.2.L43 2008
622′.3385—dc22

                                                                    2007046020

British Library Cataloguing-in-Publication Data
A catalogue record for this book is available from the British Library.

ISBN: 978-0-7506-8280-0

For information on all Gulf Professional Publishing
publications visit our Web site at www.books.elsevier.com

TABLE OF CONTENTS

INTRODUCTION

1.1 INTRODUCTION

Simply put, liquid loading of a gas well is the inability of the produced gas to remove the produced liquids from the wellbore. Under this condition, produced liquids will accumulate in the wellbore, leading to reduced production and shortening of the time until the well no longer will produce.

This book deals with the recognition and operation of gas wells experiencing liquid loading. It will present materials on methods and tools to enable you to diagnose liquid loading problems and indicate how to operate your well more efficiently by reducing the detrimental effects of liquid loading on gas production.

This book will serve as a primer to introduce the majority of the possible and most frequently used methods that can help produce gas wells when liquids start becoming a problem. Be aware that liquid loading can be a problem in both high- and low-rate wells, depending on the tubular sizes, the surface pressure, and the amount of liquids being produced with the gas.

In this book you will discover:

- How to recognize liquid loading when it occurs
- How to model gas well liquid loading
- How to design your well to minimize liquid loading effects
- What tools are available to assist you in design and analysis of gas wells for liquid loading problems
- The best methods of minimizing the effects of liquids in lower velocity gas wells and the advantages and disadvantages of the best methods

- How and why to apply various artificial lift methods for liquid removal
- What should be considered when selecting a lift method for liquids removal

1.2 MULTIPHASE FLOW IN A GAS WELL

To understand the effects of liquids in a gas well, we must understand how the liquid and gas phases interact under flowing conditions.

Multiphase flow in a vertical conduit is usually represented by four basic flow regimes as shown in Figure 1-1. A flow regime is determined by the velocity of the gas and liquid phases and the relative amounts of gas and liquid at any given point in the flow stream.

At any given time in a well's history, one or more of these regimes will be present.

- **Bubble Flow**. The tubing is almost completely filled with liquid. Free gas is present as small bubbles, rising in the liquid. Liquid contacts the wall surface and the bubbles serve only to reduce the density.
- **Slug Flow**. Gas bubbles expand as they rise and coalesce into larger bubbles, then slugs. Liquid phase is still the continuous phase. The

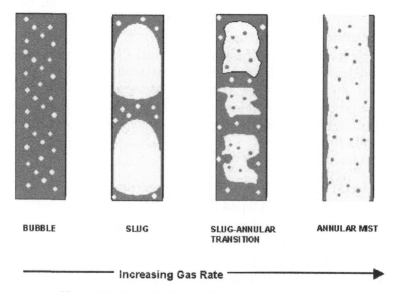

BUBBLE   SLUG   SLUG-ANNULAR   ANNULAR MIST
                                      TRANSITION

⸻⸻⸻⸻ Increasing Gas Rate ⸻⸻⸻⸻⟶

Figure 1-1: Flow Regimes in Vertical Multiphase Flow

liquid film around the slugs may fall downward. Both gas and liquid significantly affect the pressure gradient.

- **Slug-Annular Transition**. The flow changes from continuous liquid to continuous gas phase. Some liquid may be entrained as droplets in the gas. Gas dominates the pressure gradient, but liquid is still significant.
- **Annular-Mist Flow**. The gas phase is continuous and most of the liquid is entrained in the gas as a mist. The pipe wall is coated with a thin film of liquid, but pressure gradient is determined predominately from the gas flow.

A gas well may go through any or all of these flow regimes during the life of the well. Figure 1-2 shows the progression of a typical gas well from initial production to end of life. In this illustration, it is assumed that the tubing end does not extend to the mid-perforations so that there is a section of casing from the tubing end to mid-perfs.

The well may initially have a high gas rate so that the flow regime is in mist flow in the tubing but may be in bubble, transition, or slug flow below the tubing end to the mid-perforations. As time increases and production declines, the flow regimes from perforations to surface will change as the gas velocity decreases. Liquid production may also increase as the gas production declines.

Flow at the surface will remain in mist flow until the conditions change sufficiently at the surface to force the flow regime into transition

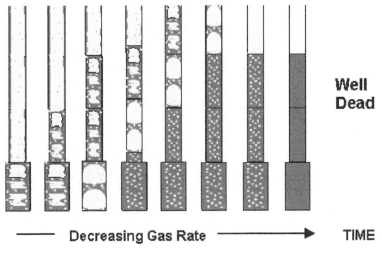

Well Dead

Decreasing Gas Rate ⟶ TIME

Figure 1-2: Life History of a Gas Well

flow. At this point, the well production becomes somewhat erratic, progressing to slug flow as gas rate continues to decline. This will often be accompanied by a marked increase in the decline rate. Note that the flow regime further downhole may be in bubble or slug flow, even though the surface production is in stable mist flow.

Eventually, the unstable slug flow at surface will transition to a stable, fairly steady production rate again as the gas rate declines still further. This occurs when the gas rate is too low to carry liquids to surface and simply bubbles up through a stagnant liquid column.

If corrective action is not taken, the well will continue to decline and eventually log off. It is also possible for the well to continue to flow for a long period in a loaded condition, producing gas up through the liquids with no liquids coming to the surface.

1.3 WHAT IS LIQUID LOADING?

When gas flows to surface, the gas will carry the produced liquids to the surface if the gas velocity is high enough. A high gas velocity generates a mist flow pattern in which the liquids are finely dispersed in the gas. This results in a low percent by volume of liquids present in the tubing (i.e., low liquid "holdup") or production conduit and, as a result, a low pressure drop due to the hydrostatic component of the flowing fluids.

According to the EIA (http://tonto.eia.doe.gov/dnav/ng/ng_prod_wells_s1_a.htm, http://tonto.eia.doe.gov/dnav/ng/hist/n9011us2A.htm) the US gas well count was 448,641 in 2006 with the annual gas production being 17,942,493 MMscf. According to these figures, the average gas rate per well is about 110 Mscf/D. Considering the critical rate for liquid loading for 2 3/8's inch ID tubing with 100 psia on the wellhead is about 300 Mscf/D, it would appear that many gas wells are liquid loaded. However loading is not limited to low rate producers as large diameter tubing wells load at a much higher rate. Comparably there are about 500,000 oil wells prodcucing about 10 bpd (http://www.eia.doe.gov/emeu/aer/pdf/pages/sec5_7.pdf).

A well flowing at a high gas velocity can have a high pressure drop due to friction, but for higher gas rate wells, the component of the pressure drop due to accumulated liquids in the conduit is relatively low. This is discussed and quantified in greater detail in the body of the book.

As the velocity of the gas in the production conduit drops with time, the velocity of the liquids carried by the gas decreases even faster. As a result, liquids begin to collect on the walls of the conduit, liquid slugs begin to form, and eventually liquids accumulate in the bottom of the well, adding to the percent of liquids in the conduit while the well is flowing. The presence of more liquids in the production conduit while the well is flowing can slow or even stop gas production altogether.

Very few gas wells produce completely dry gas. Liquids can accumulate in a well through a variety of mechanisms. Often gas wells produce liquids directly into the wellbore. In some cases, both hydrocarbons (condensate) and water can condense from the gas stream as the temperature and pressure change during travel to the surface. Fluids can also come into the wellbore as a result of coning water from an underlying zone.

Although most methods used to dewater gas wells do not depend on the source of the liquids, this is not always the case. A remediation method should consider the source of the liquid loading to be successful. If, for example, a remediation method is planned that addresses condensation only, then it must be verified that this is indeed the source of the liquid loading to ultimately be successful.

1.4 PROBLEMS CAUSED BY LIQUID LOADING

Liquid loading can lead to erratic, slugging flow and decreased production. The well may eventually die if the liquids are not continuously removed. Often, as liquids accumulate in a well, the well simply produces at a lower rate than expected.

If the gas rate is high enough to remove most or all of the liquids, the flowing tubing pressure at the formation face and production rate will reach a stable equilibrium. The well will produce at a rate that can be predicted by the reservoir inflow production relationship (IPR) curve (see Chapter 4).

If the gas rate is too low, the pressure gradient in the tubing becomes large due to the liquid accumulation, resulting in increased pressure on the formation. As the back-pressure on the formation increases, the rate of gas production from the reservoir decreases and may drop below the critical rate required to remove the liquid. More liquids will accumulate in the wellbore and the increased bottomhole pressure will further reduce gas production and may even kill the well.

Late in the life of a well, liquid may stand over the perforations with the gas bubbling through the liquid to the surface. In this scenario, the gas is producing at a low but steady rate with little or no liquids coming to the surface. If this behavior is observed with no knowledge of past well history, one might assume that the well is not liquid loaded but only a low producer.

All gas wells that produce some liquids, whether in high or low permeability formations, will eventually experience liquid loading with reservoir depletion. Even wells with very high gas-liquid ratios (GLR) and small liquid rates can load up if the gas velocity is low. This condition is typical of very tight formation (low permeability) gas wells that produce at low gas rates and have low gas velocities in the tubing. Some wells may be completed and produce considerable gas through large tubulars, but may be liquid loaded from the first day of production. See [1, 2] for an introduction to loading and some discussion of field problems and solutions.

1.5 DELIQUIFYING TECHNIQUES PRESENTED

The list below [3] introduces some of the possible methods used to deliquify gas wells that are discussed in this book. These methods may be used individually or in any combination. The list is organized roughly with regard to the static reservoir pressure.

Although the list is not necessarily complete, the methods that are outlined are discussed in some detail. Specialty methods, such as using a pumping system to inject water below a packer to allow gas to flow up the casing-tubing annulus, are not listed here, but are covered in the chapters on dewatering using beam and ESP pumping systems. Depth considerations and certain economical considerations are not considered here.

The optimum deliquifying method is defined as that which is most economic for the longest period of operation. Methods successfully implemented in similar offset fields, vendor equipment availability, reliability of equipment, manpower required to operate the equipment, etc. are all important considerations that are involved in selecting the optimum method.

- **Reservoir Pressure greater than 1500 psi**
  - Evaluate best natural flow of the well.
  - Using Nodal Analysis (TM of Macco Schlumberger), evaluate the tubing size for friction and future loading effects.
  - Consider possible coiled tubing use.
  - Evaluate surface tubing pressure and seek low values for maximum production.
  - Consider annular flow or annular and tubing flow to reduce friction effects.

- **Reservoir Pressure greater than 1500 psi**
  - These medium pressure wells may still flow using relatively smaller conduits and low surface pressures to keep flow velocities above a "critical" rate.
- **Reservoir Pressure between 500 and 1500 psi**
  - Low pressure systems
  - Plunger lift
  - Small tubing
  - Reduce surface pressure for all methods
  - Gas lift
  - Regular swabbing—for short flow periods
  - Pit blow-downs—environmentally unacceptable
  - Surfactants—soap sticks down the tubing or liquids injected down tubing or casing, use of capillary strings, or backside injection of surfactants
  - Reservoir flooding to boost pressures
- **Reservoir Pressure between 150 and 500 psi**
  - Lower-pressure systems
  - Plunger lift—can operate with large tubing
  - Small tubing
  - Reduce surface pressure for all methods
  - Surfactants (soaps), sticks, cap strings, backside injection
  - Siphon strings, usually smaller diameter
  - Rod pumps on pump-off control, PCPs if severe sand
  - Gas lift
  - Intermittent gas lift
  - Jet pump or reciprocating hydraulic pump
  - Swabbing
  - Reservoir flooding
- **Very Low Pressure systems—Reservoir Pressure less than 150 psi**
  - Rod pumps
  - Plunger in some cases
  - Siphon strings
  - Reduce surface pressure for all methods
  - Intermittent gas lift, chamber lift
  - Jet pump or reciprocating hydraulic pump
  - Swabbing
  - Surfactants (soaps), sticks, cap strings, backside injection
  - Reservoir flooding

Many times when a gas well begins to drop below the critical rate (a term to be defined in detail in this book), the operator may consider using smaller tubing or turn to plunger lift, which can operate with a number of common sizes of production tubing. Plunger lift is cheaper to install than most artificial lifting methods and will take the well to low but possibly not depletion pressures.

If the well is deep and equipped with a packer, a surfactant string may be used to inject surfactants to the bottom of the tubing. Soap sticks and back side treating are common if the produced liquids are mostly water. Beam pumps can reduce the flowing bottom hole pressure by producing liquids up the tubing and gas up the annulus but cannot ultimately achieve low pressures on the formation unless the gas flows into low pressure in the casing. Sandy wells may require PCPs or gaslift. In some cases compression can be used to evaporate all the water.

Selection of the most optimum deliquification method must necessarily consider the volume of liquid that the well will produce. Economics should dictate all decisions but, unfortunately, this is not always the case. This book will aid in the selection of the most optimum and economic method to deliquify gas wells.

1.6 SOURCE OF LIQUIDS IN A PRODUCING GAS WELL

Many gas wells produce not only gas but also condensate and water. If the reservoir pressure has decreased below the dew point, condensate is produced with the gas as a liquid; if the reservoir pressure is above dew point, the condensate enters the wellbore in the vapor phase with the gas and drops out as a liquid in the tubing or separator when the pressure drops.

Produced water may have several sources:

- Water may be coned in from an aqueous zone above or below the producing zone.
- If the reservoir has aquifer support, the encroaching water will eventually reach the wellbore.
- Water may enter the wellbore from another producing zone, which could be separated some distance from the gas zone.
- Free formation water may be produced with the gas.
- Water and/or hydrocarbons may enter the wellbore in the vapor phase with the gas and condense out as a liquid in the tubing.

1.6.1 Water Coning

If the gas rate of a well is high enough, the gas may pull water production from an underlying zone, even if the well is not perforated in the water zone. Generally a horizontal well greatly reduces gradients between the gas zone and an underlying water zone, but the same phenomenon can still occur if the well is produced at very high rates, although in this case it is commonly termed *cresting* instead of *coning*.

1.6.2 Aquifer Water

Pressure support from an aquifer will eventually allow water production to reach the wellbore, giving rise to liquid loading problems.

1.6.3 Water Produced from Another Zone

It is possible for liquids to be produced into the wellbore from another zone, either with an open hole completion or in a well having several sections perforated. In some instances, this scenario can be advantageous in that water can be reinjected, by gravity or using pumps, into an underlying zone while allowing gas to flow freely.

1.6.4 Free Formation Water

From whatever the source, it is possible for water to enter the well through the perforations with the gas. This can be a result of thin imbedded layers of gas and liquid.

1.6.5 Water of Condensation

If saturated or partially saturated gas enters the well, little or no liquids will enter the wellbore through the perforations, but condensation can occur higher in the well. At any given pressure and temperature, a certain amount of water vapor, if present, will be in equilibrium with the gas. As temperature decreases or pressure increases, the amount of water vapor in equilibrium decreases and any excess water vapor will condense, creating a liquid phase. Similarly, if temperature increases or pressure decreases, free liquid water, if present, will evaporate to the vapor phase to maintain equilibrium. If condensation occurs higher in the well, it can cause a high pressure gradient in the flow string where

Water Content as Pressure Declines

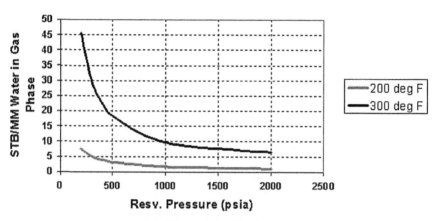

Figure 1-3: Water Solubility in Natural Gas

it occurs and also, depending on gas velocity, liquids can eventually accumulate over the perforations or pay zone.

For a given reservoir pressure and temperature, the produced gas will contain a certain amount of water vapor. Figure 1-3 shows an example of the solubility of water in natural gas in STB/MMscf. Note the rapid increase in water content as reservoir pressure declines below 500 psi.

The water will remain in the vapor phase until temperature and pressure conditions drop below the dew point. When this occurs, some of the water vapor will condense to the liquid phase. If the condensation occurs in the wellbore and the gas velocity is below the critical rate required to remove the liquid water, then liquids will accumulate in the wellbore and liquid loading will occur.

Even if the gas velocity is sufficient to remove the condensed water, corrosion problems may occur at the point in the wellbore where condensation first occurs. Condensed water can be identified as it would have a no salt content, because there is pure water in the vapor phase before condensation.

1.6.6 Hydrocarbon Condensates

Like water, hydrocarbons can also enter the well with the produced gas in the vapor stage. If the reservoir temperature is above the cricondentherm, then no liquids will be in the reservoir, but liquids can drop

Table 1-1
Water Solubility in Natural Gas

Location	P/T	Water Content (STB/MMscf)	Water Condensed in Tubing (STB/MMscf
Surface	150 psi/100 F	0.86	—
Reservoir	3500 psi/200 F	0.73	0
Reservoir	1000 psi/200 F	1.75	0.89
Reservoir	750 psi/200 F	2.22	1.36
Reservoir	500 psi/200 F	3.17	2.31
Reservoir	250 psi/200 F	6.07	5.21

out or condense in the wellbore. Then, as with water, if the gas velocity in the tubing is not high enough, the liquids can accumulate over the perforations.

Example 1-1: Water Solubility in Natural Gas

Consider the following typical example for a gas well producing initially from a reservoir at 3500 psi and 200°F producing at wellhead conditions of 150 psi and 100°F. In this example, we assume that the wellhead conditions remain constant.

As the reservoir pressure declines, the amount of water condensing out in the tubing increases. Since the gas rate will decline as the reservoir pressure decreases, we have the situation of decreasing gas rate coupled with increasing liquid production—liquid loading will inevitably occur.

1.7 REFERENCES

1. Lea, J. F. and Tighe, R. E. "Gas Well Operation with Liquid Production," SPE 11583, presented at the 1983 Production Operation Symposium, Oklahoma City, OK, February 27–March 1, 1983.

2. Libson, T. N. and Henry, J. T. "Case Histories: Identification of and Remedial Action for Liquid Loading in Gas Wells-Intermediate Shelf Gas Play," *Journal of Petroleum Technology*, April 1980, 685–693.

3. Coleman, S. B. *et al.* "A New Look at Predicting Gas Well Liquid Load-Up," *Journal of Petroleum Technology*, March 1991, 329–332.

RECOGNIZING SYMPTOMS OF LIQUID LOADING IN GAS WELLS

2.1 INTRODUCTION

Over the life of a gas well it is likely that the volume of liquids being produced will increase while the volume of gas being produced drops off. Such situations usually result in the accumulation of liquids in the wellbore until eventually the well dies or flows erratically at a much lower rate. If diagnosed early, costly losses in gas production can be avoided by implementing one of the many methods available to artificially lift the liquids from the well.

On the other hand, if liquid loading in the wellbore goes unnoticed, the liquids could also accumulate in the wellbore and the adjoining reservoir, possibly causing temporary or even permanent damage. It is vital, therefore, that the effects caused by liquid loading are detected early to prevent costly loss of production and possible reservoir damage.

This chapter is devoted to the symptoms that indicate when a gas well is having problems with liquid loading. Emphasis is placed on symptoms that are typically available in the field. Some of these are more obvious than others, but all lend themselves to more exacting methods of well analysis described in the following chapters.

Symptoms that indicate a well is liquid loading are:

- Presence of orifice pressure spikes
- Erratic production and increase in decline rate
- Tubing pressure decreases as casing pressure increases
- Pressure survey shows a sharp, distinct change in pressure gradient
- Annular heading
- Liquid production ceases

2.2 PRESENCE OF ORIFICE PRESSURE SPIKES

One of the most common methods available to detect liquid loading is that slugs of liquid begin to be produced at the wellhead. Liquids are beginning to accumulate in the wellbore and/or the flowline and are produced erratically as some of the liquids reach the surface as slugs.

This phenomenon is depicted in Figure 2-1 on a two-pen recorder showing well producing liquids normally in mist flow on the left and a well beginning to experience liquid loading problems, producing the liquids in slugs, on the right. It is recognized that two pen charts may be replaced by transducer signals on computer plots, but this is given for illustration.

When liquids begin to accumulate in the wellbore, the pressure spikes on the recorder become more frequent. Eventually, the surface tubing pressure starts to decrease due to the liquid head holding back the reservoir pressure. In addition, the gas flow begins to decline at a rate uncharacteristic of the prior production decline rate. This rapid drop in production and drop in surface tubing pressure, accompanied by the ragged two-pen recorder charts, is a sure indication of liquid loading problems. Many wells have a liquid knock-out before orifice measure-

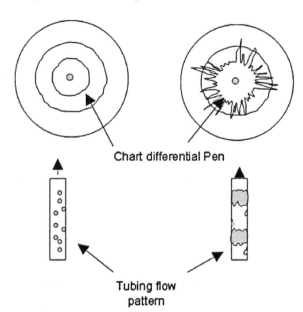

Figure 2-1: Effect of Flow Regime on Orifice Pressure Drop—Mist Flow (L) vs. Slug Flow (R) in Tubing

ments so the operator then would have to listen at the wellhead to try to determine if slugs are being produced. Also many wells now do not use the two-pen recorders, but the two-pen records shown here serve to illustrate how slugs of liquid begin to be produced by a gas well when liquid loading has commenced.

2.3 DECLINE CURVE ANALYSIS

The shape of a well's decline curve can be an important indication of downhole liquid loading problems. Decline curves should be analyzed for long periods, looking for changes in the general trend. Figure 2-2 shows two decline curves. The smooth exponential type decline curve is characteristic of normal gas-only production considering reservoir depletion. The sharply fluctuating curve is indicative of liquid loading in the wellbore and in this case is showing the well to deplete much earlier than reservoir considerations alone would indicate. Typically when decline curve trends are analyzed for long periods, wells experiencing liquid loading problems will show a sudden departure from the existing curve to a new, steeper slope. The new curve will indicate well

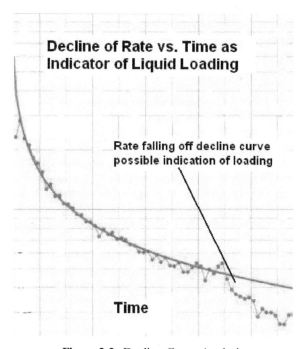

Figure 2-2: Decline Curve Analysis

abandonment far earlier then the original, providing a means to determine the extent of lost reserves as a result of liquid loading. By employing the remedial lift methods described herein, production often can be restored to the original decline curve slope.

2.4 DROP IN TUBING PRESSURE WITH RISE IN CASING PRESSURE

If liquids begin to accumulate in the bottom of the wellbore, the added pressure head on the formation has the effect of lowering the surface tubing pressure. In addition, as the liquid production increases the added liquid in the tubing being carried by the gas (liquid hold up) increases the gradient in the tubing and again provides more back pressure against the formation and reduces the surface tubing pressure.

In packer-less completions where this phenomenon can be observed, the presence of liquids in the tubing is shown as an increase in the surface casing pressure as the fluids bring the reservoir to a lower flow, higher pressure production point. As gas is produced from the reservoir, gas percolates into the tubing casing annulus. This gas is exposed to the higher formation pressure, causing an increase in the surface casing pressure. Therefore, a decrease in tubing pressure and a corresponding increase in casing pressure are indicators of liquid loading. These effects are illustrated in Figure 2-3 but the changes may not be linear with time as shown in this illustration.

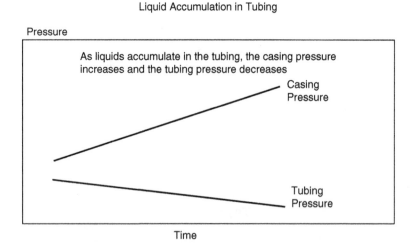

Figure 2-3: Casing and Tubing Pressure Indicators

Finally, estimates of the tubing pressure gradient can be made in a flowing well without a packer by measuring the difference in the tubing and casing pressures. In a packerless production well, the free gas will separate from the liquids in the wellbore and rise into the annulus. The fluid level in a flowing well will remain depressed at the tubing intake depth except when "heading" occurs or a tubing leak is present.

During "heading" the liquid level in the annulus periodically rises above then falls back to the tubing intake. In a flowing well, however, the difference in the surface casing and tubing pressures are an indication of the pressure loss in the production tubing. The weight of the gas column in the casing can be computed easily (see Appendix C). Comparing the casing and tubing pressure difference with a dry gas gradient for the well can give an estimate of the higher tubing gradient due to liquids accumulating or loading the tubing.

2.5 PRESSURE SURVEY SHOWING LIQUID LEVEL

Flowing or static well pressure surveys are perhaps the most accurate method available to determine the liquid level in a gas well and thereby whether the well is loading with liquids. Pressure surveys measure the pressure with depth of the well either while shut in or while flowing. The measured pressure gradient is a direct function of the density of the medium and the depth, and for a single static fluid, the pressure with depth should be nearly linear.

Since the density of the gas is significantly lower than that of water or condensate, the measured gradient curve will exhibit a sharp change of slope when the tool encounters standing liquid in the tubing. Thus the pressure survey provides an accurate means of determining the liquid level in the wellbore. If the liquid level is higher than the perforations, liquid loading problems are indicated.

Figure 2-4 illustrates the basic principle associated with the pressure survey. Note that the gas and liquid production rates can change the slopes measured by the survey, giving a higher gas gradient due to the presence of some liquids dispersed and a lower liquid gradient due to the presence of gas in the liquid. Note also that the liquid level in a shut-in gas well can be measured acoustically by shooting a liquid level down the tubing.

The fluid in the tubing in a well that produces both liquids and gases exhibits a complicated two-phase flow regime that depends on the flow rate and the amount of each constituent phase present. The flowing

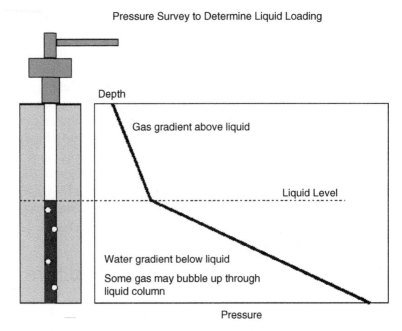

Pressure Survey to Determine Liquid Loading

Figure 2-4: Pressure Survey Schematic

pressure survey data obtained in two-phase flow is not necessarily linear as indicated earlier. When the measured pressure gradient is not linear but shows a continuously increasing pressure with depth, pressure gradient data alone is not sufficient to determine if liquid loading is in fact becoming a problem.

In these cases, it may be necessary to repeat the pressure survey at other conditions, or use techniques described later in this text to compute the gradient in smaller tubing sizes or lower surface pressures to determine if liquids are tending to accumulate. Often the pressure deflection brought about by standing liquid in the tubing can be masked by higher flow rates in small tubing. The added frictional pressure loss in these cases can "mask" the inflection point caused by the liquid interface. Large tubing usually means a lower frictional pressure loss (depends on the flow rate) and as a result typically produces a sharp deflection in the pressure survey curve.

Some wells have a tapered tubing string. In this case, a change in tubing cross-sectional flow area will cause a change in flow regime at the point where the flow area changes with a resultant change in the pressure gradient. This may appear in a gradient survey as a change in slope of the pressure-depth plot at the depth of the tubing area change and should not be confused with the gas-liquid interface at the depth of the liquid level.

Estimates of the volume of liquid production can be made by comparing the tubing pressure loss in a well producing liquids with one producing only or near dry gas. In a flowing well the bottomhole pressure (BHP) is equal to the pressure drop in the tubing (or annulus if flowing up the annulus) plus the wellhead pressure. The presence of the liquid in the production stream always increases the tubing pressure gradient. At low gas rates the proportional increase of pressure loss in the tubing due to liquids is higher than at high gas rates. The variance then allows one, with a productivity expression for gas flow from the reservoir, to see how more production is possible if the pressure increase due to liquid loading is mitigated. See Chapter 4 for illustrations of the tubing performance curve for gas with some liquids intersecting a reservoir inflow curve as a method of predicting gas well production.

2.6 WELL PERFORMANCE MONITORING

A method is presented [2] for displaying the minimum lift (and erosional gas rate) directly on the wellhead backpressure curve. These curves allow identification of when liquid loading (or erosional rates) threaten to reduce production. An overlay technique is identified whereby a minimum lift "type-curve" is generated for an entire field or a particular set of operating conditions.

2.7 ANNULUS HEADING

Some gas wells without packers establish low frequency pressure oscillations that can extend over several hours or days. These oscillations are indicative of the build up of produced liquids in the wellbore and have been reported to curtail production by over 40 percent. Figure 2-5 illustrates this oscillatory behavior for a typical packerless gas well.

The process is described by W. E. Gilbert [3]. Essentially it is a low flow rate process with a high annulus level and then later a quick high flow rate with a low annulus level that temporarily exhausts gas rapidly and wastes some of the reservoir flowing pressure since liquids are not carried with the burst of gas. Although not strictly liquid loading such as an increased concentration of liquid in the tubing or flow path, the oscillations of the liquid level and gas pressure in the annulus contribute to reduced production if this phenomenon occurs unchecked.

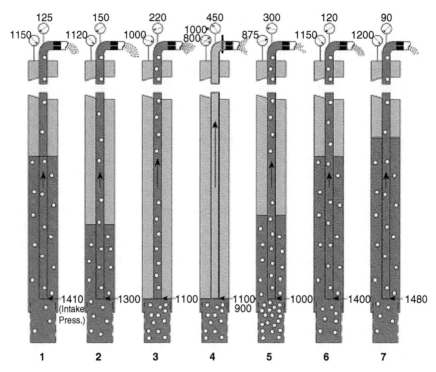

Figure 2-5: Low Frequency Pressure Oscillations in a Gas Well Producing Liquids

2.7.1 Heading Cycle without Packer

The steps of the annulus heading cycle shown in Figure 2-5 are outlined below. The cycle description begins with the annulus fluid level at the peak height.

1. Gas trickles into the annulus and slowly displaces annulus liquid into the tubing, lowering the annulus liquid level and decreasing casing pressure.
2. The well is still producing at a low rate since the tubing column is "heavy" due to the diversion of some production gas into the casing and the added production of casing liquids in the tubing as a result of the annulus gas pressure buildup. Annulus pressure is still decreasing as more annulus fluid is displaced into the tubing.
3. The pressure in the annulus continues to drop. The annulus liquid level drops to the tubing inlet as the liquid is produced out of the

annulus. Gas flows into the tubing. The weight of the tubing column is reduced since the gas from the formation is now produced up the tubing and is no longer trickling into the annulus. Liquids from the annulus are no longer being diverted into the production stream.

4. The tubing gradient drops still further due to produced gas in the flow stream, lowering the bottom hole tubing pressure and allowing dry gas from the annulus to "blow around" into the tubing. The production from the reservoir is also increased, depleting the reservoir near the wellbore much more than the other times in this cycle. For a short period, the well produces at a higher than normal rate but with relatively small amounts of liquid. Since the liquid production is low or nonexistent during the high flow rate period, the energy provided by the high gas rate is wasted as far as it is being used for the lifting of liquids.

5. The reservoir again starts to produce liquids and the gas production drops. The gas stored in the annulus is depleted and the tubing and casing annulus begin to load with liquids. As the liquid level rises in the annulus, gas also begins to percolate into the annulus. As gas is now diverted into the annulus, the gradient in the tubing increases, adding extra force against the reservoir and lowering the production rate.

6. Liquid still flows into the tubing at a higher rate than can be carried out of the tubing by the gas flow. Liquids continue to accumulate in the bottom of the well. Some gas is migrating into the casing/tubing annulus.

7. The rate of production of liquids at the surface is in balance with the rate of liquid production at the formation. Gas continues to migrate into the annulus until the annular pressure peaks, again forcing liquid into the tubing and repeating the cycle.

Note that this is not "liquid loading" in the usual sense, but is caused by an instability in the casing/tubing annulus pressures that leads to ups and downs in the production. The production of formation liquids under the cyclic behavior just described is inefficient since a portion of the cycle produces a high volume of the gas with very little lifting of the liquids. In some instances it is possible to choke back the well to control the cyclic behavior, but this method also cuts back production by increasing the average bottomhole pressure.

2.7.2 Heading Cycle with Controller

Another method used to control the oscillations while maintaining a high average flow rate is to install a downhole packer to prevent the gas from migrating into the annulus or a surface controller that monitors the pressures, preventing casing pressure buildup.

Use of a surface controller to counter the natural oscillations brought about by gas buildup in the annulus is illustrated in Figure 2-6.

1. At the start of the flowing period the tubing is opened by the rising casing pressure, which actuates the motor valve. The column of gas collected in the upper part of the tubing is produced, and the subsequent reduction of pressure ensures flow of the fluid mixture in the tubing below the gas column.
2. The tubing pressure trends downward while fluid is being displaced out of the annulus.
3. The tubing pressure then rises as annulus gas starts to break around the foot of the tubing.
4. When the casing pressure reaches the predetermined minimum, the motor valve closes the tubing outlet (or pinches it back), but flow into the well from the reservoir continues with very little decrease in

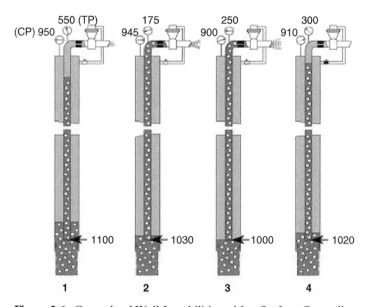

Figure 2-6: Controls of Well Instabilities with a Surface Controller

rate. This production includes both gas and liquid that flows into the annulus, effectively filling the annulus. The tubing pressure continues to rise. The casing pressure, which is directly related to the amount of gas stored in the annulus, also increases in response to gas and liquid entering the well. When the casing pressure reaches the pre-determined maximum, the cycle is repeated by opening the motor valve or opening it to a wider flow area for the gas.

By smoothing out the flow, surface controllers can be used to increase the rate of flow and extend the flowing life of wells that have reached the heading stage. This type of control produces formation liquids less effectively than pumping systems but is a good option when other lifting methods are not feasible. The use of a surface controller on heading wells can increase production and prolong the life of wells not equipped with downhole packers without expensive workovers. These controllers are applicable only for wells without packers.

2.8 LIQUID PRODUCTION CEASES

Some high rate gas wells readily produce liquids for a time and then drop off to much lower rates. As the gas production declines, the liquid production can cease. In such cases the well is producing gas at rates below the "critical" rate that can transport the liquids to the surface. The result is that the liquids continue to accumulate in the wellbore and the gas bubbles through the accumulated liquids. Depending on the accumulation of liquids and the well pressure, the well can either cease to flow or the gas can bubble up through the liquids. However, the gas rate has dropped to a value where liquids are no longer transported up the tubing.

The best method to analyze this type of well response is to calculate a minimum critical velocity in the tubing, or that minimum gas velocity required to carry liquids to the surface. If the flow is well below what is necessary to lift liquids, and especially if the flow rate is low in large diameter tubing, then the possibility of gas bubbling through accumulated liquids should be investigated.

Pumping the liquids out of the well or using coil tubing to inject N_2 may be the only solutions for this low flow rate situation.

Wireline pressure surveys can also be used to determine if there is standing liquid in the wellbore. These methods will be discussed in detail in later chapters. It is also possible to shoot an acoustically measured fluid level down the tubing if the flow does not interfere with the acoustical signals received from reflections of a pressure pulse at the surface or if the fluid shot is done quickly and periodically after shut-in of the well.

2.9 SHOOTING FLUID LEVELS ON FLOWING GAS WELLS

This section is contributed by Lynn Rowlan, Echometer.

Mr. Rowlan, BSCE, 1975, OK State University, was the recipient of the 2000 J.C. Slonneger Award bestowed by the Southwestern Petroleum Short Course Association, Inc. He has authored numerous papers for the Southwestern Petroleum Short Course, the Canadian Petroleum Society, and the Society of Petroleum Engineers.

Mr. Rowlan works as an Engineer for Echometer Company in Wichita Falls, Texas. His primary interest is to advance the technology of using the Echometer Portable Well Analyzer to analyze and optimize the real-time operation of all artificial lift production systems. He provides training and consultation in performing well analysis to increase oil and gas production, reduce failures, and reduce power consumption. He presents many seminars and gives numerous talks on the efficient operation of oil and gas wells.

As stated earlier, fluid in the bottom of the tubing indicates liquid loading. This section describes a nonintrusive method of finding a gas cut fluid level in a flowing gas well.

Shooting a fluid level down the tubing of a flowing gas well can be of benefit to the gas well operator. The most common application of an acoustic liquid level instrument is to shoot the fluid level in the casing annulus of an oil well. A less common technique is to acquire an acoustic fluid level by "shooting" down the tubing in a flowing or shut-in gas well. Analysis of the acoustic fluid levels acquired on gas wells can be used to determine (1) the amount of liquid loading on the formation, (2) the approximate gas rate into tubing, (3) the equivalent gradient of the gaseous liquid column in the tubing, and (4) the flowing bottom hole pressure at the end of the tubing and the bottom of the perforations. Fluid level instruments can be used to inexpensively identify liquid loading and determine the severity of the loading for gas wells.

When the gas well is flowing at a gas rate less than the Critical Rate, then a fluid level shot down the tubing will usually show a liquid level echo. If the gas well is flowing gas at a rate greater than the Critical Rate, then the gas/mist interface will be at the surface and the initial fluid level shot often will not show a liquid level echo in the tubing. If the gas well is shut in for an extended period of time, the high pressure gas often accumulates in the tubing and displaces all liquid out of the tubing. In a shut-in gas well the fluid level shot into the well often will show echoes at the bottom of the tubing, the perforations, and a liquid level very near the bottom of the perforations. Using a portable fluid level instrumentation permits the operator to quickly conduct a simple cost-effective test and immediately identify underperforming liquid-loaded gas wells. The information obtained from analyzing a series of fluid level shots down the tubing on a flowing gas well provides critical data in analyzing the well's performance. Flowing gas wells may be grouped into one of three different categories: (1) above critical rate, (2) below critical rate, and (3) shut-in.

2.9.1 Fluid Level—Gas Flow above Critical Rate

In the first category (gas flow rate above the Critical Rate), any liquid being produced with the gas or condensing due to temperature and pressure changes is usually uniformly distributed in the tubing. The gas velocity is sufficient to carry liquid as a fine mist or small droplets to the surface and sufficient to establish a relatively light-uniform flowing pressure gradient. At the stabilized undisturbed flowing condition, the fluid level is at the surface and the tubing is filled with a mist.

To analyze this type of gas well, the gas flow at the surface can be shut in and series of acoustic fluid level surveys should be acquired as the surface pressure increases. Analysis of these fluid level shots can be used to determine the tubing fluid gradient and the flowing bottomhole pressure. When the gas flow rate in a gas well is above the critical rate, acoustic surveys consisting of several fluid level shots has shown that a uniform light mist flowing gradient exists in the tubing string from the liquid level down to the bottom of the tubing. The gas/liquid interface pressures and height of the gaseous liquid column determined from the series of fluid level shots will usually fall along a straight line, indicating that a constant pressure gradient exists below the gas/mist liquid level interface. The change in pressure and the change in height of the gas/liquid interface from at least two fluid level measurements can be used to calculate the gradient below the fluid level. Extrapolation of the pressure at the gas/mist interface to a zero height of the gaseous liquid will give a reasonably accurate estimate of the producing bottomhole pressure.

2.9.2 Fluid Level—Gas Well Shut-in

Use of acoustic surveys to determine the static shut-in pressure is an accepted and accurate practice. Using acoustic fluid level instruments to determine static bottomhole pressure provides advantages over downhole gauges in that the equipment is compact, lightweight, and portable. Fluid level instruments are frequently used to inexpensively determine the shut-in static reservoir pressure for gas wells as opposed to traditional wire line methods, which are more intrusive and costly.

2.9.3 Fluid Level—Gas Flow below Critical Rate

An acoustic fluid level survey can be conducted to determine the depth to the fluid level and the pressure distribution in a flowing liquid loaded gas well. Generally acoustic fluid level surveys are acquired down the tubing while the well's flow at the surface is momentarily shut-in. Analysis of the acoustic fluid level surveys is used to determine the extent of liquid loading of the well and the back-pressure acting on the formation. The principal objective of the acoustic measurements in a flowing gas well is the determination of the quantity of liquid that has accumulated at the bottom of the well.

In flowing gas wells where the gas velocity is unable to lift sufficient liquid to the surface, then the liquid falls back to accumulate in the lower part of the well. A fluid level shot down the tubing will usually show a liquid level echo below the surface of the well. The flowing pressure gradient will show two values, a very light gradient (close to that of the flowing gas) above the gas/liquid interface, and a heavier gaseous liquid gradient below the gas/liquid interface. Below the liquid level the flow is characterized as net zero liquid flow with gas bubbles or slugs percolating through the liquid, and upon exiting the gaseous liquid surface the gas flows the remaining distance up the tubing to the surface.

In a liquid loaded well flowing below critical rate, the first few acoustic fluid level measurements are the most accurate in determining the gaseous liquid column gradient and the flowing bottomhole pressure. After a liquid loaded well is shut in for a period of time, surface pressure increases, and gas flow rate decreases; then the flow regime below the liquid level in the tubing is disturbed and liquids previously held up by gas flow begin to fall; the additional liquids further increase the gradient at the bottom of the tubing. Acoustic fluid level surveys acquired while the liquid is falling may result in flowing bottom hole pressures that are not accurate. When shooting fluid levels on a liquid loaded gas well, the act of shutting the flow valve at the surface of the well for a long time period or running a wire line will disturb the flow regime and can result in calculating inaccurate bottomhole pressures.

To determine the percentage of liquid below the liquid level in a flowing gas well, it is recommended that one or more fluid level measurements be undertaken shortly after stopping the flow at the surface.

2.9.4 Estimation of BHP from Fluid Level Measurement

When analyzing the acoustic data acquired from shooting a liquid loaded gas well, the gaseous column gradient below the liquid level is determined using the Echometer annular "S" curve (SPE 14254, "Acoustic Determination of Producing Bottomhole Pressure," James N. McCoy, SPE, Echometer Co.; Augusto L. Podio SPE, U. of Texas; Ken L. Huddleston, SPE, Echometer Co.). The percent liquid in the gaseous liquid column is obtained from this generalized empirical correlation (S-curve) that was developed from field data acquired from pumping oil wells. The gradients determined from the oil wells were determined under stabilized conditions, with a constant gas flow rate through the gaseous liquid column having net zero liquid flow. This correlation is applicable to stabilized flow in gas wells with some confidence as long as the liquid loaded flow regime is not disturbed. The Echometer annular S-curve does not calculate the correct gaseous column gradient after the surface valve is closed for an extended period of time. The annular S-curve does a reasonably accurate estimate of the gaseous liquid column when the liquid loaded bubble or slug flow regime is not disturbed. If a stabilized flow from the well exists, then acoustically determined bottomhole pressures are accurate.

2.9.5 Acoustic Determination of Liquid Loading in a Gas Well

Figure 2-7 plots gradients resulting from the analysis of a single fluid level shot down the tubing in a flowing liquid loaded gas well. The line

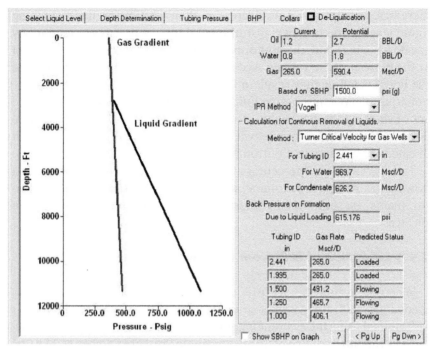

Figure 2-7: Analysis of Fluid in Tubing in Producing Gas Well

labeled Liquid Gradient displays the gradient of the gaseous liquid column determined using the Echometer S-curve. Based on the analysis of the acoustic fluid level shot there is 8551 feet of gaseous liquid in the well. The line labeled Liquid Gradient symbolizes the accumulation of fluids in the tubing, and the difference between the Liquid Gradient and the Gas Gradient line represents the "Back Pressure on Formation" at the 11261-foot tubing intake depth. The 385 psig gas/liquid interface pressure at the 2782-foot liquid level is extrapolated using the gaseous liquid gradient to the producing bottom hole pressure of 1067 psi. The S-curve gradient is used to determine an equivalent 685 feet of gas free liquid load is applying 615 psi of back pressure on the formation. This Deliquification tab shows the additional gaseous liquid pressure above the flowing Gas Gradient line.

The current 265 Mscf/D average gas flow rate is below the Turner critical rate of 969.7 Mscf/D. This critical gas flow rate is calculated at the intake pressure of the 2.441 inch internal dimension tubing.

At a glance, various tubing sizes can be evaluated to determine if the critical gas flow rate could be exceeded and whether the well will flow in a unload state with a smaller diameter velocity string. The gas rate is determined at the intersection of a simple Vogel inflow curve modeling the flow from the formation intersecting with the outflow curves for each specific tubing size. The predicted status shows that this well continues to flow at the current liquid loaded state with 2-3/8 inch tubing. If the Turner or Coleman critical rate is greater than the existing flow rate for any tubing size, then the well's predicted status stays loaded. But velocity strings with 1.5-inch internal dimension and smaller will result in a sufficiently high gas velocity for the well to flow above critical in an unloaded state. With a 1.5-inch velocity string the well would flow continuously at 491 Mscf/D, resulting in a 226 Mscf/D incremental increase in the gas production rate.

A single acoustic fluid level "shot" down the tubing in flowing gas wells can be used to determine:

- Amount of liquid in the bottom of the tubing
- Backpressure on the formation due to liquid
- Gas flow rate into the tubing
- Equivalent fluid gradient below the liquid level
- Flowing bottomhole pressure
- Feasibility of using various lift methods to remove the liquid loading
- Incremental gas flow rate if liquid loading is removed

A fluid level down the tubing can be used to confirm that a well is liquid loaded. The results from the fluid level analysis determine the back pressure on the formation due to liquid load, plus predict the incremental increase in the gas production if the well is unloaded. In a liquid loaded well the annular S-curve predicts a reasonably accurate gradient of the gaseous liquid column and is a good technique to determine flowing BHP and liquid loading.

2.10 SUMMARY

In summary, several symptoms of wells suffering from liquid loading have been illustrated. These indicators provide early warning of liquid loading problems that can hamper production and sometimes permanently damage the reservoir. These indicators should be monitored on a regular basis to prevent loss of production. Methods to analytically predict loading problems and the subsequent remedial action will be discussed in the later chapters in this book.

2.11 REFERENCES

1. Gilbert, W. E. "Flowing and Gas-Lift Well Performance," presented at the spring meeting of the Pacific Coast District, Division of Production, Los Angeles, May 1954, Drilling and Production Practice, pp. 126–157.

2. Thrasher, T. S. "Well Performance Monitoring: Case Histories," SPE 26181, presented at the SPE Gas Technology Symposium, Calgary, Alberta, Canada, June 28–30, 1993.

C H A P T E R   3

CRITICAL VELOCITY

3.1 INTRODUCTION

To effectively plan and design for gas well liquid loading problems, it is essential to be able to accurately predict when a particular well might begin to experience excessive liquid loading. In the next chapter, Nodal Analysis (Macco-Schlumberger™) techniques are presented that can be used to predict when liquid loading problems and well flow stability occur. In this chapter, the relatively simple "critical velocity" method is presented to predict the onset of liquid loading.

This technique was developed from a substantial accumulation of well data and has been shown to be reasonably accurate for vertical wells. The method of calculating a critical velocity will be shown to be applicable at any point in the well. It should be used in conjunction with methods of Nodal Analysis if possible.

3.2 CRITICAL FLOW CONCEPTS

The transport of liquids in near vertical wells is governed primarily by two complementing physical processes before liquid loading becomes more predominate and other flow regimes such as slug flow and then bubble flow begin.

3.2.1 Turner Droplet Model

It is generally believed that the liquids are both lifted in the gas flow as individual particles and transported as a liquid film along the tubing wall by the shear stress at the interface between the gas and the liquid before the onset of severe liquid loading. These mechanisms were first investigated by Turner *et al.* [1], who evaluated two correlations developed on the basis of the two transport mechanisms using a large experimental database as illustrated here. Turner discovered that liquid loading could best be predicted by a droplet model that showed when droplets move up (gas flow above critical velocity) or down (gas flow below critical velocity).

Turner *et al.* [1] developed a simple correlation to predict the so-called *critical velocity* in near vertical gas wells assuming the droplet model. In this model, the droplet weight acts downward and the drag force from the gas acts upward (Figure 3-1). When the drag is equal to the weight, the gas velocity is at "*critical*". Theoretically, at the critical velocity the droplet would be suspended in the gas stream, moving neither upward nor downward. Below the critical velocity, the droplet falls and liquids accumulate in the wellbore.

In practice, the critical velocity is generally defined as the minimum gas velocity in the production tubing required to move liquid droplets upward. A "velocity string" is often used to reduce the tubing size until the critical velocity is obtained. Lowering the surface pressure (e.g., by compression) also increases velocity.

Turner's correlation was tested against a large number of real well data having surface flowing pressures mostly higher than 1000 psi. Examination of Turner's data, however, indicates that the range of applicability for his correlation might be for surface pressures as low as 5 to 800 psi.

Two variations of the correlation were developed, one for the transport of water and the other for condensate. The fundamental equations derived by Turner were found to underpredict the critical velocity from the database of well data. To better match the collection of measured

Liquid Transport in a Vertical Gas Well

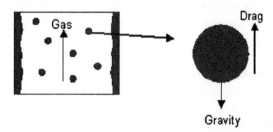

Figure 3-1: Illustrations of Concepts Investigated for Defining Critical Velocity

field data, Turner adjusted the theoretical equations for required velocity upward by 20 percent. From Turner's [1] original paper, after the 20 percent empirical adjustment, the critical velocity for condensate and water were presented as

$$V_{gcond} = \frac{4.02(45-0.0031p)^{1/4}}{(0.0031p)^{1/2}} \ ft/sec \tag{3-1}$$

$$V_{gwater} = \frac{5.62(67-0.0031p)^{1/4}}{(0.0031p)^{1/2}} \ ft/sec \tag{3-2}$$

where p = psi.

The theoretical equation from Ref. 1 for critical velocity V_t to lift a liquid (see Appendix A) is

$$V_t = \frac{1.593\sigma^{1/4}(\rho_l - \rho_g)^{1/4}}{\rho_g^{1/2}} \ ft/sec \tag{3-3}$$

where σ = surface tension, dynes/cm, ρ = density, lbm/ft^3.
Inserting typical values of:

Surface Tension	20 and 60 dyne/cm for condensate and water, respectively
Density	45 and 67 lbm/ft^3 for condensate and water, respectively
Gas Z factor	0.9

$$V_{t,condensate} = \frac{1.593(20)^{1/4}(45-.00279P/Z)^{1/4}}{(.00279P/Z)^{1/2}} = \frac{3.368(45-.00279P/Z)^{1/4}}{(.00279P/Z)^{1/2}}$$

$$V_{t,water} = \frac{1.593(60)^{1/4}(67-.00279P/Z)^{1/4}}{(.00279P/Z)^{1/2}} = \frac{4.43(67-.00279P/Z)^{1/4}}{(.00279P/Z)^{1/2}}$$

Inserting $Z = 0.9$ and multiplying by 1.2 to adjust to Turner's data gives:

$$V_{t,condensate} = \frac{4.043(45-.0031P)^{1/4}}{(.0031P)^{1/2}}$$

$$V_{t,water} = \frac{5.321(67-.0031P)^{1/4}}{(.0031P)^{1/2}}$$

Turner [1] gives 4.02 and 5.62 in his paper for these equations.

These equations predict the minimum critical velocity required to transport liquids in a vertical wellbore. They are used most frequently at the wellhead with P being the flowing wellhead pressure. When both water and condensate are produced by the well, Turner recommends using the correlation developed for water because water is heavier and requires a higher critical velocity.

Gas wells having production velocities below that predicted by the preceding equations would then be less than required to prevent the well from loading with liquids. Note that the actual volume of liquids produced does not appear in this correlation and the predicted terminal velocity is not a function of the rate of liquid production.

3.2.2 Critical Rate

Although critical velocity is the controlling factor, one usually thinks of gas wells in terms of production rate in SCF/d rather than velocity in the wellbore. These equations are easily converted into a more useful form by computing a critical well flow rate. From the critical velocity V_g, the critical gas flow rate q_g, may be computed from:

$$q_g = \frac{3.067 P V_g A}{(T+460)Z} \; MMscf/D \tag{3-4}$$

where

$$A = \frac{(\pi)d_{ti}^2}{4 \times 144} \; ft^2$$

T = surface temperature, °F
P = surface pressure, psi
A = tubing cross-sectional area
d_t = tubing ID, inches

Introducing the preceding into Turner's [1] equations gives the following:

$$q_{t,condensate}(MMscf/D) = \frac{.0676 P d_{ti}^2}{(T+460)Z} \frac{(45-.0031P)^{1/4}}{(.0031P)^{1/2}}$$

$$q_{t,water}(MMscf/D) = \frac{.0890 P d_{ti}^2}{(T+460)Z} \frac{(67-.0031P)^{1/4}}{(.0031P)^{1/2}}$$

These equations can be used to compute the critical gas flow rate required to transport either water or condensate. Again, when both liquid phases are present, the water correlation is recommended. If the actual flow rate of the well is greater than the critical rate computed by the preceding equation, then liquid loading would not be expected.

3.2.3 Critical Tubing Diameter

It is also useful to rearrange the preceding expression, solving for the maximum tubing diameter that a well of a given flow rate can withstand without loading with liquids. This maximum tubing is termed the *critical tubing diameter*, corresponding to the minimum critical velocity. The critical tubing diameter for water or condensate is shown here as long as the critical velocity of gas, V_g, is for either condensate or water.

$$d_{ti}, inches = \sqrt{\frac{59.94 q_g (T+460)Z}{P V_g}}$$

3.2.4 Critical Rate for Low Pressure Wells—Coleman Model

Recall that these relations were developed from data for surface tubing pressures mostly greater than 1000 psi. For lower surface tubing pressures, Coleman et al. [2] has developed similar relationships for the minimum critical flow rate for both water and liquid. In essence the Coleman et al. formulas (to fit their new lower wellhead pressure data, typically less than 1000 psi) are identical to Turner's equations but without the Turner [1] 1.2 adjustment to fit his data.

With the same data defaults given above to develop Turner's equations, the Coleman et al.[2] equations for minimum critical velocity and flow rate would appear as:

$$V_{t,condensate} = \frac{3.369(45-.0031P)^{1/4}}{(.0031P)^{1/2}}$$

$$V_{t,water} = \frac{4.434(67-.0031P)^{1/4}}{(.0031P)^{1/2}}$$

$$q_{t,condensate}(MMscf/D) = \frac{.0563Pd_{ti}^2}{(T+460)Z}\frac{(45-.0031P)^{1/4}}{(.0031P)^{1/2}}$$

$$q_{t,water}(MMscf/D) = \frac{.0742Pd_{ti}^2}{(T+460)Z}\frac{(67-.0031P)^{1/4}}{(.0031P)^{1/2}}$$

However, if the original equations of Turner were used, the coefficients would be 4.02 and 5.62 both divided by 1.2 to get the Coleman equations, so there can be some confusion. The concern is that even if some slight errors in the Turner development are present, the equations with the coefficients have been used with success, and the question is "are the original coefficients better than if they are corrected"?

Example 3-1: Calculate the Critical Rate Using Turner *et al.* and Coleman *et al.*

Well surface pressure = 400 psia
Well surface flowing temperature = 120° F
Water is the produced liquid
Water density = 67 lbm/ft³
Water surface tension = 60 dyne/cm
Gas gravity = 0.6
Gas compressibility factor for simplicity = 0.9
Production string = 2-3/8 inch tubing with 1.995 in ID, A = .0217 ft²
Production = .6 MMscf/D

Critical Rate by Coleman et al. [2]

Calculate the gas density:

$$\rho_g = \frac{M_{air}\gamma_g P}{R(T+460)Z} = \frac{28.97\gamma_g P}{10.73(T+460)Z} = 2.7\frac{0.6\times400}{580\times.9} = 1.24\,lbm/ft^3$$

$$V_g = \frac{1.593\sigma^{1/4}(\rho_l-\rho_g)^{1/4}}{\rho_g^{1/2}} = \frac{1.59360^{1/4}(67-1.24)^{1/4}}{1.24^{1/2}} = 11.30\,ft/sec$$

$$q_{t,water} = \frac{3.067PAV_g}{(T+460)Z} = \frac{3.067 \times 400 \times .0217 \times 11.30}{(120+460) \times 0.9} = .575 \; MMscf/d$$

Critical Rate by Turner et al. [1]

Since the Turner and Coleman variations of the critical rate equation differ only in the 20 percent adjustment factor applied by Turner for his high pressure data, then

$$V_g = 1.2 \times 11.30 = 13.56 \; ft/sec$$

$$q_{t,water} = 1.2 \times 0.575 = 0.690 \; MMscf/d$$

For this example, the well is above critical considering Coleman (.6 > .575 MMscf/D) but below critical (.6 < .69 MMscf/D) according to Turner. We would say it is above critical since the more recent lower wellhead pressure correlation of Coleman *et al.* [2] says it is flowing above critical.

This example illustrates that the more recent Coleman *et al.* [2] relationships require less flow to be above critical when analyzing data with lower wellhead pressures. Also the example shows that the relationships require surface tension, gas density at a particular temperature, and pressure including use of a correct compressibility factor and gas gravity. If these factors are not taken into account for each individual calculation, then the approximate equation may be used. For this example, the approximate Coleman equation gives

$$\begin{aligned} q_{t,water} &= \frac{.0742Pd_{ti}^2}{(T+460)Z} \frac{(67-.0031P)^{1/4}}{(.0031P)^{1/2}} \\ &= \frac{.0742 \times 400 \times 1.995^2}{(120+460) \times 0.9} \frac{(67-.0031 \times 400)^{1/4}}{(.0031 \times 400)^{1/2}} = .579 \; MMscf/d \end{aligned}$$

and is very close to the previously calculated 0.575 MMscf/D.

3.2.5 Critical Flow Nomographs

To simplify the process for field use, the following simplified chart from Trammel [6] can be used for both water and condensate production. To use the chart, enter with the flowing surface tubing pressure (see the dotted line) at the bottom x-axis for water and top axis for condensate. Move upward to the correct tubing size then either left for water or right for condensate to the required minimum critical flow rate.

Example 3-2: Critical Velocity from Figure 3-2 [6]

200 psi well head pressure
2-3/8 inch tubing, 1.995 inch ID

What is minimum production according to the Turner equations?

The example indicated by the dotted line shows that for a well having a well head pressure of 200 psi and 2-3/8 inch tubing, the flow rate must be at least ≈586 Mscf/D (actually 577 calculated) or liquid loading will likely occur.

A similar chart was developed by Coleman *et al.* [2] using the Turner correlation for flowing well head pressures below about 800 psi. Note only one set of curves are represented on this chart to be used for both

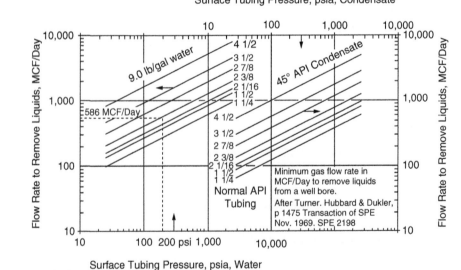

Figure 3-2: Nomograph for Critical Rate for Water or Condensate (after [6]) for a Constant Z = 0.8, Temperature of 60° F, and the Original Turner Assumptions of Surface Tension of σ = 20 dynes/cm for Condensate, and 60 dynes/cm for Water, ρ = 45 lbm/ft³ for Condensate and 67 lbm/ft³ for Water, and Gas Gravity = 0.6

water and condensate. The chart is used in the same manner as the above chart with no distinction between water and condensate. If water and condensate are present, the more conservative water coefficients are used anyway.

The Coleman *et al.* [5] correlation would then be applicable for flowing surface tubing pressures below about 800 psi and the Turner chart (or Turner correlation) for surface tubing pressures above about 800 psi. The dividing line between using Turner or Coleman might best be obtained from experience or even a blend of the two from 500 to 1000 psi.

The chart of Figure 3-4 is another way of looking at critical velocity. It was prepared using a routine calculating actual Z factor (gas compressibility) at each point but still depends on fluid properties and temperatures. For this 60 dyne/cm for surface tension, 67 lbm/ft^3, gas gravity of 0.6 and 120° F were used.

Example 3-3: Critical Velocity with Water: Use Turner's [1] Equations with Figure 3-4

100 psi wellhead pressure
2-3/8 inch tubing, 1.995 inch ID
Read from Figure 3-4 a required rate of about 355 Mscf/D.
Compare to the simplified Turner equations using Z = 0.9 for
  simplicity.

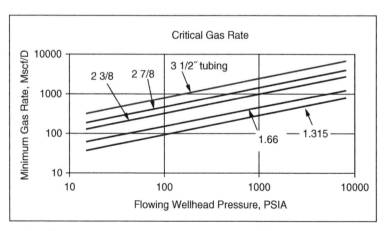

Figure 3-3: The Exxon Nomograph for Critical Rate [2] (for lower surface tubing pressures)

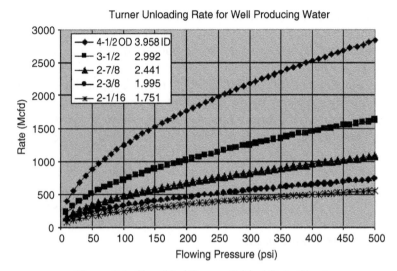

Figure 3-4: Simplified Turner Critical Rate Chart

$$V_{t,water} = \frac{5.32(67 - .0031P)^{1/4}}{(.0031P)^{1/2}}$$
$$= \frac{5.32(67 - .0031 \times 100)^{1/4}}{(.0031 \times 100)^{1/2}} = \frac{5.32 \times 2.86}{.557} = 27.22 \, ft/\sec$$

$$q_g = \frac{3.067PV_gA}{(T+460)Z} = \frac{3.067 \times 100 \times 27.22 \times .0217}{580 \times .9} = 0.346 \, MMscf/D$$

In this case the difference between the calculations and reading from chart can be attributed to that fact that the chart was calculated using actual Z factors and not an assumed value of 0.9.

Using one of the critical velocity relationships, the critical rate for a given tubing size vs. tubing diameter can be generated as in Figure 3-5 where a surface pressure of 200 psi and surface temperature of 80° F is used. (In this case, specific liquid and gas properties were used in the critical flow equations rather than the typical values given above.) This type of a presentation provides a ready reference for maximum tubing size given a particular well flow rate.

A large tubing size may exhibit below critical flow and a smaller tubing size may indicate that the velocity will increase to be above critical. Tubing sizes approaching and less than 1 inch, however, are not

Min. Qg for Unloading Gas Well

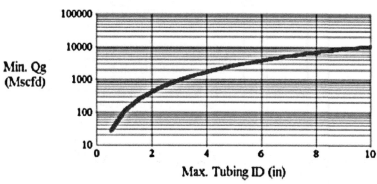

Figure 3-5: Critical Rate vs. Tubing Size (200 psi and 80° F) from Maurer Engineering, PROMOD program. Use this type of presentation with critical velocity model desired

generally recommended as they can be difficult to initially unload due to the high hydrostatic pressures exerted on the formation with small amounts of liquid. It is difficult to remove a slug of liquid in a small conduit. See also Bizanti [4] for pressure, temperature, diameter relationships for unloading and Nosseir *et al.* [5] for consideration of flow conditions leading to different flow regimes for critical velocity considerations.

3.3 CRITICAL VELOCITY AT DEPTH

Although the preceding formulas are developed using the surface pressure and temperature, their theoretical basis allows them to be applied anywhere in the wellbore if pressure and temperature are known. The formulas are also intended to be applied to sections of the wellbore having a constant tubing diameter. Gas wells can be designed with tapered tubing strings, or with the tubing hung off in the well far above the perforations. In such cases, it is important to analyze gas well liquid loading tendencies at locations in the wellbore where the production velocities are lowest.

For example, in wells equipped with tapered strings, the bottom of each taper size would exhibit the lowest production velocity and thereby be first to load with liquids. Similarly, for wells having the tubing string hung well above the perforations, the analysis must be performed using

Liquid Transport in a Vertical Gas Well

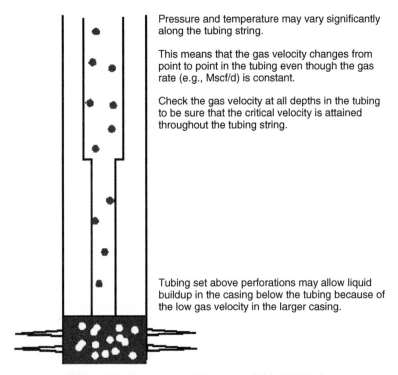

Pressure and temperature may vary significantly along the tubing string.

This means that the gas velocity changes from point to point in the tubing even though the gas rate (e.g., Mscf/d) is constant.

Check the gas velocity at all depths in the tubing to be sure that the critical velocity is attained throughout the tubing string.

Tubing set above perforations may allow liquid buildup in the casing below the tubing because of the low gas velocity in the larger casing.

Figure 3-6: Completions Effects on Critical Velocity

the casing diameter near the bottom of the well since this would be the most likely location of the initial liquid buildup. In practice, it is recommended that liquid loading calculations be performed at all sections of the tubing where diameter changes occur. In general for a constant diameter string, if the critical velocity is acceptable at the bottom of the string, then it will be acceptable everywhere in the tubing string.

In addition, when calculating critical velocities in downhole sections of the tubing or casing, downhole pressures and temperatures must be used. Minimum critical velocity calculations are less sensitive to temperature, which can be estimated using linear gradients. Downhole pressures, on the other hand, must be calculated by using flowing gradient routines (perhaps with Nodal Analysis, Macco Sclumberger™) or perhaps a gradient curve. Bear in mind that the accuracy of the critical velocity prediction depends on the accuracy of the predicted flowing gradient.

Critical Flow Rate - Pressure with Gray (Mod)

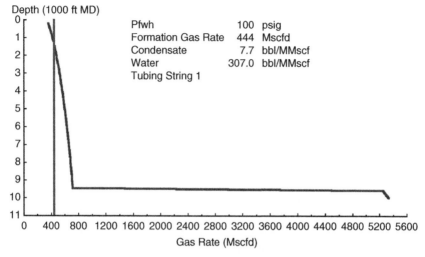

Figure 3-7: Critical Velocity with Depth

Figure 3-7 shows critical rate calculated using the Gray correlation. The vertical line is the actual rate. The blue line is the required critical rate for the tubing and the casing on the bottom. Note the well is predicted to be just above critical rate at the surface but the rest of the tubing is below critical and as usual, well below critical for the casing flow. Normally the required rate is maximum at the bottom of the tubing but for high pressure, high temperature (unusual for most loaded gas wells) the critical may be calculated to be maximum at surface conditions.

Guo *et al.* [7] present a kinetic energy model and show critical rate and velocity at downhole conditions. They mention that Turner underpredicts the critical rate. They mention the controlling conditions are downhole.

3.4 CRITICAL VELOCITY IN HORIZONTAL WELL FLOW

In inclined or horizontal wells the preceding correlations for critical velocity cannot be used. In deviated wellbores, the liquid droplets have very short distances to fall before contacting the flow conduit rendering the mist flow analysis ineffective. Due to this phenomenon, calculating gas rates to keep liquid droplets suspended and maintain mist flow in horizontal sections is a different situation than for tubing. Fortunately, hydrostatic pressure losses are minimal along the lateral section of the well and only begin to come into play as the well turns vertical where critical flow analysis again becomes applicable.

Another, less understood effect that liquids could have on the performance of a horizontal well has to do with the geometry of the lateral section of the wellbore. Horizontal laterals are rarely straight. Typically, the wellbores "undulate" up and down throughout the entire lateral section. These undulations tend to trap liquid, causing restrictions that add pressure drop within the lateral. A number of two phase flow correlations that calculate the flow characteristics within undulating pipe have been developed over the years and, in general, have been met with good acceptance. Once such correlation is the Beggs and Brill method [6]. These correlations have the ability to account for elevation changes, pipe roughness and dimensions, liquid holdup, and fluid properties. Several commercially available nodal analysis programs now have this ability.

A rule of thumb developed from gas distribution studies suggests that when the superficial gas velocity (superficial gas velocity = total in-situ gas rate/total flow area) is in excess of ≈14 fps, then liquids are swept from low lying sections as illustrated in Figure 3-8.

Upon examination, this is a conservative condition and requires a fairly high flow rate. Bear in mind, however, when performing such calculations that the velocity at the toe of the horizontal section can be substantially less than that at the heel.

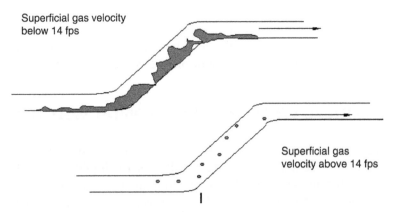

Figure 3-8: Effects of Critical Velocity in Horizontal/Inclined Flow

3.5 REFERENCES

1. Turner, R. G., Hubbard, M. G., and Dukler, A. E. "Analysis and Prediction of Minimum Flow Rate for the Continuous Removal of Liquids from Gas Wells," *Journal of Petroleum Technology*, Nov. 1969. pp. 1475–1482.

2. Coleman, S. B., Clay, H. B., McCurdy, D. G., and Norris, H. L. III. "A New Look at Predicting Gas-Well Load Up," *Journal of Petroleum Technology*, March 1991, pp. 329–333.

3. Trammel, P. and Praisnar, A. "Continuous Removal of Liquids from Gas Wells by use of Gas Lift," *SWPSC*, Lubbock, Texas, 1976, 139.

4. Bizanti, M. S. and Moonesan, A. "How to Determine Minimum Flowrate for Liquid Removal," *World Oil*, September 1989, pp. 71–73.

5. Nosseir, M. A. *et al.* "A New Approach for Accurate Predication of Loading in Gas Wells Under Different Flowing Conditions," SPE 37408, presented at the 1997 Middle East Oil Show in Bahrain, March 15–18, 1997.

6. Beggs, H. D. and Brill, J. P. "A Study of Two-Phase Flow in Inclined Pipes," *Journal of Petroleum Technology*, May 1973, 607.

7. Guo, B., Ghalambor, A., and Xu, C. "A Systematic Approach to Predicting Liquid Loading in Gas Wells," SPE 94081, presented at the 2995 SPE Production and Operations Symposium, Oklahoma City, Ok., 17–19, April 2005.

SYSTEMS NODAL ANALYSIS

4.1 INTRODUCTION

A typical gas well may have to flow against many flow restrictions in order for the produced gas to reach the surface separator.

The gas must flow through the reservoir rock matrix, then through the perforations and gravel pack, possibly through a bottomhole standing valve, then through the tubing, possibly a subsurface safety valve, through the surface flowline, and flowline choke to the separator.

Each of these components will have a flow dependent pressure loss. A change in any of the well restrictions will affect the well production rate. To determine overall well performance, all components of the well must be considered as a unit or total system.

One useful tool for the analysis of well performance is system Nodal Analysis. Nodal Analysis divides the total well system into two subsystems at a specific location called the *nodal point*. One subsystem considers the inflow from the reservoir, through possible pressure drop components and to the nodal point. The other subsystem considers the outflow system from some pressure on the surface down to the nodal point. For each subsystem, the pressure at the nodal point is calculated and plotted as two separate, independent pressure-rate curves.

The curve from the reservoir to the nodal point is called the *inflow curve*, and the curve from the separator to the nodal point is called the *outflow curve*. The intersection of the inflow and outflow curves is the predicted operating point where the flow rate and pressure from the

---

two independent curves are equal. The inflow and outflow curves are illustrated in Figure 4-1.

Although the nodal point may be located at any point in the system, the most common position is at the mid-perforation depth inside the tubing. With this nodal point, the inflow curve represents the flow from the reservoir through the completions into the tubing, and the outflow curve represents the flow from the node to a surface pressure reference point (e.g., separator); summing pressure drops from the surface to the node at the mid-perforations depth.

The Nodal Analysis method employs single or multiphase flow correlations, as well as correlations developed for the various components of reservoir, well completion, and surface equipment systems to calculate the pressure loss associated with each component in the system. This information then is used to evaluate well performance under a wide variety of conditions that will lead to optimum single well completion and production practices. It follows that nodal analysis is useful for the analysis of the effects of liquid loading on gas wells.

First, Nodal Analysis will be used to analyze the effects of various tubing sizes on the ability of gas wells to produce reservoir liquids. Rough estimates of the onset of liquid loading problems are possible, and examples will be given to illustrate the beneficial effects of reducing tubing size to increase the gas flow velocity in the tubing, thereby improving the efficiency of the liquid transport process.

Second, Nodal Analysis is used to clarify the detrimental effects of excessive surface production tubing pressure. Increased surface

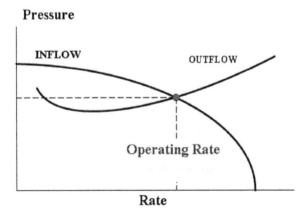

Figure 4-1: System Nodal Analysis

pressure adds backpressure on the reservoir at the sand face. The added backpressure reduces gas production and lowers the gas velocity in the tubing, that again reduces the efficiency with which the liquids are transported to the surface.

4.2 TUBING PERFORMANCE CURVE

The outflow or tubing performance curve (TPC) shows the relationship between the total tubing pressure drop and a surface pressure value, with the total liquid flow rate. The tubing pressure drop is essentially the sum of the surface pressure, the hydrostatic pressure of the fluid column (composed of the liquid "hold up" or liquid accumulated in the tubing and the weight of the gas), and the frictional pressure loss resulting from the flow of the fluid out of the well. For very high flow rates there can be an additional "acceleration term" to add to the pressure drop but the acceleration term is usually negligible compared to the friction and hydrostatic components. The frictional and hydrostatic components are shown by the dotted lines in Figure 4-2 for a gas well producing liquids. Duns and Ros [1] and Gray [2] are examples of correlations used for gas well pressure drops that include liquid "hold up" effects.

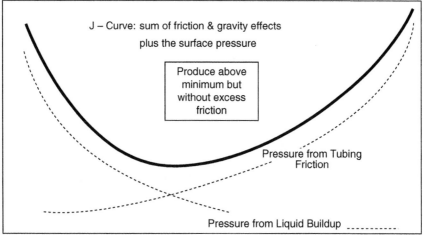

Figure of Tubing Performance Curve with labels: "Tubing Performance Curve", "Flowing tubing bottom", "J – Curve: sum of friction & gravity effects plus the surface pressure", "Produce above minimum but without excess friction", "Pressure from Tubing Friction", "Pressure from Liquid Buildup"

Figure 4-2: Tubing Performance Curve

Notice that the TPC passes through a minimum. To the right of the minimum, the total tubing pressure loss increases due to increased friction losses at the higher flow rates. The flow to the right of the minimum is usually in the mist flow regime that effectively transports small droplets of liquids to the surface.

At the far left of the TPC the flow rate is low and the total pressure loss is dominated by the hydrostatic pressure of the fluid column brought about by the liquid hold up, or that percent of the fluid column occupied by liquid. The flow regime exhibited in the left-most portion of the curve is typically bubble flow, characteristically a flow regime that allows liquids to accumulate in the wellbore.

Slightly to the left of the minimum in the TPC, the flow is often in the slug flow regime. In this regime liquid is transported to the surface periodically in the form of large slugs. Fluid transport remains inefficient in this unstable regime as portions of the slugs "fall-back" as they rise and must be lifted again by the next slug. Fall-back and relifting the liquids results in a higher producing bottomhole pressure.

It is common practice to use the TPC alone, in the absence of up-to-date reservoir performance data, to predict gas well liquid loading problems. It is generally believed that flow rates to the left of the minimum in the curve are unstable and prone to liquid loading problems. Conversely, flow rates to the right of the minimum of the tubing performance curve are considered to be stable and significantly high enough to effectively transport produced liquids to the surface facilities.

Understandably, this method is inexact but is useful to predict liquid loading problems in the absence of better reservoir performance data. Therefore you can just select the flow rate you are measuring currently and see if it is in a favorably predicted portion of the TPC or not, regardless of having the reservoir inflow curve.

With reservoir performance data, however, intersections of the tubing outflow curve and the reservoir inflow curve allow a prediction of where the well is flowing now and into the future if reservoir future IPR curves can be generated.

4.3 RESERVOIR INFLOW PERFORMANCE RELATIONSHIP (IPR)

In order for a well to flow, there must be a pressure differential from the reservoir to the wellbore at the reservoir depth. If the wellbore pressure is equal to the reservoir pressure, there can be no inflow. If the wellbore pressure is zero, the inflow would be the maximum possible—the Absolute Open Flow (AOF). For intermediate wellbore pressures, the inflow will vary. For each reservoir, there will be a unique relationship between the inflow rate and wellbore pressure.

Figure 4-3 shows the form of a typical gas well IPR curve. The IPR curve is often called the deliverability curve.

4.3.1 Gas Well Backpressure Equation

The equation for radial flow of gas in a well perfectly centered within the well drainage area with no rate dependent skin is

$$q_{sc} = \frac{.000703 k_g h (P_r^2 - P_{wf}^2)}{\mu Z T \ln\left(\left(.0472 \frac{r_e}{r_\omega}\right) + S\right)} \tag{4-1}$$

where:

q_{sc} = gas flow rate, Mscf/D
k_g = effective permeability to gas, md
h = stratigraphic reservoir thickness (perpendicular to the reservoir layer), ft
P_r = average reservoir pressure, psia
P_{wf} = flowing wellbore pressure at the mid-perforation depth, psia
μ_g = gas viscosity, cp
Z = gas compressibility factor at reservoir temperature and pressure
T = reservoir temperature, °R
r_e = reservoir drainage radius, ft
r_w = wellbore radius, ft
S = total skin

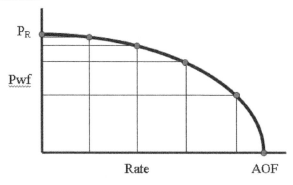

Figure 4-3: Typical Reservoir IPR Curve

Equation 4-1 can be used to generate an inflow curve of gas rate vs. P_{wf} for a gas well if all the preceding data is known. However, often the data required to use this equation are not well known, and a simplified equation is used to generate an inflow equation for gas flow that utilizes well test data to solve for the indicated constants.

$$q_{SC} = C(P_r^2 - P_{wf}^2)^n \qquad (4\text{-}2)$$

where

q_{SC} = gas flow rate, in consistent units with the constant C
   n = a value that varies between about 0.5 and 1.0. For a value of
         0.5, high turbulence is indicated and for a value of 1.0, no
         turbulence losses are indicated.

This equation often is called the *backpressure* equation with the radial flow details of Equation 4-1 absorbed into the constant C. The exponent n must be determined empirically. The values of C and n are determined from well tests. At least two test rates are required, since there are two unknowns, C and n, in the equation, but four test rates are recommended to minimize the effects of measurement error.

If more than two test points are available, the data can be plotted on log-log paper and a least squares line fit to the data, to determine n and C.

Taking the log of Equation 4-2 gives

$$\log(q_{SC}) = \log(C) + n\log(P_r^2 - P_{wf}^2) \qquad (4\text{-}3)$$

On a log-log plot of rate vs. $(P_r^2 - P_{wf}^2)$, n is the slope of the plotted line and $\ln(C)$ is the Y-intercept, the value of q when $(P_r^2 - P_{wf}^2)$ is equal to 1.

For two test points, the n value can be determined from the equation

$$n = \frac{\log(q_2) - \log(q_1)}{\log(P_r^2 - P_{wf}^2)_2 - \log(P_r^2 - P_{wf}^2)_1} \qquad (4\text{-}4)$$

This equation may also be used for more than two test points by plotting the log-log data as described and picking two points from the best-fit line drawn through the plotted points. Values of the gas rate, q, and the corresponding values of $P_r^2 - P_{wf}^2$ can be read from the plotted line at the two points corresponding to the points 1 and 2 to allow solving for n.

Once n has been determined, the value of the performance coefficient C may be determined by the substitution of a corresponding set of values for q and $P_r^2 - P_{wf}^2$ into the backpressure equation. (See more detail in Appendix C.)

If pseudo-stabilized data can be determined in a convenient time, then this equation can be developed from test data easily. Pseudo-steady state indicates that any changes have reached the boundary of the reservoir, but practically it means that for wells with moderate to high permeability, pressures and rates recorded appear to become constant with time. If the well has very low permeability, then pseudo-stabilized data may be nearly impossible to attain, and then other means are required to estimate the inflow of the gas well. Rawlins and Schellhardt [3] provide more information on using the backpressure equation. (For more details on the backpressure equation, see Appendix C.)

In truth many operators do not find the time or the expense involved with testing low pressure gas wells worthwhile. Instead, for loading analysis they use the critical rate correlations and examine the decline curves. However for sizing compression and tubing size, it is advantageous to have an IPR for the well. If one knows the approximate shut-in pressure of a well, then a flowing bottom-hole pressure can be calculated as a point on the IPR and if using the backpressure equation, with an assumed value of the n, then an IPR can be constructed with calculations and without testing. Better success is obtained with this approach if done before the well is loaded, however.

4.3.2 Future IPR Curve with Backpressure Equation

For predicting backpressure curves at different shut-in pressures (at different times), the following approximation from Fetkovitch [4] can be used for "future" inflow curves.

$$q = C\left(\frac{P_r}{P_{ri}}\right)(P_r^2 - P_{wf}^2)^n \tag{4-5}$$

where:

q = current gas rate
C = coefficient consistent with the gas rate and pressure units
P_r = average reservoir pressure, at current time, psia
P_{wf} = flowing current well bore pressure, psia
P_{ri} = initial average reservoir pressure use to determine C and n, psia

4.4 INTERSECTIONS OF THE TUBING CURVE AND THE DELIVERABILITY CURVE

Figure 4-4 shows a tubing performance curve intersecting a well deliverability inflow curve (inflow performance curve—IPR). The figure shows the tubing curve intersecting the inflow curve in two places. Stability analysis shows that the intersection between points C and D is stable whereas the intersection between A and B is unstable and, in fact, will not occur.

For example, if the flow rate strays to point D, then the pressure from the reservoir is at D but the pressure required to maintain the tubing

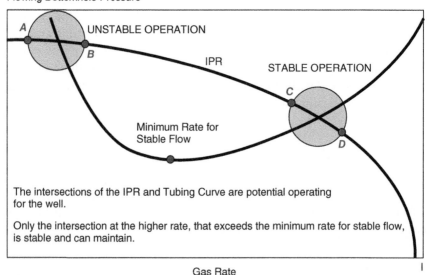

Tubing J-Curve and Flow Stability

Flowing Bottomhole Pressure

A
UNSTABLE OPERATION
B
IPR
STABLE OPERATION
C
Minimum Rate for
Stable Flow
D

The intersections of the IPR and Tubing Curve are potential operating for the well.

Only the intersection at the higher rate, that exceeds the minimum rate for stable flow, is stable and can maintain.

Gas Rate

Figure 4-4: Tubing Performance Curve in Relation to Well Deliverability Curve

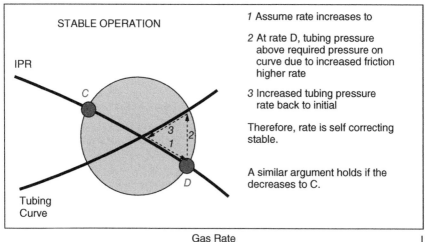

Figure 4-5: Stable Flow

flow is above D. The added backpressure against the sand face of the reservoir then decreases the flow back to the point of stability where the two curves cross. Similarly, if the flow temporarily decreases to point C, the pressure drop in the tubing is decreased, decreasing the pressure at the sand face, prompting an increase in flow rate back to the equilibrium point. Note also that the stable intersection between C and D is to the right of the minimum in the tubing curve. When the intersection of the tubing performance curve and the IPR curve occurs to the right of the minimum in the J-curve, the flow tends to be more stable, and stable vs. erratic flow almost always means more production.

If, on the other hand, the flow happens to decrease to point A, the pressure on the reservoir is increased due to an excess of fluids accumulating in the tubing. The increase in reservoir pressure decreases the flow further, thus increasing the pressure on the reservoir further until the well dies. Similarly, if the well flows at point B, the increased pressure against the reservoir reduces the flow, which again increases the pressure drop in the tubing until ultimately the well again dies. The crossing point on the left side, therefore, uniformly tends toward zero flow, consistent with the minimum point of the tubing performance curve.

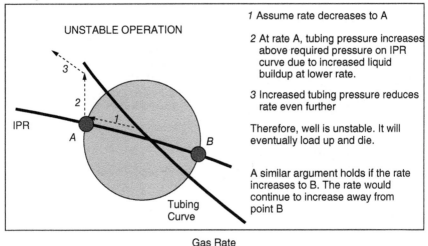

Tubing J-Curve and Flow Stability

Flowing Bottomhole Pressure

UNSTABLE OPERATION

1 Assume rate decreases to A

2 At rate A, tubing pressure increases above required pressure on IPR curve due to increased liquid buildup at lower rate.

3 Increased tubing pressure reduces rate even further

Therefore, well is unstable. It will eventually load up and die.

A similar argument holds if the rate increases to B. The rate would continue to increase away from point B

IPR

A

B

Tubing Curve

Gas Rate

Figure 4-6: Unstable Flow

Thus the crossing point of the IPR and the tubing performance curve, to the right of the minimum of the J-curve, represents a stable flow condition where liquids are effectively transported to the surface, and that to the left of the minimum represents unstable conditions where the well loads up with liquids and dies.

4.5 TUBING STABILITY AND FLOWPOINT

Another way to summarize unstable flow is presented in Figure 4-7. Here the difference between the two curves is the difference between the flowing bottom-hole pressure and the flowing tubing surface pressure. The apex of the bottom curve is called the *flowpoint*. Greene [5] provides additional information on the flowpoint and also for more information on gas well performance, as well as fluid property effects on the AOF of the well.

The reasons for the flow rates below the flowpoint not being sustainable are explained at each rate by the slopes of the inflow and outflow curves as shown earlier. From Greene [5], "a change in the surface pressure is transmitted downhole as a similar pressure change, but a compatible inflow rate in the same direction as the pressure change does not exist. The result is an unstable flow condition that will either kill the

Tubing J-Curve and Flow Stability
The "Flowpoint"

Flowing Bottomhole Pressure

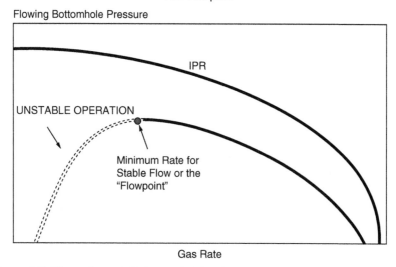

IPR

UNSTABLE OPERATION

Minimum Rate for
Stable Flow or the
"Flowpoint"

Gas Rate

Figure 4-7: Flowpoint or Minimum Stable Flow Rate for Gas Well with Liquids Production

well or, under certain conditions, move the flow rate to a compatible position above the flowpoint rate."

This discussion on stable rates is not the same as flowing below the critical rate as discussed in Chapter 3. There the mechanism is flowing below a certain velocity in the tubing that permits droplets of liquid to fall and accumulate in the wellbore instead of rising with the flow. Discussed here is the interaction of the tubing performance with the inflow curve and reaching a point where the well will no longer flow in a stable condition.

However, the instability is brought on by regions of tubing flow where liquid is accumulating in the tubing due to insufficient gas velocity, so although the arguments are dissimilar, the root causes of each phenomenon are similar. Many Nodal programs will plot the "critical" point on the tubing performance curve. Often this point is on the minimum of the tubing curve or to the right of the minimum in the tubing curve.

When analyzing a gas well, check for Nodal intersections that are "stable" and check for critical velocity at the top and bottom of the well. For instance if selecting tubing size, choose one that that will allow the well to flow above the critical flow rate and also to be stable from Nodal Analysis, for a more complete analysis.

4.6 TIGHT GAS RESERVOIRS

A possible exception to this stability analysis is the tight gas reservoir. A tight gas reservoir generally is defined as one where the reservoir permeability is less than 0.01md. Tight gas reservoirs having low permeability have steep IPR relationships and react to changes in pressure very slowly. A possible tight gas inflow curve is shown in Figure 4-8. This figure shows that the right-most crossing of the tubing performance curve and the IPR might be to the left of the minimum of the J-curve. The above "slope" arguments would lead the conclusion that the right-most intersection is unstable but the well is flowing to the left of the minimum in the tubing performance curve. For tight gas wells, pseudo-steady state data is usually impossible to obtain to get a good inflow curve, and often just using the critical velocity concept is the best tool to analyze liquid loading.

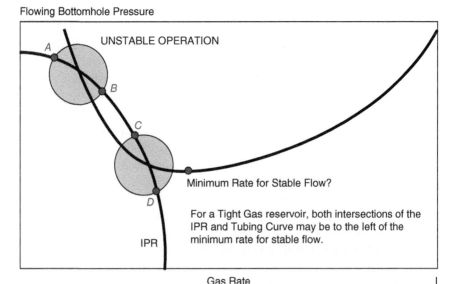

Figure 4-8: Tight Gas Well Modeled with Nodal Intersections

4.7 NODAL EXAMPLE – TUBING SIZE

From the preceding analysis, it is clear that size (diameter) of the production tubing can play an important role in the effectiveness with which the well can produce liquids. Larger tubing sizes tend to have lower frictional pressure drops due to lower gas velocities that in turn lower the liquid carrying capacity. Smaller tubing sizes, on the other hand, have higher frictional losses but also higher gas velocities and provide better transport for the produced liquids. Chapter 5 provides additional information on sizing the tubing following this introductory example.

In designing the tubing string, it then becomes important to balance these effects over the life of the field. To optimize production it may be necessary to reduce the tubing size later in the life of the well.

Figure 4-9 shows tubing performance curves superimposed over two IPR curves. For the higher pressure IPR curve, C, D, and E tubing curves would perform acceptably, but D and E would have more friction and less rate than would the tubing performance curve C. Curves A and B may be intersecting to the left of the minimum in the tubing performance curve; this is thought to generate unstable flow.

For the low pressure or "future" IPR curve, curves A, B, C, and D are all showing intersection below the minimum for the tubing performance curves and as such would not be good choices. Tubing performance curve E, the smallest tubing, performs acceptably for the low pressure IPR curve. Curve E could intersect a little lower on the low pressure IPR curve, and does not, in this case, because of a fairly high (400 psi) surface tubing pressure.

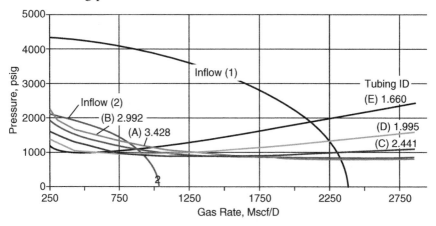

Figure 4-9: Effect of Tubing Size on Future Well Performance

4.8 NODAL EXAMPLE—SURFACE PRESSURE EFFECTS: USE COMPRESSION TO LOWER SURFACE PRESSURE

Frequently the production sales line pressure dictates the surface pressure at the wellhead, which may be beyond the control of the field production engineer. Some installations, however, have compressor stations near the sales line to maintain low pressures at the wellhead while boosting pressure to meet the levels of the sales line. Other methods to lower surface pressure are available to the engineer or technician. This section demonstrates the effects of lowering the wellhead pressure to enhance production and better lift the produced liquids to the surface. Chapter 6 provides more information on the use of compression following this introductory example.

Figure 4-10 shows various tubing performance curves plotted against an IPR curve. The TPC curves or the J-curves are all computed using the same tubing size but with various tubing surface pressures.

Note that reducing the surface pressure has the effect of lowering the tubing performance curve. Lower pressures are beneficial until the steep portion of the gas deliverability curve is reached and then production returns diminish. For instance, the drop in surface pressure from 100 to 50 psi shows only a small gain in production because the deliverability curve is steep in this portion of the curve near the maximum flow rate or the AOF.

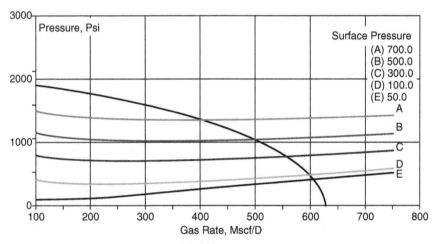

Figure 4-10: Effect of Surface Tubing Pressure on Well Performance (2 3/8's to 19,000 ft)

Reductions in the surface wellhead pressure can be implemented by:

- compression
- larger or "twinned" flowlines
- elimination of small lines, bends, tees, elbows, chokes, or choke bodies at the surface
- reduced separator pressure
- eductors

4.9 SUMMARY NODAL EXAMPLE OF DEVELOPING IPR FROM TEST DATA WITH TUBING PERFORMANCE

This summary problem shows developing the gas IPR (inflow performance relationship) from test data and intersecting the IPR with a calculated tubing performance curve for a well. The object is to analyze if the well is or soon will be in any danger of liquid loading problems.

Example Problem 4.9.1

Calculate the inflow curve from test data and intersect with tubing performance data.
Given data:

Reservoir Pressure, $\bar{P}_r$	3500 psia
2-7/8 inch tubing	2.441 inch ID
Depth	12000 ft
Water production	60 Bbls-water/MMscf
Water Sp. Gr.	1.03

(Note: many operators do not know what the disposed water volume actually is)

Gas gravity, γ_g	0.65
Tsurf	120°F
BHT	70°F
Psurf	300 psia

From flow-after-flow testing (Appendix C) the following pseudo-steady state data are available:

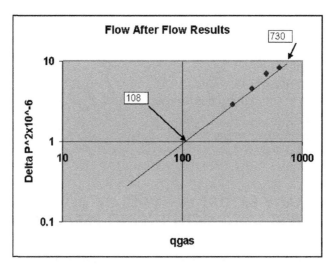

Figure 4-11: Log-Log Plot of Flow-After-Flow Tests

Gas Rate, qg, Mscf/D	P_{wf}, Psiap	$(\bar{p}_r^2 - P_{wf}^2)^2$ 10⁶, psia
263	3170	2.911
380	2897	4.567
497	2440	7.006
640	2150	8.338

Solving for the *n* and *C* value for the backpressure equation:

$$n = \frac{\Delta \log(q_g)}{\Delta \log(\bar{p}_r^2 - p_{wf}^2)} = \frac{\log 730 - \log 108}{\log 10^7 - \log 10^6} = \frac{2.863 - 2.033}{1} = 0.83$$

$$C = \frac{q}{(\bar{p}_r^2 - p_{wf}^2)^n} = \frac{730}{(10 \times 10^6)^{.83}} = 1.13 \times 10^{-3} \frac{\text{Mscf/D}}{\text{psia}^{2n}}$$

The inflow equation is then:

$$q_g = 1.13 \times 10^{-3} (3600^2 - P_{wf}^2)^{.83}$$

Using data from the gas backpressure equation, a conventional IPR equation can be plotted. Using a computer program, a tubing performance curve can be calculated using the Gray correlation (described in Appendix C). Plotting both the inflow and the tubing performance curve (outflow curve) gives the following plot. The pressure is the sum of the tubing surface pressure and the tubing pressure drop.

Tubing performance data using the Gray correlation:

Qg, Mscf/D	Pwf, psia
0	0
338	1200
524	900
666	865
772	835
846	838
890	841
905	844
1000	849
1200	855
1400	890

The following plot shows the intersection of the tubing performance curve at about 846 Mscf/D and the intersection is to the right of the minimum in the tubing curve so this should be a stable situation. However, the minimum in the tubing curve is close to the inflow curve, so further declines in the reservoir may lead to an unstable flow rate. The surface pressure is high for this well, so a reduction in the surface tubing pressure would tend to allow flow further into the future if pressure declines; however, since the tubing curve is in the near vertical area of the IPR, a reduction in the surface tubing pressure would not increase flow much.

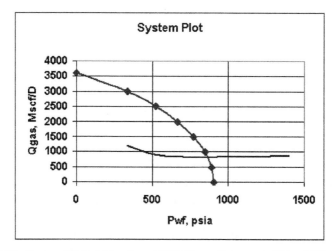

Figure 4-12: System Plot of Developed IPR Curve Intersected Liquid/Gas Producing Tubing Performance Curve

The critical rate according to Coleman is calculated as follows with z assumed as 0.9:

$$q_{t,water}(MMscf/D) = \frac{.0742Pd_{ti}^2}{(T+460)Z} \frac{(67-.0031P)^{1/4}}{(.0031P)^{1/2}}$$
$$= \frac{.0742 \times 300 \times 2.44^2(67-.0031 \times 300)^{1/4}}{580 \times .9 \times (.0031 \times 300)^{1/2}}$$
$$= 0.751 MMscf/D$$

The system plot shows a stable situation, and according to the Coleman *et al.* critical rate, the well is flowing above the critical rate of about 751 Mscf/D also.

4.10 CHOKES

Many operators say never to use chokes on loading gas wells. Wells that are not loaded simply have production reduced (for various reasons) when using chokes. However despite the fact that critical rate shows that lower pressures without the use of chokes requires less rate to be above critical, there are many reports of loaded gas wells having beneficial effects from using chokes.

Figure 4.12 shows that on nodal plots that using a choke at the surface (or downhole) on a loaded well will bring the well performance combined with choke performance to a stable curve. However, in Figure 4.12, the well is still below critical when it is choked to a stable Nodal outflow curve.

There are reports of loaded gas wells that have to be flowed intermittently to get them to flow and if flowed continuously, they will load and die completely. Again this is contrary to the concepts of critical rate.

Case History

The well was drilled toward the end of 2005 and in the beginning of 2006. It is completed in the Cotton Valley and Travis Peak formations. The interval is over 3,000′. It is operated by a major.

By necessity the well was put to a higher pressure separator. The well was not strong enough to flow against the back pressure from the longer line and the well died. Strangely enough, it was found once we shut in the well and brought it back on with a slight choke, the production

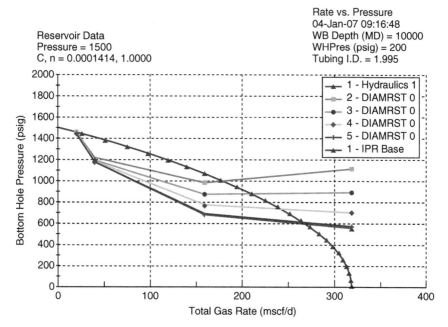

Reservoir Data
Pressure = 1500
C, n = 0.0001414, 1.0000

Rate vs. Pressure
04-Jan-07 09:16:48
WB Depth (MD) = 10000
WHPres (psig) = 200
Tubing I.D. = 1.995

Legend:
1 - Hydraulics 1
2 - DIAMRST 0
3 - DIAMRST 0
4 - DIAMRST 0
5 - DIAMRST 0
1 - IPR Base

Bottom Hole Pressure (psig)

Total Gas Rate (mscf/d)

Figure 4-13: Effects of a Choke on a Liquid Loaded Gas Well

became more stable. This has been seen on some other wells in this field once the production has fallen below the critical rate. The choke seems to flatten the production curve rather than having to shut in the well and flowing intermittently.

From the flow rate report, the tank was lost around the 4th of June. The well was shut in and the operator tried to bring it back on to the tank battery. The well came on, lasted a day or so, and then died. The well was shut in again and brought back on with a slight choke. The tubing pressure can be seen coming up to around 450 psi. The well then began to flow at a more constant rate.

Recently, the well has been put on a tank back on location and by dropping the tubing pressure back down and we are flowing.

So it seems that use of a choke on a loaded well may allow it to flow constantly instead of intermittently. Is this a situation for more rate than by intermittent flow? It would seem so but there is currently a lack of data in this regard.

In summary, however, do not put a choke on a well above critical unless you want less rate. Do not put a choke on a well with artificial lift (unless very special cases exist) or you will see less rate, especially

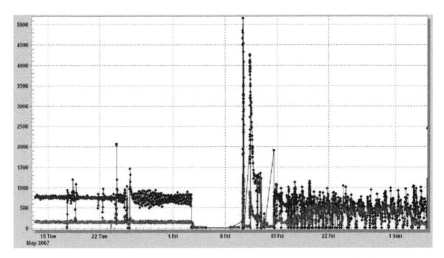

Figure 4-14: Loaded Gas Well Performing with Surface Choke

with plunger lift and gas lift in general. However, there may be a window where the operator can extend the well with continuous production as opposed to intermittent production by using a choke before artificial lift is required.

4.11 MULTIPHASE FLOW FUNDAMENTALS

R P Sutton, Marathon
Robert P. Sutton is a Senior Technical Consultant for Marathon Oil Company. He has BS and MS degrees in Petroleum Engineering from Marietta College and University of Louisiana at Lafayette. Since joining Marathon in 1978, Rob has worked in the areas of reservoir and PVT simulation and developed Marathon's in-house Nodal Analysis computer program.

4.11.1 Fundamentals of Multiphase Flow in Wells

Many correlations have been developed over the years to evaluate the pressure drop resulting from the multiphase flow of fluids in a vertical or deviated well. The following list provides a reasonable cross-section of these methods in order of their publication date.

Poettmann and Carpenter [6]	1952
Baxendell and Thomas [7]	1961
Fancher and Brown [8]	1963
Duns and Ros [1]	1963
Hagedorn and Brown [9]	1965
Orkiszewski [10]	1967
Aziz, Govier, and Fogarasi [11]	1972
Beggs and Brill [12]	1973
Griffith, Lau, Hon, and Pearson [13]	1973
Chierici, Ciucci, and Sclocchi [14]	1974
Cornish [15]	1976
Gray [2]	1978
Mukherjee and Brill [16]	1979
Reinicke, Remer, and Hueni [17]	1984
Ansari [18]	1988
Kaya [19]	1998

All these methods evaluate the pressure drop resulting from the flow of condensate, gas, and water using the following general equation. This equation identifies the pressure drop resulting from changes in elevation, friction, and kinetic energy (acceleration) as the fluids flow through the tubing string.

$$\frac{dP}{dL} = \frac{\left(\rho_m \cos\theta + \frac{f\rho_m v^2}{2gd} + \frac{\rho_m v dv}{gdL}\right)}{144} \tag{4-6}$$

where

dP = pressure drop, psi
dL = length, ft
dv = change in velocity, ft/sec
ρ_m = mixture density, lb/ft^3
θ = angle, degrees from vertical
f = Moody friction factor
v = velocity, ft/sec
g = gravitational constant, 32.17 ft/sec^2
d = pipe internal diameter, ft

The various multiphase flow correlations differ primarily in the way flow patterns are calculated and the resulting liquid holdup is evaluated.

Brill and Mukherjee [20] provide a comprehensive discussion of the more recently published methods, and Brown *et al.* [21] provide background on many of the older methods.

The application of these methods to problems of gas wells producing free water yield a wide range of results. An example is presented to better illustrate this point.

4.11.2 Application of Multiphase Flow Correlations for a Well Producing Gas and Free Water

An offshore gas well was tested with the following results:

Gas Rate, MCFPD	Water Rate, STBPD	FWHP, psia	FWHT, °F	FBHP, psia	FBHT, °F
1300	37	133	75	397	161

A flowing pressure survey was run in this well and the following data was recorded:

Depth (TVD) feet	Pressure psia
0	133
999	163
1969	191
2802	211
3575	234
4350	264
4797	281
5172	299
5620	318
5979	337
6441	358
6867	379
6991	389
7127	397

The well produces gas with a gravity of 0.65 and a water gravity of 1.0. The low specific gravity of the water indicates that produced water could be condensed water. However, a check of the water vapor content of the gas at reservoir conditions reveals a vapor content of 1.8 BBLS/ MMCF (about 2 BWPD) so free water is produced from the reservoir. With the directional profile of the well provided next, calculate the pressure profile in the well and compare to the measured. The tubing size is 2.441 in.

MD, ft	TVD, ft
0	0
2000	1968
4000	3573
6000	5171
7000	5978
8000	6818
8357	7127

An evaluation of the calculated pressure traverse in the well compared with the measured data is shown in Figure 4-15.

The calculated bottomhole pressure ranges from 246 to 1328 psia compared with the measured pressure of 397 psia, highlighting the need to properly select appropriate multiphase flow correlations suitable for a gas well producing water. The following list of correlations are more suited to gas wells with free liquid production.

Hagedorn and Brown	1965
Cornish	1976
Gray	1978
Reinicke, Remer, and Hueni	1984
Ansari	1988
Kaya	1998

For low rate gas well evaluations where liquid loading can be an issue in the well, the Cornish method is unsuitable as it was proposed for high rate flow where there is little or no liquid fall back. A closer look at the results shows the method of Reinicke *et al.* to best model the observed pressure drop in this example well.

The method developed by Reinicke *et al.* was developed to model multiphase flow behavior in gas wells producing water, so this selection is reasonable.

Another procedure for validating multiphase flow pressure drop correlations is to examine their behavior using the calculated tubing performance curves. The methods chosen for evaluating gas-water flow are presented in Figure 4-17. Although this analysis still confirms the Reinicke *et al.* method, additional information is provided to analyze the performance of the well. The shape of the curve in this presentation can be used to infer flow stability. Rate intervals where the curve exhibits a zero or negative slope indicate unstable flow. In this instance the

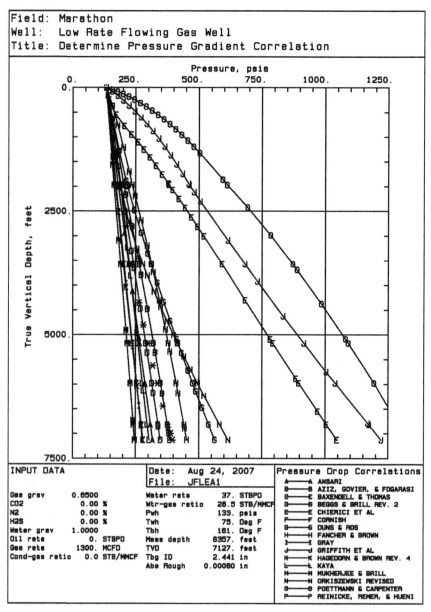

Field: Marathon
Well: Low Rate Flowing Gas Well
Title: Determine Pressure Gradient Correlation

Figure 4-15: Comparison of Pressure Gradient Calculations with Measured Pressure Profile

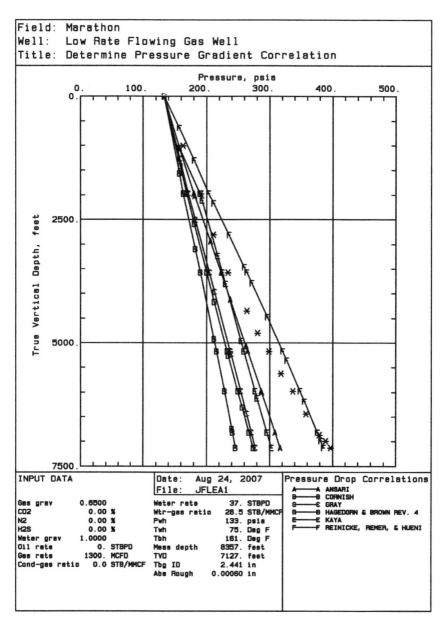

Figure 4-16: Comparison of Pressure Gradient Calculation Methods for Gas Wells Producing Water

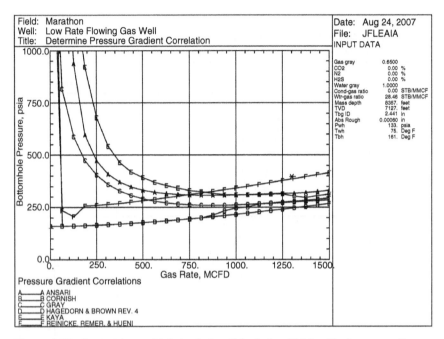

Figure 4-17: Comparison of Methods for Calculating Tubing Performance Curves against Measured Data

gas velocity is insufficient to effectively lift liquids from the well. The higher density liquids then have a tendency to accumulate in the well and liquid loading issues follow. The rate at which flow becomes unstable typically differs as determined by the multiphase flow correlations. Methods such as Cornish and the method of Hagedorn and Brown do not effectively evaluate this condition; they evaluate a no-slip condition where there is no liquid fall back in the well. These are shortcomings in these methods. Other methods tend to identify a range of rates where flow is predicted to become unstable. Additional information from the field (wellhead pressure charts) or other analysis techniques (critical flow velocity) can be utilized to confirm unstable flow conditions.

The analysis of this well can be extended by examining the critical velocity necessary to unload the well. Turner's [22] method originally was proposed to evaluate the critical velocity using wellhead conditions. Generally, this is a reasonable choice unless the well's flowing pressure gradient is dominated by the produced liquid or the geometry of the well changes. Using the Reinicke *et al.* method to determine the pressure profile for this example, the flow velocity and Turner critical velocity are determined in the well. The results are depicted in the following graph.

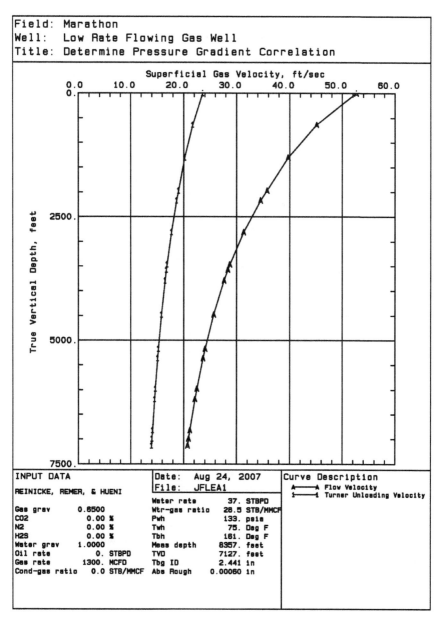

Figure 4-18: Flow Velocity and Critical Velocity Profile for Well with Constant Geometry

As long as the evaluated flow velocity (curve A) is greater than the critical velocity (curve 1), the well should not experience liquid loading issues and stable flow should be observed. For low pressure operations, industry experience has shown the Coleman [23–26] method to be applicable. This method results in velocity values 20 percent lower than those determined by Turner.

Many wells exhibit a change in flow geometry in the well. A common example of this occurs when the tubing string is landed above the perforation interval in the well. In this instance, the velocity profile is significantly altered below the tubing. If the tubing in the referenced example had been set 500 feet above the perforations, the bottom portion of the well would have had to flow through a 7-in casing with an internal diameter of 6.184 in.

In this scenario, the well rate results in sufficient velocity to lift the liquids in the tubing string. However, the flow velocity in the casing is below the critical velocity and liquids can then begin to accumulate in the wellbore.

Wells frequently produce at low rates with flow velocities well below the critical velocity. Liquids accumulate in the well and the gas is then produced through this static liquid column. The result is additional backpressure against the reservoir as shown in Figure 4-20. This additional pressure drop reduces gas production from the reservoir. Furthermore, the liquid can imbibe into the reservoir in the near well area, which reduces the reservoir's ability to produce gas.

Sutton *et al.* [27] evaluated six methods to predict the pressure drop resulting from the accumulation of liquid in the well. They found the procedure developed by Hasan and Kabir [28] to be best suited for calculating the pressure loss through an aerated liquid column in a gas well producing below the critical velocity. Laboratory experiments to replicate static liquid columns in wells have been conducted to enhance the understanding of this phenomenon. Figure 4-21 shows the results of an experiment in which air flow through 4-in casing with 2-in tubing landed above the entry point to the flow loop. Air is injected into the flow loop at a rate yielding the critical velocity in the tubing portion of the flow loop. Any liquid reaching the tubing is produced from the flow loop; however, the bulk of the liquid remains stationary in the 4-in casing where the gas velocity is reduced.

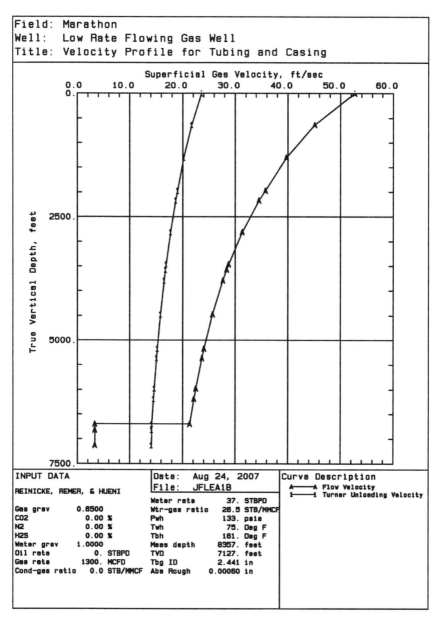

Field: Marathon
Well: Low Rate Flowing Gas Well
Title: Velocity Profile for Tubing and Casing

Figure 4-19: Flow Velocity and Critical Velocity Profile for Well with Changing Geometry

76   *Gas Well Deliquification*

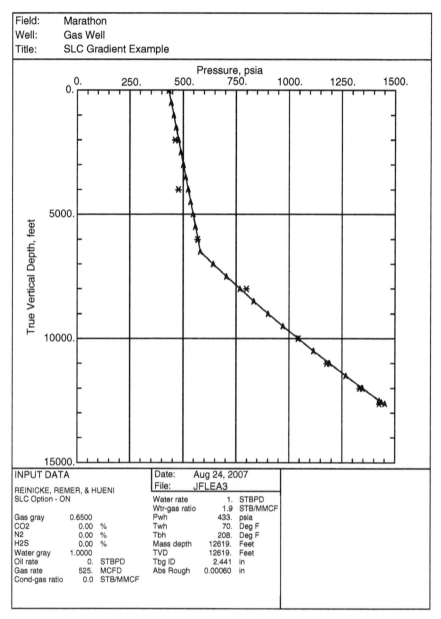

Figure 4-20: Effect of Static Liquid Column on Pressure Gradient Profile

Figure 4-21: Laboratory Experiment of a Static Liquid Column (courtesy Richard Christiansen and Colorado School of Mines)

4.12 SUMMARY

Systems Nodal Analysis can be used to study the effects of a wide variety of conditions on the performance of gas wells. The effects of tapered tubing strings, perforation density and size, formation fluid properties, and fluid production rates are just a few of the many parameters that the technique can analyze. Only a few sample problems varying tubing size and surface pressure with different inflow expressions are shown here.

Use Nodal Analysis to examine the effects of variables that you have control of such as number of perforations, perhaps surface pressure, and tubular sizes if designing a well or considering tubing resize. For liquid loading, look for intersections of the tubing curve with the inflow curve to be to the right of the minimum in the tubing curve for stability. Look for flow rates that will allow the well to flow above the critical rate at the surface and at points downhole as well.

4.13 References

1. Duns, H., Jr. and Ros, N. C. J. "Vertical Flow of Gas and Liquid Mixtures in Wells," Proc. Sixth World Pet. Congress, 1963, 451.

2. Gray, H. E. "Vertical Flow Correlation in Gas Wells," API User's Manual for API 14, *Subsurface Controlled Subsurface Safety Valve Sizing Computer Program*, Appendix B, June, 1974.

3. Rawlins, E. L. and Schellhardt, M. A. "Back Pressure Data on Natural Gas Wells and Their Application to Production Practices," Bureau of Mines Monograph 7, 1935.

4. Fetkovitch, M. J. "The Isochronal Testing of Oil Wells," SPE Paper No. 4529, 48th Annual Fall Meeting of SPE of AIME, Las Vegas, Nevada, Sept. 30–Oct 3, 1973.

5. Greene, W. R. "Analyzing the Performance of Gas Wells," presented the annual SWPSC, Lubbock, Texas, April 21, 1978.

6. Poettmann, F. H. and Carpenter, P. G. "The Multiphase Flow of Gas, Oil, and Water Through Vertical Flow Strings with Application to the Design of Gas Lift Installations," *Drill. and Prod. Prac.,* API, 1952, 257–317.

7. Baxendell, P. B. and Thomas, R. "The Calculation of Pressure Gradients in High Rate Flowing Wells," *J. Pet. Tech.*, Oct. 1961, 1023–1028.

8. Fancher, G. H., Jr. and Brown, K. E. "Prediction of Pressure Gradients for Multiphase Flow in Tubing," *Soc. Pet. Eng. J.*, March, 1963, 59–69.

9. Hagedorn, A. R. and Brown, K. E. "Experimental Study of Pressure Gradients Occurring During Continuous Two Phase Flow in Small Diameter Vertical Conduits," *J. Pet. Tech.*, April, 1965, 475–484.

10. Orkiszewski, J. "Predicting Two Phase Pressure Drops in Vertical Pipe," *J. Pet. Tech.*, June 1967, 829–838.

11. Aziz, K., Govier, G. W., and Fogarasi, M. "Pressure Drop in Wells Producing Oil and Gas," *J. Cdn. Pet. Tech.*, July–Sept., 1972, 38–48.

12. Beggs, H. D. and Brill, J. P. "A Study of Two Phase Flow in Inclined Pipes," *J. Pet. Tech.*, May 1973, 607–617.

13. Griffith, P., Lau, C. W., Hon, P. C., and Pearson, J. F. "Two Phase Pressure Drop in Inclined and Vertical Pipes," *Tech. Report No. 80063 81*, Mass. Inst. Technol., Aug. 1973.

14. Chierici, G. L., Ciucci, G. M., and Sclocchi, M. "Two Phase Vertical Flow in Oil Wells Prediction of Pressure Drop," *J. Pet. Tech.*, Aug. 1974, 927–938.

15. Cornish, R. E. "The Vertical Multiphase Flow of Oil and Gas at High Rates," *J. Pet. Tech.*, July 1976, 825–831.

16. Mukherjee, H. and Brill, J. P. "Liquid Holdup Correlations for Inclined Two Phase Flow," *J. Pet. Tech.*, May 1983, 1003–1008.

17. Reinicke, K. M., Remer, R. J., and Hueni, G. "Comparison of Measured and Predicted Pressure Drops in Tubing for High Water Cut Gas Wells," *SPEPE*, Aug. 1987, 165–177.

18. Ansari, A. M., Sylvester, N. D., Sarica, C., Shoham, O., and Brill, J. P. "A Comprehensive Mechanistic Model for Upward Two-Phase Flow in Well-bores," *SPE Prod. Fac.*, May 1994, 143–152.

19. Kaya, A. S., Sarica, C., and Brill, J. P. "Mechanistic Modeling of Two-Phase Flow in Deviated Wells," *SPE Prod. Fac.*, Aug. 2001, 156–165.

20. Brill, J. P. and Mukherjee, H. *Multiphase Flow in Wells*, Society of Petroleum Engineers, Richardson, TX, 1999.

21. Brown, K. E. and Beggs, H. D. *The Technology of Artificial Lift Methods, Vol. 1*, PennWell Publishing Co, Tulsa, OK, 1977, Chapt. 2.

22. Turner, R. G., Hubbard, M. G., and Dukler, A. E. "Analysis and Prediction of Minimum Flow Rate for the Continuous Removal of Liquids from Gas Wells," *J. Pet. Tech.*, Nov. 1969, 1475–1482.

23. Coleman, S. B., Hartley, B. C., McCurdy, D. G., and Norris III, H. L. "A New Look at Predicting Gas-Well Load-up," *J. Pet. Tech.*, Mar. 1991, 329–333.

24. Coleman, S. B., Hartley, B. C., McCurdy, D. G., and Norris III, H. L. "Understanding Gas-Well Load-Up Behavior," *J. Pet. Tech.*, Mar. 1991, 334–338.

25. Coleman, S. B., Hartley, B. C., McCurdy, D.G., and Norris III, H. L. "The Blowdown-Limit Model," *J. Pet. Tech.*, Mar. 1991, 339–343.

26. Coleman, S. B., Hartley, B. C., McCurdy, D. G., and Norris III, H. L. "Applying Gas-Well Load-Up Technology," *J. Pet. Tech.*, Mar. 1991, 344–349.

27. Sutton, R. P., Cox, S. A., Williams, E. G., Jr., Stoltz, R. P., and Gilbert, J. V. "Gas Well Performance at Subcritical Rates," paper SPE 80887 presented at the SPE Production and Operations Symposium, Oklahoma City, OK, March 23–25, 2003.

28. Hasan, A. R., Kabir, C. S., and Rahman, R. "Predicting Liquid Gradient in a Pumping-Well Annulus," *SPEPE*, Feb. 1988, 113–120.

SIZING TUBING

5.1 INTRODUCTION

As seen in Chapters 3 and 4, the size of the flow conduit through which the gas is produced (this could be the tubing or the casing-tubing annulus or simultaneous flow up the casing-tubing annulus and the tubing) determines the performance of velocity or siphon strings or just determines how well and for how long the production tubing will produce the well.

The basic concept of tubing design is to have a large enough tubing diameter such that excessive friction will not occur and a small enough tubing such that the velocity is high and liquid loading will not occur. The objective is to design a tubing installation meeting these requirements over the entire length of the tubing string or flow conduit. Also it is desired to meet these requirements for as long as possible into the future before another well configuration may be required.

The concepts needed to properly size and evaluate a tubing change-out have been described in Chapter 4 using Nodal concepts and in Chapter 3 using critical velocity concepts. Both of these concepts should be considered when sizing tubing to reduce liquids loading. A well decline curve is also desirable to help in deciding if liquid loading is a problem and to properly post-evaluate the installation of smaller tubing.

5.2 ADVANTAGES AND DISADVANTAGES OF SMALLER TUBING

The reason to run smaller tubing is to increase the velocity for a given rate and sweep the liquids out of the well and the tubing. In general, faster velocity reduces the liquid holdup (% liquid by volume in the tubing) and lowers the flowing bottomhole pressure attributed to gravity effects of the fluids in the tubing. However, tubing too small for the production rate can cause excess friction and require a larger flowing bottomhole pressure.

There are many other methods of deliquifying a gas well, and tubing design must be compared to other possible methods before making a final decision. For instance, Plunger Lift will be shown in Chapter 7 to work better, in general, in larger tubing. Therefore you may reach a time in the life of the well when you must decide if you want to install and operate with smaller tubing or install Plunger Lift to reduce liquid loading in the future.

There are some pros and cons of smaller tubing that should be evaluated before proceeding in this direction. Some of the disadvantages are:

1. Pressure bombs, test tools, and coiled tubing cannot be run in the smaller strings. This is especially true in 1.05, 1.315, and 1.66, and even in 1.9 inch OD tubing. This makes small diameter tubing unpopular with field personnel.
2. If you change to a smaller tubing today, then later you may have to downsize to even smaller tubing. There may be cases where using, for instance, plunger lift could last longer into the future of the well without significant changes in the hardware. It is critical to evaluate the longevity of a smaller tubing design using Nodal Analysis or by comparison to the history of similar installations.
3. If the small tubing becomes loaded, then you cannot swab the tubing and may not even be able to nitrogen lift it. One-inch tubing is especially bad about loading up and is hard to get started flowing again. Figure 5-1 shows how small tubing requires more pressure to support a given volume of fluid. The same volume of fluid that may be negligible in larger tubing can be significant in small tubing.

5.3 CONCEPTS REQUIRED TO SIZE SMALLER TUBING

To resize tubing, we need the reservoir inflow from a reservoir model or an IPR curve obtained from well test data. Then we have the Nodal concepts of generating a tubing curve for various sizes of tubing and obtaining some information from the shape of the tubing curve. Also we have the concept of critical flow and we want the velocity in the tubing to be greater than critical velocity so the holdup or percent by volume of liquids in the tubing will be greatly reduced.

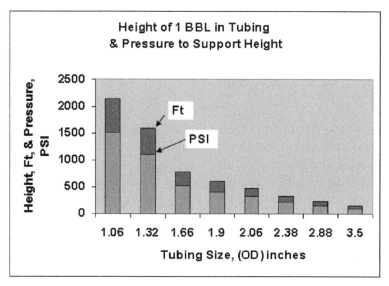

Figure 5-1: Effects of Constant Amount of Liquid Standing in Various Tubing Sizes

Example 5-1

To illustrate these concepts, let's look at an example for various tubing sizes. Consider a well with these conditions:

Well Depth	10,000 ft
Bottomhole Temperature	180°F
Surface Flowing Temperature	80°F
Surface Flowing Pressure	100 psig
Gas Gravity	0.65
Water Gravity	1.02
Condensate Gravity	57 API
Water Rate	2 bbl/MMscf
Condensate Rate	10 bbl/MMscf
Reservoir Pressure	1000 psia
Reservoir Backpressure n	1.04
Tubing	various
Reservoir Backpressure C	0.002 Mscf/D/psi^{2n}

The flowing bottomhole pressure for each tubing is calculated (using the Gray correlation, see Appendix C) for a range of gas production rates and plotted on the same graph with the reservoir inflow curve (Inflow Performance Relationship, IPR) in Figure 5-2.

From Figure 5-2 we see that:

- The 1 in, the 1.25 in, and the 1.5 in ID tubing strings are acceptable because the minimum in the tubing or "outflow" curves is to the left of the expected intersection point with the IPR, or the point where they are calculated to flow.
- The 1.75 in ID tubing curve is very flat at the intersection point and we cannot be sure that the minimum is to the left of the intersection point of the IPR curve.
- The 1.995 in ID curve definitely has the minimum somewhere to the right of the intersection point with the IPR curve.

We conclude from the Nodal plot that the 1.5 in curve looks like a good design. The 1.75 in performance is questionable and as the reservoir declines further, the 1.75 in curve would definitely not be a good choice. So from this analysis, the best design for the most production would be the 1.5 in ID tubing for current conditions.

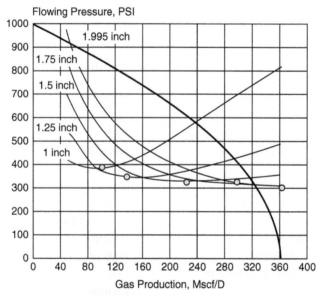

Figure 5-2: Tubing Performance vs. Tubing ID: Critical Rates Plotted on Tubing Curves

5.3.1 Critical Rate at Surface Conditions

Now let's check the critical rate for each tubing ID for tubing size. Since the surface pressure is low, we will use the Coleman *et al.* [2] findings for lower surface pressure wells that modify the original Turner [1] formulas. Since both water and condensate are present, we will conservatively use the water equation.

The surface critical gas rate required for water is calculated from Equation 5-1 using Z = 0.9 and tabulated in

$$Qgas, water = \frac{14.33PA(67-.0031P)^{1/4}}{TZ(0.0031P)^{1/2}} \tag{5-1}$$

The critical rates from surface pressures in Table 5-1 are plotted in Figure 5-2 as dots on the corresponding tubing curve. We can compare the critical rates for each tubing size with the flow rates predicted from the Nodal solution at the intersection point of each tubing curve with the IPR.

- The critical rate for the 1 in ID tubing is to the right of the minimum in the tubing curve but not close to the larger intersection of the tubing curve/inflow curve.
- The critical rate for the 1.25 in tubing is perhaps a little to the left of the minimum in the tubing curve but still to the left of the intersection.
- The critical rate for the 1.5 in tubing is just to the right of the minimum in the tubing curve but still to the left of the intersection.
- The critical rate for the 1.75 in tubing is to the left of the minimum in the curve but still to the left of the intersection.
- The critical rate for the 1.995 in curve is to the left of the minimum in the curve, but it is to the right of the intersection. We previously stated not to use the 1.995 in curve since the intersection with the deliverability curve is to the left of the minimum in the tubing curve,

Table 5-1
Critical Rates vs. Tubing Size for Figure 5-2 Using Surface Pressure

Tubing ID (in)	1.000	1.250	1.500	1.750	1.995
Flow Area (in^2)	.785	1.226	1.766	2.400	3.120
$Q_{critical}$ (Mscf/D)	88.2	142.2	204.9	278.5	362.0

but the critical velocity also says not to try to produce at this intersection, since the critical rate of 346 Mscf/D is larger.

5.3.2 Critical Rate at Bottomhole Conditions

The previous analysis of critical rates used the well flowing surface pressure to calculate the critical rate at surface conditions. A similar analysis can be done at the bottomhole pressure conditions.

Using the Nodal solution pressure (bottomhole pressure at the Nodal intersections), the Nodal solution rate can be calculated. If the critical rate calculated at the Nodal solution pressure is less than the Nodal solution rate, then the Nodal solution rates are acceptable; if not, then the critical velocity condition is violated.

From Table 5-2 the biggest tubing that has enough rate (above critical) at the bottom of the tubing is the 1.50 in tubing. The larger ID tubings would have velocity at the bottom of the tubing less than the critical.

Note that the calculation of critical rate at bottomhole conditions will depend somewhat on the particular method used to calculate the tubing curves. Multiphase flow correlations are developed for a range of fluid properties and tubing sizes that may not match your well conditions exactly. Different multiphase flow correlations can often result in drastically different flowing gradients. The most significant difference between correlations is usually in regard to how each calculates the beginning of the turn up or liquid loading at low rates. Thus, it is imperative to use a method appropriate for your well.

For lower rate gas wells with moderate liquids production, the Gray correlation is quite good to predict the tubing J-curve and is recommended unless you have specific data that indicates otherwise. Gray was used for the tubing curves in Figure 5-2.

Table 5-2
Critical Rates Needed at Nodal Intersections Compared to Nodal Rates

Tubing ID (in)	1.000	1.250	1.500	1.750	1.995
Nodal Solution Pressure (psia)	585	435	355	335	335
Nodal Solution Rate (Mscf/D)	220	275	320	325	325
Critical Rate for Nodal Solution Pressure (Mscf/D)	167	226	294	388	505

The best way to ensure a good flowing bottomhole calculation is to measure the actual flowing bottomhole pressure and the associated well production rate and compare the different calculation methods to the measured data. Some software allows the user to adjust the calculations slightly to better match actual well data.

5.3.3 Summary of Tubing Design Concepts

To summarize, when redesigning a tubing string:

- Check the Nodal analysis for stability.
- Compare the Nodal solution rate to the critical velocity requirement at the top of the tubing.
- Compare the Nodal solution rate to the critical velocity at the bottom of the flow string. For a constant diameter string, if the velocity is acceptable at the bottom of the string, then it will be acceptable.
- Ensure that the flow correlation used to calculate the Nodal solutions is appropriate for your well conditions by comparison to some measured data if available.

In this example, the critical velocity at the bottom of the tubing limits the choices to the 1.5 in ID tubing or smaller when considering critical velocity.

5.4 SIZING TUBING WITHOUT IPR INFORMATION

In the previous analysis, we used Nodal Analysis to evaluate different tubing options. This is the best way to design a tubing string provided that you have a good IPR curve. But you do not have to have an accurate representation of the reservoir or IPR curve, nor do you have to run a reservoir model to make choices on the tubing size.

If you know where the well is flowing now, you can calculate the tubing curves for the current tubing string and see if you are currently flowing to the right or the left of the minimum in the tubing J-curve. If you are flowing to the left of the minimum in the tubing curve, you can investigate different tubing sizes and generate curves where you would be expected to flow to the right of the minimum curve. You can make these evaluations without having a reservoir curve (IPR curve) or running a reservoir model.

If you do have a reservoir IPR curve to work with and the tubing curves intersect and match actual rates, then you can have more confidence in the results. But the reservoir curve is not necessary to analyze stability and critical velocity requirements.

Critical rates typically are evaluated at surface conditions. However, you can also calculate the downhole flowing pressures and enter the critical rate correlations for downhole conditions as shown in Table 5-2. You should especially make these calculations if you have any larger diameter flow paths such as casing flow up to the entrance to the tubing. However it is almost certain that for wells on the verge of loading, which flow up the casing, they will be well below the critical velocity. However, if the length of casing flow from perforations to the tubing intake is not too long, then even if it is flowing below critical, the net additional pressure drop may not be too large. A Nodal program that can model flow string diameter changes with depth can analyze this situation. You should check downhole critical flow in any case as a precaution even if the tubing size is constant down to the perforations. A critical rate that is acceptable at the bottom of the tubing means that it will be acceptable for the rest of the tubing or conduit.

5.5 FIELD EXAMPLE 1—RESULTS OF TUBING CHANGE-OUT

Regardless of the precautions listed in this section, there are many success stories related to coiled tubing and smaller tubing installations as illustrated earlier. Still, economics must be considered and one must be careful to consider whether a velocity string is the best method for long term results. Other methods should be investigated to see if they would provide similar or greater rate benefits and perhaps require fewer modifications to the well over a period of time.

5.6 FIELD EXAMPLE 2—RESULTS OF TUBING CHANGE-OUT

Dowell/Schlumberger published results from several case histories of using smaller tubing. A summary of some cases is shown in Table 5-4, from the Dowell/Schlumberger report [4].

A typical production chart from the Dowell/Schlumberger report is reproduced in Figure 5-3. Clearly, dramatic improvements can be made by the proper and timely installation of coiled tubing in wells that are experiencing liquid loading.

Table 5-3
Additional Field Results from Installation of Smaller Tubing
(after Wesson [3])

		Typical Velocity String Results			
Well	Production String Size (In.)	Perforation Depth (Ft)	Initial Production	Velocity String Size	Post Production (In.)
1	2-7/8″	8,200	40 Mcfd	1-1/4″	500 Mcfd
			4 BLPD		8 BLPD
2	2-7/8″	12,600	80 Mcfd	1-1/4″	200 Mcfd
			1-2 BLPD		10 BLPD
3	2-7/8″	13,000	50 Mcfd	1-1/4″	350 Mcfd
			2 BLPD		10 BLPD
4	2-7/8″	13,300	140 Mcfd	1-1/4″	300 Mcfd
			3 BLPD		6 BLPD
5	2-7/8″	13,300	Dead	1-1/4″	250 Mcfd with soap injection
6	2-7/8″	11,380	150 Mcfd	1-1/4″	155 Mcfd
			6 BOPD		12 BOPD
7	3-1/2″	11,860	8 Mcfd	?	255 Mcfd
			2 BLPD		26 BLPD
8	2-7/8″	11,850	25 Mcfd	1-1/4″	419 Mcfd
			4 BOPD		19 BOPD
9	2-7/8″	11,365	350 Mcfd	3865′-1-1/4″	450 Mcfd
			50 BWPD	7500′-1-1/2″	115 BWPD
10	2-7/8″	9,475	167 Mcfd	1-1/4″	533 Mcfd
			2 BOPD		5 BOPD
11	2-7/8″	9,415	167 Mcfd	1-1/4″	367 Mcfd
			No liquid		2 BLPD
12	2-7/8″	16,250	100 Mcfd	1-1/2″	425 Mcfd
			No Liquid		3 BLPD
13	3-1/2″	14,900	440 Mcfd	4900′-1-3/4″	750 Mcfd
			35 BWPD	10000′-2″	50 BWPD
14	3-1/2″	12,938	250 Mcfd	4538′-1-1/2″	575 Mcfd
			1 BWPD	8400′-1-3/4″	2.5 BWPD

5.7 PRE- AND POST-EVALUATION

One additional method of evaluation of a prospect for smaller tubing is the decline curve. This does not determine what size tubing should be used, but it does show if the production is sharply dropping off and it can be due to liquid loading. From the previous discussions of Nodal and Critical Velocity concepts, you can then analyze the well to see if decline drop-offs are due to liquid loading. Or you might run a pressure

Table 5-4
Coiled Tubing Installation Results

Well Name	Tbg OD"	CT OD"	Feet to Perfs	BHP (psi)	Mscf/D Prior CT	Mscf/D With CT	NPV ($)	Time to Pay-out Days	Other Information and Benefits
Well #1	3½	1¼	12,700	1,400	220	340	28,200	34	Pulled Compressor Rates stabilize after job
Well #2	2⅞	1½	14,200	750	400	390	11,406	119	Good example—Rates stabilize after job
Well #3	2⅞	1½	15,360	1,200	200	400	6,900	90	Simplified well ops after job. Good prod. response
Well #4	2⅞	1¼	13,500	3,400	185	175	23,266	98	Delay compressor Inst. rates stabilize after job
Well #5	2⅞	1½	9,430	1,100	625	550	14,345	123	Delay compressor Inst. rates stabilize after job
Well #6	2⅞	1¼	12,390	1,700	370	360	11,917	124	Delays compressor Inst. Simplifies well ops after job
Well #7	3½	1½	12,580	500	450	560	23,141	96	Rates stabilize well
Well #8	2⅞	1¼	12,600	400	80	180	13,534	132	Well almost dead before job. Rates stabilize
Well #9	2⅞	1½	16,380	1,400	280	170	-5,499	DNPO	CT well above perfs (270' of 5")
Well #10	3½	1½	14,250	400	450	280	-32,000	DNPO	Gauge sheets did not show need for CT. CT above perfs (160')
Summary			13,339	1,225	326	341	9,521	102	

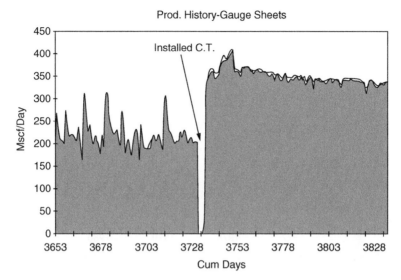

Prod. History-Gauge Sheets

Figure 5-3: Example of Rate Change after Coiled Tubing Installation[4]

bomb in the well to see if the well has liquid loading near the bottom of the well. Early remedial action will eliminate some problems, and any actions taken later will not show quite as dramatic an effect on production.

Also be sure to check for holes in the tubing before making any well evaluation. This is especially true if the well has no packer because some liquids can fall back and reload the tubing. If there is a packer, then a hole will just allow the casing to pressure up, which could be a casing integrity problem or corrosion problem.

Figure 5-4 shows results of a study for a 10,000′ well (after Weeks [5]).

Note that in late 1978–early 1979, a sharp production decline is evident. It was determined that this sharp decline was due to liquids loading. A small string of coiled tubing was installed and the shallower decline curve was measured after installation.

Note that the production did not increase that much immediately after the installation of the small diameter tubing. If the decline curve information was not available, then the installer of the small diameter tubing might have thought that the installation did not cause a very favorable production response from the well. The point is that after installation of small tubing (or even plunger, gaslift, foam, or other methods) the production may or may not increase very much. However,

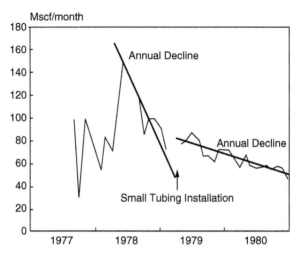

Figure 5-4: Example of Slope Change in Decline Change after Coiled Tubing Installations [5]

if the steep decline curve is arrested and a flatter decline with less interference from liquids loading is achieved, then the installation is a success and more recoverable reserves will be a consequence.

Therefore always try to get decline curve information before installation of small tubing (or other methods of dewatering) and then keep post installation decline curve data to properly evaluate the installation.

Figure 5-5 shows the completion corresponding to the decline curve in Figure 5-4.

Although 1-inch coiled tubing was (initially) successful in this case history, use 1-inch string with caution. When the tubing is this small, an intermittent slug of liquid can load the tubing and it can be difficult or impossible to get the string unloaded again.

There is one other caution to consider when viewing decline curves as indicated by the decline curve of Figure 5-6. At first glance, this curve appears to be a fairly normal decline curve without a sharp break in the curve to a steeper decline. Therefore one might conclude that this well has no liquid loading problems.

However, after some diagnosis, this well, even though it has a smooth downward decline, was found to be liquid loaded. In fact it was liquid loaded from the first day of production. It has a smooth decline because it is always liquid loaded and did not show the characteristic change in decline rate from no loading to liquid loading conditions. Since the

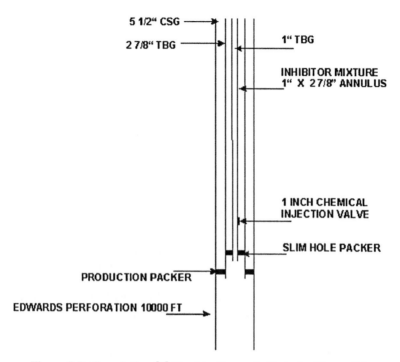

Figure 5-5: Completion [5] Used to Generate Data for Figure 5.4.

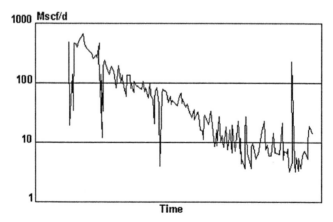

Figure 5-6: Rate vs. Time: Well That Is Liquid Loaded [3]

decline rate change is not observed, it was assumed that the well was not liquid loaded. This well is capable of producing a higher rate on a shallower decline curve than the one shown earlier.

5.8 WHERE TO SET THE TUBING

It is recommended to set the tubing at the top of the pay but not past the top one-third of pay. If the tubing is set too deep, liquid could collect over the perforations during a shut-in. When the well is brought back on production, the relatively large liquid volume in the casing-tubing annulus must be displaced into the tubing, making the well difficult, if not impossible, to flow due to a high fluid level in the tubing. Also if the tubing end is set below the perforations, then pressure buildup during shut-in cannot push liquids below the tubing end or near the tubing end since there is no place for the liquids to enter the formation.

5.9 HANGING OFF SMALLER TUBING FROM THE CURRENT TUBING

Sometimes tubing is landed high (Figure 5-7) and the flow though the casing below the tubing end is most likely well below critical, creating an extra pressure drop due to liquid loading of the casing.

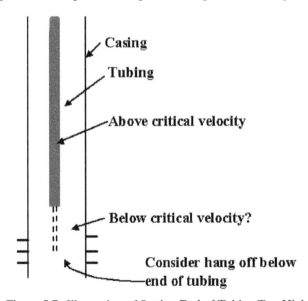

Figure 5-7: Illustration of Setting End of Tubing Too High

If wells are completed with considerable casing flow before you come up to the tubing intake or if there is a very large pay interval and the tubing is currently set at the top of the pay, it may be beneficial to hang off a section of smaller tubing from the end of the current tubing end to a deeper well depth. There are at least two systems [6,7] that will allow you to hang off a smaller tubing from the end of the current tubing.

The two tools are the double grip hydraulic set or wireline set packer for suspending coiled tubing [7]. The packer can be set hydraulically (Figure 5-8) or by a charge using an electric line—set similar to the

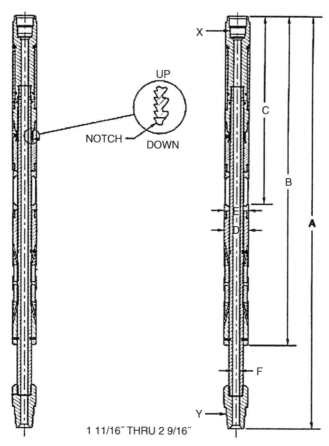

Figure 5-8: A Hydraulic Set Packer [7] That Can Be Run Inside Existing Tubing to Hang of a Smaller Section of Coiled Tubing to Eliminate Areas of Flow Below Critical Velocity Below High Set Tubing

Baker Model D packer. For the hydraulically set packer, there is a stinger torn disconnecting the CT above the packer after it is set.

5.10 SUMMARY

Use of smaller tubing can be successful. It is usually successful at rates of several hundred Mscf/D as opposed to smaller rates where plunger lift might be used. It has fewer problems if the tubing installed is well above 1 in ID, considering limitations on tools that can be run and methods used to possibly unload the well.

- Size the tubing using Nodal Analysis for stability and critical rate for minimum velocity and use the conservative higher rates indicated for a particular tubing size.
- Look at critical rate both at surface and downhole.
- Be sure that the tubing size selected flows above critical velocity from top to bottom and that the tubing is landed such that no large tubulars below the tubing bottom contribute to liquid loading.
- Do not land the tubing below the perforations but rather at the top or in the top one-third of the pay to avoid large liquid slugs on startup.
- A packer will avoid annulus pressure cycling, but planning ahead for possible plunger lift would dictate a completion without a packer.
- Even if the well is producing what seems to be an acceptable rate, check for liquid loading.
- Be cautious with velocity strings as you can easily design for too much friction on the smaller diameter side and design for downhole liquid loading on the larger diameter side. Remember it may not be as permanent a solution as other methods. Try to project performance to future well conditions.
- Analyze any changes in tubing size not only from the immediate rate that is obtained, but also from the slope of the new decline curve before conclusions are reached on a new completion.

5.11 REFERENCES

1. Turner, R. G., Hubbard, M. G., and Dukler, A. E. "Analysis and Prediction of Minimum Flow Rate for the Continuous Removal of Liquids from Gas Wells," *J. Pet. Tech.*, Nov. 1969, 1475–1482.

2. Coleman, S. B., Clay, H. B., McCurdy, D. G., and Norris, H. L. III. "A New Look at Predicting Gas-Well Load Up," *J. Pet. Tech.*, March 1991, 329–333.

3. Wesson, H. R. "Coiled Tubing Velocity /Siphon String Design and Installation," 1st Annual Conference on Coiled Tubing Operations & Slimhole Drilling Practices, Adams Mark Hotel, Houston, Texas, March 1–4, 1993.

4. WIS Solutions on "Coiled Tubing Velocity Strings: A Simple, Yet Effective Tool for the Future Technology," Schlumberger, Dowell.

5. Weeks, S. G. "Small Diameter Concentric Tubing Extends Economical Life of High Water-Sour Gas Edwards Producers," SPE 10254, presented at the 56th Annual Fall Technical Conference and Exhibition of the SPE of AIME, San Antonio, Texas, Oct. 5–7, 1961.

6. Campbell, J. A. and Bays, K. "Installation of 2 7/8-in. Coiled-Tubing Tailpipes in Live Gas Wells," OTC 7324 presented at the 25th Annual OTC in Houston, Texas, USA, May 3–6, 1993.

7. Petro-Tech Tools, Inc., Houston, Texas.

CHAPTER 6

COMPRESSION

Larry Harms

Larry Harms is a Principal Production Optimization Engineer for Cono-coPhillips. Larry has over 25 years of experience in the application of compression to optimize production. He has conducted numerous training courses for operations, maintenance, and engineering personnel on compression, production optimization, and gas well deliquification.

6.1 INTRODUCTION

Compression is crucial to all gas well production as it is the primary means to transport gas to market. Compression is also vital to deliquification, lowering wellhead pressure and increasing gas velocity. The lower bottomhole producing pressure from deliquifying wells and lowering surface pressures with compression can result in substantial production and reserves increases. These increases can range from a few percent to many times the current production. This uplift requires investment for the compressor and associated equipment as well as operating costs for the maintenance and power to continue running the compressor. However, many times compression can be the most economical way to keep wells deliquified, providing higher production rates at lower pressures.

Compressing associated gas in oil wells often is seen as a simple "rate acceleration" project that seldom has good economics. The argument has been made successfully that compressing gas reservoirs exposes a significantly larger portion of the original gas in place (OGIP) to production and it actually adds significant reserves. This phenomena is very pronounced in CBM and other adsorbtion reservoirs. Without compression, recovery from the San Juan Basin "fairway" (see Chapter 13) would have been under 50 percent. With compression the recovery will be closer to 95 percent of OGIP, a change that approaches 6 TCF increased production.

The process of choosing how to apply compression and the proper equipment to achieve the desired pressures and rates is important in optimizing results. Fortunately Systems Nodal Analysis can be used effectively to help in the process of evaluating wells and compression equipment.

Compression and reduced surface pressure is usually the first tool used in the life of a gas well to keep it deliquified and sometimes the only artificial lift method used, but compression can also be used to increase the effectiveness of other artificial lift deliquification methods including foamers, gas lift, beam pumping, ESPs, and velocity strings. When applying compression or any deliquification method it is important to insure that downstream equipment has sufficient capacity to insure uplift to the overall production.

There are many different types of compressors, each of which has its own operating ranges, efficiencies, strengths, and weaknesses. A majority of the applications for gas well deliquification involve the use of reciprocating or screw compressors.

6.2 COMPRESSION HORSEPOWER AND CRITICAL VELOCITY

As was noted in Chapter 4, critical velocity is directly proportional to the surface pressure. A reduction in surface pressure requires energy (horsepower). Compression horsepower is related to the ratio of the discharge and suction pressures in psia commonly known as the compression ratio.

Table 6-1 shows the horsepower required to compress gas at different pressures using a multistage reciprocating compressor to pipeline conditions of 1000 psig. Note that as suction pressures are reduced and compression ratios increase, the amount of horsepower increases dramatically. Also note in Table 6-1 that the amount of fuel gas that would be required to drive a natural gas engine to power the compressor is almost 6 percent of the gas being compressed at 0 psig (14.7 psia) even with efficient equipment.

Table 6-1
Compression Horsepower and Fuel Gas

Suction, psig	Suction, psia	Discharge, psia	Compression Ratio	Horsepower/ MMCFD	% Fuel Gas Required
0	14.7	1014.7	69.0	309	5.9%
10	24.7	1014.7	41.1	253	4.9%
25	39.7	1014.7	25.6	216	4.2%
50	64.7	1014.7	15.7	181	3.5%
125	139.7	1014.7	7.3	130	2.5%
300	314.7	1014.7	3.2	75	1.4%

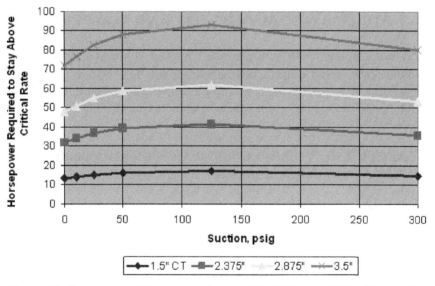

Figure 6-1: Compression Horsepower Required for Different Tubing Sizes to Stay Above the Critical Rate

By combining the amount of horsepower required at a given pressure with the critical velocity or rate (see Chapter 3) required to keep a well deliquified, it is possible to identify the minimum amount of horsepower required to keep any well deliquified. This is shown in Figure 6-1 for different tubing sizes assuming again a 1000 psig pipeline pressure and a reciprocating compressor.

A similar evaluation can be done for any specific well to determine the horsepower that will be required to keep the well deliquified based solely on well fluids, surface pressures, and compressor performance. However, this evaluation shows only the minimum horsepower required and neglects the well's performance. Also neglected are the limits on the specific compressor's performance. In order to assure that the well will respond as desired it is necessary to include the well's performance characteristics as well as the specific compressor's performance.

Considering the fuel gas used as well as the increased size and cost of the compression equipment needed to keep gas flowing above the critical rate, it is important to closely scrutinize the economics of using compression to keep a well unloaded.

It is therefore very useful to apply System Nodal Analysis to evaluate the current potential uplift and future results expected from compression.

6.3 SYSTEMS NODAL ANALYSIS AND COMPRESSION

Systems Nodal Analysis tools (see Chapter 4) are ideally suited to evaluating the effect of reducing the surface tubing pressure using compression. The following is an example of how Systems Nodal Analysis can be used to evaluate a specific well and compressor options.

Example 6-1: Wellhead Compressor Option Evaluation Using System Nodal Analysis

Well Depth—10,000 ft.
Bottom hole Temperature—180° F
Surface Flowing Temperature—80° F
Surface Flowing Pressure—125 psig
Gas Gravity—.65
Water Gravity—1.03
Condensate Gravity—57 API
Reservoir Pressure—600 psia
Reservoir Backpressure n—.97
Reservoir Backpressure C—.0015 Mscf/D/psi2
C and n used in "Back Pressure Equation for Gas Flow", q, gas = C $(Pr^2-Pwf^2)^n$
Pr is reservoir shut in or average pressure, psia
Pwf is producing pressure at perforation mid-point, psia
Tubing Flow Correlation—Gray

The example well has loaded up and will not flow continuously but averages about 200 MCFD from intermittent flow periods. Systems Nodal Analysis for this well is shown in Figure 6-2. The solution point indeed falls to the left of the inflection point, which represents the minimum rate for stable flow (see Figure 4-8).

In addition, Figure 6-2 shows that the well can become stable with reduced surface pressure. One way to more easily see this is to plot the solution points from the Systems Nodal Analysis well prediction and the critical rate calculated for the tubing size and fluid in the well per Coleman as shown in Figure 6-3.

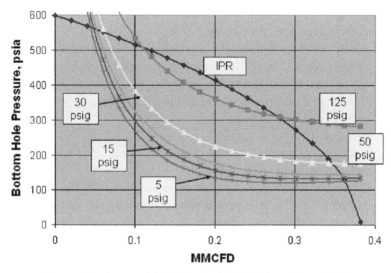

Figure 6-2: Systems Nodal Analysis for Well in Example 6-1

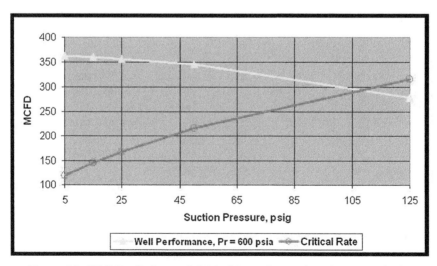

Figure 6-3: Well Prediction and Critical Rate Comparison for Example 6-1

It can be seen in Figure 6-3 that the wellhead pressure must be reduced to about 105 psig to become stable; however, additional flow and stability can be obtained by reducing the pressure further.

A wellhead compressor can provide this reduction. Figure 6-4 shows the well's performance at 600 psia reservoir pressure and the critical rate along with the performance curves for two different wellhead compression units. Additionally the well's performance at 400 psia reservoir pressure is shown. It is assumed that the compressor would have a 125 psig discharge pressure into the current gathering system and no pressure drop in the suction piping.

Inspection of these curves shows that either unit would have the ability to deliquify the well with an increase in rate. Compressor A, the larger unit, would result in a wellhead pressure of about 6 psig at a rate of 362 MCFD, and compressor B would provide a rate of 360 MCFD at a wellhead pressure of 13 psig.

As Section 6.2 showed, the increase of fuel gas must be considered at the lower pressures, and indeed in this case reducing the pressure to 6 psig instead of 13 psig results would burn more fuel gas than the 2 MCFD production increase, resulting in a net reduction of rate from reducing the wellhead pressure an additional 7 psig. Of course the natural gas engine driving the compressor can be slowed down to control the capacity and optimize the performance and fuel gas use. In addition the larger Compressor A is more expensive to purchase or rent.

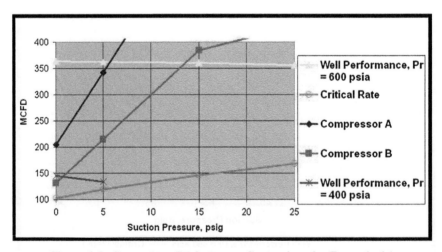

Figure 6-4: Compressor Performance Comparison for Example 6-1

Since either compressor will work to deliquify the well and increase production, and the smaller compressor B should cost less and consume less fuel gas, it seems to be the obvious choice. However future performance must be considered. As the reservoir depletes will compressor B still be the best choice to keep the well unloaded, and what reduction in reservoir pressure can be expected before the well begins to have loading problems again?

In order to investigate this, use the well performance curve at a reservoir pressure of 400 psia in Figure 6-4. This shows that Compressor B still has enough capacity to keep the well deliquified to deplete the reservoir to at least 400 psia and slightly below. Additional runs dropping the reservoir pressure to 380 psia shows that this is the point where there is not enough productivity for the well to stay unloaded and it can be expected to stop flowing steadily again. This fluctuating flow situation can be particularly difficult for gas engine driven wellhead compressors unless they have sufficient bypass capacity and fuel gas supply to run through the low flow periods (see Section 6.11).

In order to do a more thorough evaluation of this situation, an Integrated Production Model [1] (IPM), which incorporates a material balance model of the reservoir, as well as the wellbore, compressor performance curve, and surface facilities can be used to predict flow streams for different cases, allowing an economic comparison.

The well in Example 6-1 can be considered a tight (low permeability) well so the main value from the compression is in keeping the well unloaded with small uplifts seen from the incremental lowering of wellhead pressure past this point. There can be substantial differences when evaluating higher permeability wells.

6.4 THE EFFECT OF PERMEABILITY ON COMPRESSION

The goal of compression can be expanded from keeping the well deliquified to including a significant acceleration component in higher permeability/productivity wells. Also high productivity wells do not need compression to stay deliquified until significantly lower reservoir pressures. These differences can be very important to optimizing the effect of compression on different productivity gas wells.

This is shown in Table 6-2, where Wells L and H are identical except for the difference in permeability and stimulation type.

Systems Nodal Analysis shows that lower permeability Well L can be expected to experience liquid loading at a substantially higher reservoir

Table 6-2
Comparison of Well's Permeability and Compression Uplift

	Well L	Well H
Perm., md	0.2	2
Reservoir thickness, ft.	100	100
Skin	−3	0
Depth, ft.	7000	7000
Tubing Diameter, in.	2.875	2.875
Surface Pressure, psig	500	500
Critical Rate, MCFD	900	900
Reservoir Pressure at Critical Rate, psia	1500	870
Increase from drop to 100 psig surface pressure, psig	200	1100

pressure. Also if compression is put on Well L just as the well reaches the critical rate in order to prevent liquid loading, a modest accompanying uplift of about 200 MCFD is the expected result. Well H can be expected to provide a much higher 1100 MCFD uplift even though it is at a much lower reservoir pressure. This indicates that it might be worthwhile to install compression sooner on this well depending on the economics of accelerating the production.

Since higher permeability wells are so sensitive to pressure, it is imperative that these wells be monitored continuously for liquid loading and their performance anticipated and optimized. Compression is often the primary or most important artificial lift method on these type wells.

A dependable rule of thumb is that high productivity, high cumulative production, low pressure wells make the best compression candidates.

That is not to say that compression is unimportant for tighter gas wells. Especially on fairly tight gas wells that have been allowed to liquid load and produced intermittently and/or to flow in bubble flow at low average flow rates for a long period of time, large changes in producing bottom hole pressure can occur with compression, making it attractive to reduce surface pressures on these wells also [2].

In all cases IPM can be a very useful tool to determine the increased reserves and acceleration component over time for different compression options in order to optimize the value of compression.

6.5 PRESSURE DROP IN COMPRESSION SUCTION

Because the goal of compression is to transmit the suction pressure of the compressor all the way to the bottom of the well, anything that causes a pressure drop between the wellhead and the compressor is undesirable. Surface restrictions increase the horsepower used and/or result in a reduction in uplift or quicker liquid loading because the pressure reduction at the wellhead is reduced. In extreme cases surface restrictions may completely choke the flow, resulting in a very small reduction of pressure at the wellhead even though the compressor has a low suction pressure.

Examples of surface restrictions include small diameter piping, pipe elbows, chokes, orifice meters, and suction control valves. Although all these restrictions may not be eliminated, they can be minimized.

As an example, a well with a 2 in. flow line produces 390 MCFD up 2.875 in. tubing. This well is below the critical rate and analysis shows the rate can be increased to 650 MCFD if the tubing pressure is dropped to 17 psig. The pressure drop through the flow line was measured at 21 psi for the current average rate conditions. Pressure drop calculations show that it would be impossible to reach the target conditions with the current flow line used as the compressor suction due to choked flow. Increasing the pipeline size to 3 in would result in 6 psi pressure drop at the expected 650 MCFD and 17 psig compressor suction, but the best economic decision is to install a 4 in flow line with an expected pressure drop less than 2 psi. System Nodal Analysis as well as IPM can be helpful in modeling these situations. Using the actual pressure drop as a basis for the model can improve the accuracy of the results.

A suction control valve is installed upstream of compressors to prevent horsepower and mechanical limits from being reached. Unfortunately good control requires some pressure drop to occur. When choosing these valves lower pressure drop designs such as V-Ball or butterfly valves may be appropriate to achieve good control with minimal pressure drop.

Discharge piping restrictions are not as large a factor as the suction piping because they have smaller effects on the compression ratio and thus horsepower. However, discharge piping systems should always be analyzed to determine the expected effect and whether changes to reduce pressure loss can be economically justified.

6.6 WELLHEAD VERSUS CENTRALIZED COMPRESSION

Because there will always be some pressure drop in piping, no matter the size, between a well and the compressor suction, minimizing the distance helps to minimize the pressure drop. The ability to have low losses on the suction explains why wellhead compression is always potentially more efficient on a hydraulic basis than centralized compression.

Offsetting this is the increased fuel and mechanical efficiencies along with reduced capital cost per horsepower of installed compression that can be obtained with a few larger horsepower centralized units when compared with many smaller wellhead units. Also costs to operate and maintain larger centralized units is less per unit of compression capacity.

However, as has been shown, when tight gas wells or wells in one operating area with significant contrast in productivity are involved, customizing the compression to the individual well can be attractive to optimize rates, fuel, and operating cost.

In most cases with multiple gas wells being gathered in the same system, a combination of wellhead compression and centralized compression will provide optimum economic results.

6.7 DOWNSTREAM GATHERING AND COMPRESSION'S EFFECT ON UPLIFT FROM DELIQUIFYING INDIVIDUAL GAS WELLS

A frequent subject of debate when deliquifying gas wells in a field (or actually when any production increasing project is done) is whether there has been a true uplift in not only the individual well but the overall field's production. This is an area of considerable concern when using any of the deliquification methods. In order to insure that true uplift is seen from deliquifying an individual well there must be adequate gathering piping size and compression capacity to result in minimal pressure increases on the other wells in the system. This can be confirmed using a Systems Nodal Analysis of the complete system but should be examined with at least a simple piping and compressor analysis to assure that the method of deliquifying the well will result in the expected overall uplift demanded to provide economic success.

Consider the following example; a well has loaded up in the 1000 psig gathering system and needs to be put into the low pressure system to unload and increase flow. A check of the compression system shows that there is adequate compression horsepower to put the well on compression at 85 psig, which will unload the well. Upon switching the well to the low pressure system it is found that wellhead pressure dropped only to 180 psig, but the well unloaded with an uplift of 500 MCFD. Unfortunately the increase in pressure on the downstream gathering system resulted in more than a 500 MCFD loss from other wells on this same system so that overall production actually decreased. In this case the well was put back into the high pressure system until an additional flow line could be used to reduce the effect of putting the high pressure well into the low pressure system in the future.

6.8 COMPRESSION ALONE AS A FORM OF ARTIFICIAL LIFT

Sometimes compression is the most economical, lowest risk choice as the sole artificial lift method. This may be true in wells that are sand producers, making them prone to operating problems with capillary strings, plungers, or pumps and sand cleanouts with the intermittent flow from plunger lift. Also mechanical problems in the tubulars such as holes or restrictions may make it very expensive and risky to install other artificial lift types. In some cases the available expertise in the area to operate plunger lift or pumping systems may be insufficient dictating only compression be used.

The substantial risk of reducing the well's productivity during any work-over operations that may be required to run other artificial lift methods may also swing the best choice to staying with compression only. This can be particularly true in low pressure, high productivity reservoirs in which the reduction of flow area from the insertion of additional equipment in the well will also potentially limit the rate due to pressure drop. At very low pressures the extra water vapor carrying capacity of the gas or even evaporation can also be helpful. However, in many cases lower wellhead pressures with compression is an aid to the other types of artificial lift.

6.9 COMPRESSION WITH FOAMERS

Compression can aid foamers because reducing the wellhead pressure also reduces the density of the gas and increases agitation (see Chapter 8). When significantly reducing wellhead pressures, checks on the effect of the lower pressures on foam quality, foam stability, foamer performance, and rate should be conducted.

6.10 COMPRESSION AND GAS LIFT

Gas lift and compression almost always are used concurrently as the high pressure gas used for lift gas injection usually is provided by the discharge of a compressor (see Chapter 11). There is an optimum combination of wellhead pressure, gas lift injection pressure, gas lift injection rate, and compression requirements that can be found for any specific well. Many times this optimum is at a low wellhead pressure (below 100 psig), which allows reduced lift gas injection rates.

A limit to the advantage of lower wellhead pressures in gas lift is seen when the velocity and friction pressure drop become too large a factor. Systems Nodal Analysis can be very helpful in these evaluations.

6.11 COMPRESSION WITH PLUNGER LIFT SYSTEMS

Plunger lift is one artificial lift system that can benefit greatly from compression. The basics of plunger lift are discussed in Chapter 7 where it is pointed out that lower pressure at the wellhead is very desirable. Both gas powered and electric compressors have been shown to have application in plunger lift installations.

Figure 6-5 shows a simple schematic of a plunger lift installation equipped with a surface compressor. The compressor is switched on to lower the wellhead pressure when the well is opened while the plunger is arriving and during the after flow period. The compressor is then shut

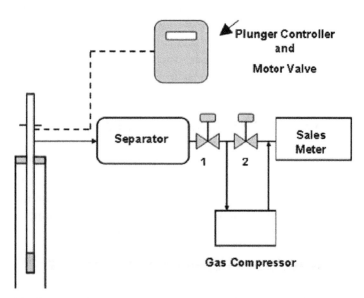

Figure 6-5: Compression System Installation with Plunger Lift (Phillips and Listiak [3])

down or in full recycle while the well is shut in to build up pressure to bring the plunger up for the next cycle.

Plunger lift is sometimes thought of as an intermittent gas lift system where the lift gas is provided by the reservoir. Sometimes the reservoir gas is supplemented taking gas from the compressor discharge or other external gas and injecting into the well annulus, allowing the plunger to be cycled more frequently.

Figure 6-6 shows production data for a plunger lift installation equipped with compression. The initial production followed a fairly steep decline until it was put on compression, at which time the oil and gas increased markedly then fell off. A sustained uplift was achieved after the plunger lift was installed four months later and operated in conjunction with the compressor.

Morrow [4] further discusses compression with plunger lift.

6.12 COMPRESSION WITH BEAM PUMPING SYSTEMS

In a beam pump system, the liquid production is governed primarily by the downhole stroke, the number of strokes per minute (SPM), and the pump size. Lowering the casing head pressure (CHP) on pumping well producing liquids up the tubing and gas up the annulus will bring

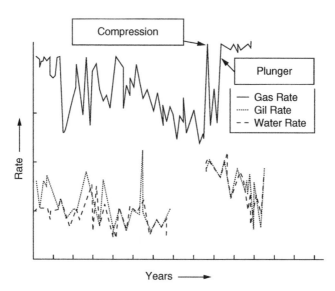

Figure 6-6: Performance Improvement Using Plunger Lift and Compression (Phillips and Listiak [3])

up the liquid level in the annulus and allow you to pump faster if desired. If the pump is set below the perforations and there is adequate liquid over the pump but all liquid is below the perforations, then the producing pressure is a function of the CHP. This emphasizes that pumping liquids lowers the producing pressure, but if the CHP remains high the producing pressure on the perforations is always a little higher than the CHP. This shows that pumping *and compression* are needed for low producing pressures on the formation. Lowering the pressure on the tubing will allow the liquid to be removed with less pumping system horsepower.

Figure 6-7 compares the effect of three wellhead pressures on the liquid level in the tubing casing annulus. Note that these assume a constant production rate and flowing bottom pressure. The figure shows that as the CHP is reduced, the liquid level in the casing/tubing annulus is raised substantially. Conversely, high wellhead pressure puts the liquid level in the annulus low in the well near the pump intake. If the CHP is too high and annulus fluid level too low, there is a distinct possibility of incomplete pump fillage and lower pump efficiency. This can be overcome by putting the pump below the perforations if possible. Otherwise a low but adequate fluid level is needed. As reservoir pressures deplete, lower annulus fluid levels over time are expected with a fixed wellhead surface pressure.

Therefore, one way to insure proper pump fillage and more efficient pump operation is to lower the surface casing pressure by compression.

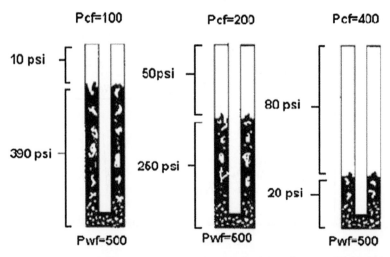

Figure 6-7: Pressure Relations on a Pumping Well with a Gaseous Fluid Column (McCoy *et al.* [5])

If the liquid level was initially low and pump fillage was not complete, compression will have the effect of increasing the well's production simply from higher pump efficiency.

In addition, since there is a higher liquid level in the annulus and better pump fillage with lower surface pressure, the unit can also be run at higher speeds and/or longer stroke lengths to thus lower the fluid level, which lowers the producing bottom hole pressure, resulting in higher production rates of liquids and gas. As shown in Section 6.4, this increase can be considerable if the well's productivity is high.

Increased pump fillage can also reduce pump failures and other associated failures, especially if "fluid pound" is avoided. See Chapter 10 for a complete discussion on gas separation and effects on the beam pump system.

6.13 COMPRESSION WITH ESP SYSTEMS

In general, electric submersible pumps (ESPs) operate with a fixed pressure between the pump intake and the pump discharge (see Chapter 12). This translates to a fixed pressure increase between the well's bottomhole flowing pressure and the wellhead pressure. Therefore, lowering the surface wellhead pressure on a typical ESP installation proportionally lowers the flowing pressure. The lower bottomhole flowing pressure increases production and/or lowers the power demand of the unit.

However again if the CHP is still high, even with all the fluid off the perforations (pump below perfs with motor cooling options), the producing pressure will never be lower than the CHP, emphasizing the need for compression on the casing.

6.14 TYPES OF COMPRESSORS

There are a number of compressor types that are used to lower the pressure on entire fields of gas wells, or to lower the pressure on individual gas wells.

For single well applications, the following list of compressor types may be used:

- Reciprocating
- Liquid injected rotary screw
- Liquid ring

- Sliding vane
- Rotary lobe
- Reinjected rotary lobe

Some of the following descriptions are taken from Thomas [6].

6.14.1 Rotary Lobe Compressor

Refer to Figure 6-8.

- Low cost per CFM.
- Air cooled.
- Approximately 2.0 compression ratio.
- Small amounts of liquid ingestion are acceptable.
- High displacement is achievable (50–12,000 cfm).
- Power frame supporting bearings, gears, shafts.

6.14.2 Liquid Injected Rotary Screw Compressor

Refer to Figure 6-9.

- Higher cost per cfm.
- Liquid injected.
- Approximately 10-20 compression ratio.
- Medium displacement.

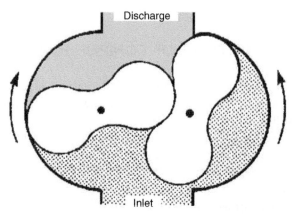

Figure 6-8: Elements of Rotary Lobe Compressor

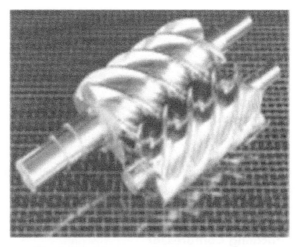

Figure 6-9: Elements of a Screw Compressor

- Power frame required.
- Requires seal oil cooling system.
- Requires gas/oil separator.
- Liquid ingestion dilutes seal oil.
- If not specifically designed for vacuum and low discharge pressure operation, then:
  - Questionable mechanical seals
  - High back pressure valve
  - High separator velocities
- Can handle very high compression ratios in one stage of compression as the oil absorbs the majority of the heat of compression. Excellent for very low suction pressure even down to vacuum. Oil cooling system required.
- Except for gear amplification, very few wearing parts, which provides very high reliability.
- Mechanical and adiabatic efficiency is high if unit is run at design conditions.
- Efficiency suffers if unit is run too far off of design conditions or if multiple stages are used.
- Discharge pressure limited by manufacturing choices, often the maximum pressure is less than 300 psig.
- Oil can become contaminated with heavy hydrocarbons and other liquids causing operational problems. Selection of proper oil type is absolutely critical. Test oil frequently for fines content.

6.14.3 Liquid Ring Compressor

- Medium cost per cfm.
- Liquid injected.
- Approximately 4.0 compression ratio.
- High displacement.
- Power frame required.
- Requires seal liquid cooling system (normally oil).
- Requires gas/liquid separator.
- Large amounts of liquid ingestion possible (but water will contaminate oil system requiring replacement of seal fluid).
- Generates about 25 psi delta pressure.

6.14.4 Reciprocating Compressor

Refer to Figure 6-10.

Figure 6-10: 10 HP Gas Engine Drive Reciprocating Compressor Package; Operating Conditions: Ps0/Pd50 Psig @40 Mscf/D

- High cost per cfm.
- Air or liquid cooled.
- Approximately 4.0 maximum compression ratio per stage.
- Low displacement/power frame.
- No amount of liquid ingestion allowed.
- Valve losses greatly affect compression ratio and volumetric efficiency but can have the highest efficiency.
- Most flexible of all compressors in that it can handle varying suction and discharge pressures and still maintain high mechanical and adiabatic efficiency within temperature and mechanical design limits.
- Overall compression ratio is dependent only on discharge temperature and rod load rating of frame. Units can be 2 staged (or even 3+ staged) to produce very high discharge pressures with low suction pressure.
- Level of knowledge required for maintaining unit can easily be obtained. Good engine mechanics can be good compressor mechanics.
- Potentially high operating expense and downtime due to compressor valve maintenance. This valve maintenance is highly dependent on gas quality (solid and liquid contamination), which can be a problem with well head compression.
- Not as efficient with very low suction pressures (vacuum).

6.14.5 Reinjected Rotary Lobe Compressor

- Low cost per CFM.
- Air cooled.
- Approximately 4.0 compression ratio (high vacuum).
- Small amounts of liquid ingestion are acceptable.
- High displacement is achievable (50–12,000 cfm).
- Power frame supporting bearings, gears, shafts.
- Requires intercooler.

6.14.6 Sliding Vane Compressor

- Medium cost per cfm.
- Liquid cooled (jacket).
- Approximately 3–4.5 compression ratio.
- Two stage units with higher compression ratios available.
- Medium displacement/power frame.
- Requires external oil lubrication system.

- Once through lubrication-oil leaves with gas.
- Tolerates no liquid ingestion.
- Low capital and expense cost unit, very simple operation.
- Simple design makes for easy and high availability (depending on as quality).
- Useful in VRU service.
- Bearings isolated from sour gas. Separate lube system.
- Blades wear on interior case, so compressor life is heavily dependent on gas quality and contaminants. Blades can get stuck in the case if many solids are present in the gas.
- Limited to lower discharge pressure and lower volume applications.

6.15 GAS JET COMPRESSORS OR EJECTORS

Gas jet compressors, or ejectors, are classified as thermocompressors and are in the same family as jet pumps, sand blasters, and air ejectors. They use a high-pressure gas for motive power. Ejectors using gas can impart up to two compression ratios; using liquid they can generate higher ratios if cavitation can be avoided.

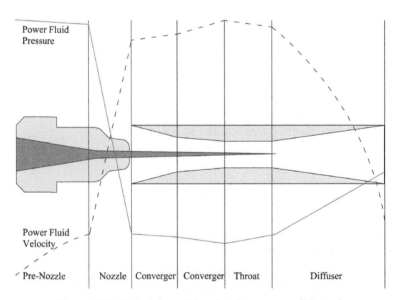

Figure 6-11: Principles of Gas Jet Compressor (Ejector)

The ejector, or gas jet compressor, operates on the Bernoulli principle as illustrated in Figure 6-11. The high pressure motive fluid enters the nozzle and is accelerated to a high velocity/low pressure at the nozzle exit. The wellhead is exposed to the low pressure at the nozzle exit through the suction ports and is mixed with the motive fluid at the entrance to the throat. Momentum transfer between the motive and produced fluids in the throat and velocity decrease in the diffuser increases the pressure to the discharge pressure.

Eductors have potential advantages including:

- No moving parts
- Low maintenance/high reliability
- Easy to install, operate, and control
- Can handle liquid slugs
- Low initial cost/payback time usually short
- Nozzle sizes can be changed to meet changing well conditions

Figure 6-12 shows an ejector in actual field service. One successful configuration uses a flooded screw compressor to pull the tubing/casing annulus down to 8–10 psig. A portion of the gas discharged by the compressor is used to drive an ejector to pull the tubing down to 1–5 psig.

Figure 6-12: Ejector Installed on a Wellhead [7]

The exhaust of the ejector is combined with the casing gas and sent to the compressor.

If high pressure fluid is available (e.g., from a nearby high pressure gas well) to power the ejector, then it is advantageous to utilize this wasted energy with an eductor to lower surface pressure on a lower rate well to prevent liquid loading.

The principal disadvantage of eductors is that they have a higher hp/MMCF requirement than other technologies (i.e., they have lower mechanical efficiency). This lower efficiency can often be offset by extremely low capital cost. For example, a well was limited to 600 MCF/d with 9 psig wellhead pressure by the 2-stage reciprocating compressor installed. The compressor had plenty of horsepower to move more gas but the piping and cylinder configuration did not allow lower pressures. Replacing the 300 hp compressor with a different machine would be very expensive so an ejector was added between the wellhead and the compressor to compress the full stream—basically adding a compression stage. This ejector lowered the wellhead pressure to −5 psig with 9 psig discharge pressure (atmospheric pressure at this site is 11 psia so the ejector developed 3.3 compression ratios) and increased the well's production to 900 MCF/d. The efficiency of the ejector is only 46 percent, but it reduced the capital outlay required by more than an order of magnitude.

6.16 OTHER COMPRESSORS

Other types of compressors continue to be developed or adapted for application on gas wells including multiphase pumps, which can act as compressors and scroll compressors. There is also an attempt being made to develop a down-hole gas compressor.

Discussion of centrifugal compressors, which are installed on all the largest gas lift, gas plant, and transmission applications, has not been included as these are used infrequently in normal gas field deliquification service.

6.17 SUMMARY

Compression can help a liquid loading well by increasing the gas velocity to equal or exceed the critical unload velocity, and also lowers pressure on the formation for more production by lowering the wellhead flowing pressure.

Because of the differing response that can be expected from different types of wells it is important that the compressor type and size be matched to the well. Systems Nodal Analysis can be a helpful tool to accomplish this.

Compression often is used on a field-wide basis to lower the gathering system pressure; however, for any compressor the amount of pressure reduction that can be transmitted back to the wellhead must be taken into account for optimal results.

Compression can be used as a primary artificial lift method or to aid the other types of artificial lift to different degrees.

There are many types of compressors that can be successfully applied to help deliquify gas wells. The key to attaining the best economic success in deliquifying gas wells is to match the compressor to the well's performance.

6.18 REFERENCES

1. Harms, L. K. "Installing Low-Cost, Low-Pressure Wellhead Compression on Tight Lobo Wilcox Wells in South Texas: A Case History," paper SPE 90550 presented at the 2004 SPE Annual Technical Conference and Exhibition, Houston, TX, Sept. 26–29, 2004.

2. Harms, L. K. "Better Results Using Integrated Production Models for Gas Wells," paper SPE 93648 presented at the 2005 SPE Production and Operations Symposium, Oklahoma City, OK, April 17–19, 2005.

3. Phillips, D. and Listiak, S. "Plunger Lifting Wells with Single Wellhead Compression," presented at the 43rd Southwestern Petroleum Short Course, Lubbock, Texas, April 23–25, 1996.

4. Morrow, S.J. and Aversante, O. L. "Plunger-Lift, Gas Assisted," 42nd Annual Southwestern Petroleum Short Course, Lubbock, Texas, April 19–20, 1995.

5. McCoy, J.N., Podio, A. L., and Huddleston, K. L. "Acoustic Determination of Producing Bottomhole Pressure," SPE Formation Evaluation, September 1988, 617–621.

6. Thomas, F. A. "Low Pressure Compressor Applications," presentation at the 49th Annual Liberal Gas Compressor Institute, April 4, 2001.

7. Simpson, D. A. "Use of an Eductor for Lifting Water," BP Forum on Gas Well De-Watering, Houston, TX, May 5, 2002.

PLUNGER LIFT

7.1 INTRODUCTION

Plunger lift is an intermittent artificial lift method that uses only the energy of the reservoir to produce the liquids. A plunger is a free-traveling piston that fits snugly within the production tubing and depends on well pressure to rise and solely on gravity to get back to the bottom of the well. Figure 7-1 illustrates a typical plunger lift installation.

Plunger lift operates in a cyclic process with the well alternately flowing and shut-in. During the shut-in period with the plunger on bottom, gas pressure accumulates in the annulus and liquids have mostly already accumulated in the well during the last portion of the flow period. Liquids accumulate in the bottom of the tubing, and the plunger falls through the liquids to the bumper spring to await a pressure buildup period. The pressure of the annulus gas depends on the shut-in time, reservoir pressure, and permeability. When the annulus pressure increases sufficiently, the motor valve is opened to allow the well to flow. The annulus gas expands into the tubing, lifting the plunger and liquids to the surface.

Conventional plunger, being the most common, has as part of the cycle a shut-in period where the plunger can fall and pressure can build in the formation and the casing. Continuous flow plunger has only a brief shut-in period to allow the plunger to fall out of the lubricator, and then flow commences as the plunger falls against the flow. Complete satisfactory cycles with a shorter shut-in period will result in more production. The two-piece plunger is one such plunger, but there are other plungers with valves and caged balls and seats

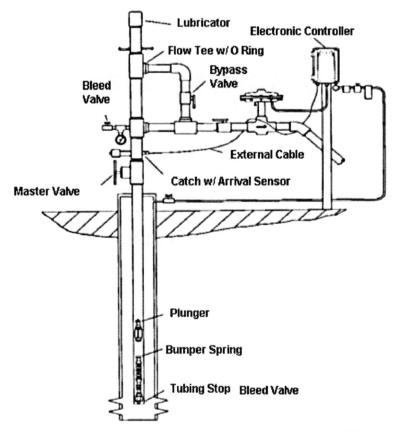

Figure 7-1: Typical Conventional Plunger Lift Installation [1]

that allow flow to bypass the seal mechanism when falling against flow.

The reservoir is allowed to produce gas until the production rate decreases to some value near the critical rate and liquids begin to accumulate in the wellbore. The well is then closed and the plunger falls back to the bumper spring, first through gas and then some accumulated liquid.

The pressure buildup period follows. Then using the gas pressure that has been allowed to build up in the annulus, the well is opened to production again, bringing the liquids and plunger to the surface. With the plunger at the surface, the well remains open and the gas again is allowed to flow until production rates begin to fall. The well is closed in and the plunger falls to bottom, repeating the cycle.

Figure 7-2 shows an approximate depth-rate application chart where plunger is shown to be feasible in the region of lower rates and depths identified by the curve. This is an approximate chart, as plunger lift has been operated successfully to depths of 20,000 feet.

A plunger lift system is relatively simple and requires few components. A typical plunger lift installation as in Figure 7-1 would include the following components:

- A downhole bumper spring, which is wirelined into the well to allow the plunger to land more softly downhole
- A plunger free to travel the length of the tubing
- A wellhead designed to catch the plunger and allow flow around the plunger
- A controlled motor valve that can open and close the production line
- A sensor on the tubing to sense arrival of the plunger
- An electronic controller that contains logic to decide how the cycles of flowing production and time of well shut-in period are determined for best production

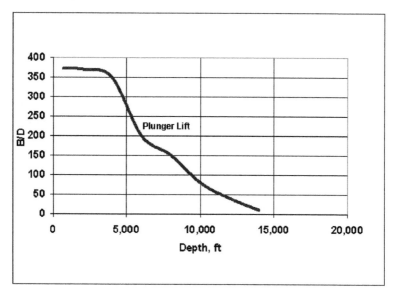

Figure 7-2: Approximate Depth-Rate Application Chart for Conventional Plunger Lift

7.2 PLUNGERS

Figure 7-3 shows some typical plungers that were tested to provide data for developing plunger lift system models [1]. These shown are typical but do not include all types of plungers available to the industry.

In this figure, the plungers are identified from left to right as:

1. Capillary plunger, which has a hole and orifice through it to allow gas to "lighten the liquid slug above the plunger."
2. Turbulent seal plunger with grooves to promote the "turbulent seal."
3. Brush plunger used especially when some solids or sand is present.
4. Another type of brush plunger.
5. Combination grooved plunger with a section of "wobble washers" to promote sealing.
6. Plunger with a section of turbulent seal grooves and a section of spring-loaded expandable blades. Also a rod can be seen that will open/close a flow-through path through the plunger depending on whether it is traveling down or up.
7. Plunger with two sections of expandable blades with a rod to open flow-through plunger on down stroke.
8. Mini- plunger with expandable blades.

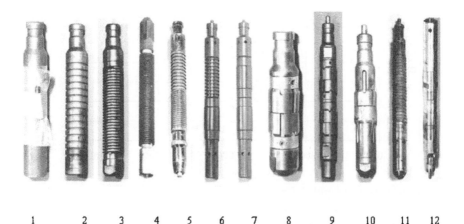

| 1 | 2 | 3 | 4 | 5 | 6 | 7 | 8 | 9 | 10 | 11 | 12 |

Figure 7-3: Various Types of Plungers

9. Another with two sections of expandable blades and a rod to open flow-through passage during plunger fall.
10. Another with expandable blades and a rod to open a flow through passage during the plunger fall and close it during the plunger rise.
11. Wobble washer plunger and a rod to open flow passage during the plunger fall.
12. Expandable blades with a rod to open a flow-through passage on the plunger fall that could fall against the flow and operate as continuous flow.

Several of these plungers have a push rod to open a flow passage through the plunger to allow flow through the plunger when falling to increase the fall velocity. When the plunger arrives at the surface, the push rod forces the flow passage open for the next fall cycle. When the plunger hits on bottom, the rod is pushed upward to close the flow passage for the next upward cycle.

The brush plunger was found in testing to show the best seal for gas and liquids, but it typically wears sooner than other plungers. The brush plunger is the only plunger that will run in wells, making a trace of sand or solids. Plungers with the spring-loaded expandable blades showed the second best sealing mechanism and they do not wear nearly as fast as the brush plunger.

7.3 PLUNGER CYCLE

Conventional plunger lift operates on a relatively simple cycle as illustrated in Figure 7-4. Figure 7-5 shows in more detail the casing and tubing and bottomhole pressures throughout one complete plunger cycle. The numbers on top of Figure 7-4 labeling the steps of the cycle are also provided on the figure for clarity.

1. The well is closed and pressure in the casing is building. When the pressure is enough to lift the plunger and the liquids to the surface at a reasonable velocity ($\approx$750 fpm) against the surface pressure, the surface tubing valve will open.
2. The valve opens and the plunger and liquid slug rise. The gas in the annulus expands into the tubing to provide the lifting pressure. Also the well is producing some during the rise time to add to the energy required to lift the plunger and liquid.

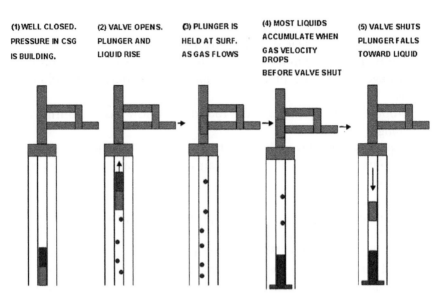

| (1) WELL CLOSED. PRESSURE IN CSG IS BUILDING. | (2) VALVE OPENS. PLUNGER AND LIQUID RISE | (3) PLUNGER IS HELD AT SURF. AS GAS FLOWS | (4) MOST LIQUIDS ACCUMULATE WHEN GAS VELOCITY DROPS BEFORE VALVE SHUT | (5) VALVE SHUTS PLUNGER FALLS TOWARD LIQUID |

Figure 7-4: Simplified Pictorial Illustrations of Plunger Cycle Events

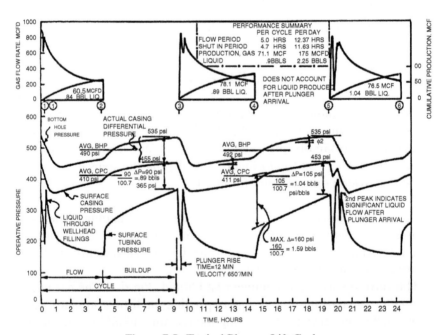

Figure 7-5: Typical Plunger Lift Cycle

3. The liquid reaches the surface and travels down the flowline. The plunger is held at the surface by pressure and flow. The gas is allowed to flow for some time.
4. The flow velocity begins to decrease and liquids begin to accumulate in the bottom of the well. The casing pressure begins to rise some, indicating a larger pressure drop in the tubing. If flow is allowed to continue too long, a "too large" liquid slug will accumulate in the bottom of the well, requiring a high casing buildup pressure to lift it.
5. The valve is shut. The plunger falls. The liquids are at the bottom of the well for the most part. The plunger will hit the bottom and the cycle will repeat.

The cycles continue and may be adjusted according to different schemes that may be programmed into the various controllers available.

7.4 PLUNGER LIFT FEASIBILITY

Field testing of various artificial lift methods to determine their applicability can be costly. Although plunger lift is a relatively inexpensive technique (possibly $4000 for a minimum installation), additional equipment options can add to the initial costs. Also downtime for installation, adjustments to see if the plunger installation will perform, and adjustments to optimize production add to the costs.

To alleviate these costs, methods have been developed to predict whether plunger lift will work in advance of the installation, under particular well operational conditions. These methods vary in rigor as well as accuracy but historically have proven to be useful tools when predicting the feasibility of the plunger lift method.

There are several screening procedures that can be used to determine if plunger lift will work for a particular set of well conditions.

7.4.1 GLR Rule of Thumb

The crudest of these procedures is a simple rule of thumb that states that the well must have a gas/liquid ratio (GLR) of 400 scf/bbl for every 1000 ft of lift or some value that is fairly close to the 400 approximate value (this corresponds to approximately 233 m^3 gas/ (m^3 liquid for every 1000 m depth)).

Example 7-1

Will plunger lift work for a 5000 foot well producing a GLR of 500 scf/bbl?

Applying the rule of thumb of 400 sc/bbl for each 1000 ft of lift, the required GLR is:

GLR, required = 400 scf/ (bbl-1000 ft) × 5 = 2000 scf/bbl

However, the actual producing GLR is 500 scf/bbl, so this well is not a candidate for plunger lift, according to this rule.

Although useful, this approximate method can give false indications when the well conditions are close to that predicted by the rule of thumb. Due to its simplicity, the simple rule method neglects several important considerations that can determine plunger lift's applicability. This rule of thumb, for instance, does not consider the reservoir pressure and resultant casing build-up operating pressure that can play a pivotal role in determining the feasibility of plunger lift. Well geometry, specifically whether or not a packer is installed, can also determine if plunger lift is feasible.

7.4.2 Feasibility Charts

To get around some of the shortcomings of the GLR rule-of-thumb requirement, charts from Beeson *et al.* [2] have been developed that provide a more accurate means for determining the applicability of plunger lift. These are shown in Figures 7-6 and 7-7, which examine the feasibility of plunger lift for 2-3/8 in. and 2-7/8 in. tubing, respectively.

With reference to the charts, the horizontal x-axis lists the net operating pressure. The net operating pressure is the difference in the casing build-up operating pressure and the separator or line pressure to which the well flows when opened.

The casing build-up pressure represents a casing pressure to which the well builds to within a reasonable operating period of time. Since this time dictates the time permitted for each plunger cycle, reasonable time suggests a matter of a few hours rather than days or weeks.

Although the line pressure used in the net operating pressure is more straightforward, some special considerations deserve mentioning. The line pressure used to enter the chart must be the flowing wellhead pressure. Often, if the separator is located a significant distance from the well, and particularly if the two are connected through a small diameter flow line, the line pressure might build when the well is allowed to flow. For example, if the separator pressure is 100 psi, the line pressure might build to 200 psi at the wellhead when the well comes on as the liquid slug is forced into the small diameter line. Therefore, the proper use of Figures 7-6 and 7-7 requires some judgment on the part of the design engineer. The vertical y-axis of the charts is simply the required minimum produced gas/liquid ratio in scf/bbl.

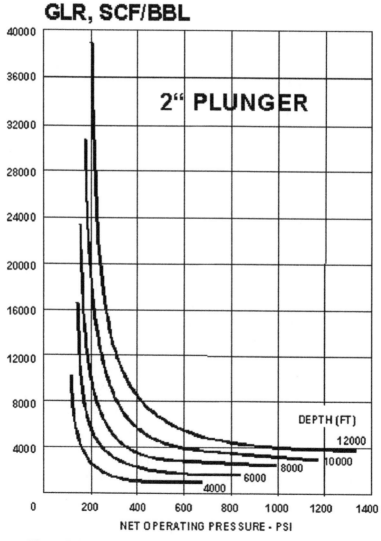

Figure 7-6: Feasibility of Plunger Lift for 2-3/8's Inch Tubing [2]

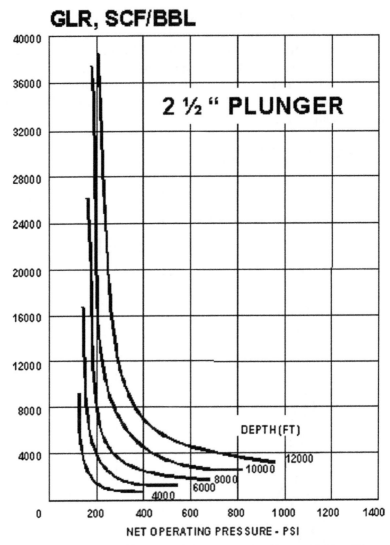

Figure 7-7: Feasibility of Plunger Lift for 2-7/8's Inch Tubing [2]

Use the figures by entering the x-axis with the net operating pressure. Track vertically upward to the intersection with the well depth. Then track horizontally to the y-axis and read the minimum produced GLR required to support plunger lift.

If the well's measured producing GLR is greater than or equal to that given by the chart, then plunger lift will likely work for the well.

If the measured GLR of the well is close to the value given by the charts, the well may or may not be a candidate for plunger lift. Under those conditions, the accuracy of the charts requires that other means be employed to determine the applicability of plunger lift. The following example illustrates the use of the charts shown in Figures 7-6 and 7-7.

Example 7-2

A given well is equipped with 2-3/8 in. tubing (a 2-inch plunger, approximately). Is this well a good candidate for plunger lift?
Operational data:

Casing build-up pressure	350 psi
Line or separator pressure	110 psi
Well GLR	8500 scf/bbl
Well depth	8000 ft

Use Figure 7-6 to determine whether plunger lift will work for this well.

Net operating pressure = (Casing build-up pressure − Line pressure)
= 350 − 110 = 240 psi

Entering Figure 7-6 shows that at a depth of 8000 ft, the well is required to produce a GLR (gas-liquid ratio) of about 8000 scf/bbl to maintain plunger lift.

The example well has a measured GLR of 8500 scf/bbl and therefore is a likely plunger lift candidate. Note pressure, gas rate, and depth are accounted for from this chart.

Comparing Figure 7-6 with Figure 7-7 suggests that there is an advantage to using the larger diameter tubing. As the tubing diameter increases, however, the likelihood that the plunger loses the liquid on the upstroke (due to liquid fallback around the plunger) and comes up dry increases. If the plunger comes up dry, the plunger (a large metal object) will impact the well head with great force, possibly causing damage. Because of this and other reasons, plunger lift is not as common with 3½ in. tubing and especially larger tubing sizes. On the way up the well liquid is lost from above to below and gas is lost from below to above the plunger.

Also note that the casing size is not listed with Figure 7-6 or Figure 7-7. Since the casing volume is used to store the pressured gas used to bring the plunger to the surface, the casing size is important. In general, the bigger the casing size, the smaller the required casing build-up pressure to lift the plunger and liquid. From the reference [2] it is unclear if the figures were developed using 51/2 in casing data or using 7 in casing data, or both.

7.4.3 Maximum Liquid Production with Plunger Lift

Figure 7-8 [3] helps to evaluate the effect of liquid production rate on the feasibility of using plunger lift. This figure shows the maximum possible liquid production rate that plunger lift will tolerate for a given depth and tubing size. The chart tubing size versus depth in feet is on the x-axis and the maximum allowable liquid production in bbls/day for plunger lift on the y-axis.

MAXIMUM FLUID PRODUCTION FOR PLUNGER PUMP

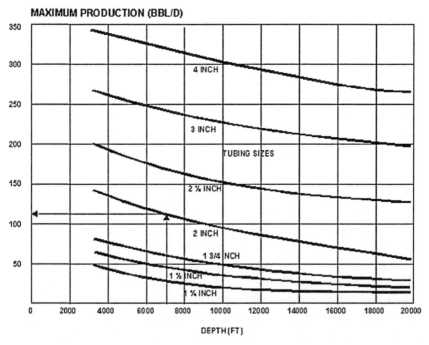

Figure 7-8: Liquid Production Estimate for Plunger Lift [3]

The chart generally is used by entering the x-axis with the well depth. Then track vertically upward to the given tubing size. Finally read horizontally to the left and find the maximum allowable liquid production rate for the use of plunger lift on the y-axis.

Example 7-3

A given well is 7000 ft deep and is to be produced by plunger lift through 2-inch tubing. What is the maximum liquid that can be produced?

Entering Figure 7.6 with the depth of 7000 ft and 2-inch tubing gives the maximum production by plunger lift of about 110 bbls/day. (The process is shown in the figure with the arrows.)

7.4.4 Plunger Lift with Packer Installed

Although some installations have employed plunger lift systems successfully in wells having packers installed, packerless completions are highly preferred. In the event that the well does have a packer installed, perforation of the tubing above and near the packer, allowing the casing annulus to accommodate gas storage, can drastically improve the efficiency of the plunger lift system. However, packer liquid may have to be drained from the well annulus before going on production, perhaps by setting a plug below and bailing the liquid out of the well.

Some wells, however, have sufficient reservoir pressure and gas flow to produce liquids with plunger lift even with a packer. When a packer is installed in the well, Figure 7-9 can be used to estimate whether the well conditions are sufficient to support a plunger lift system.

This figure plots two curves that represent the upper limit of conditions required for plunger lift for the cases with and without packers installed in the completion. These are plotted against the GLR on the x-axis and the well depth in feet on the y-axis. If the intersection of the well's GLR and depth falls on or below the respective curve, then plunger lift will likely work for the well. This figure clearly demonstrates the adverse effects that packers have on plunger lift installations.

GAS REQUIREMENT FOR OPERATING PLUNGER PUMP

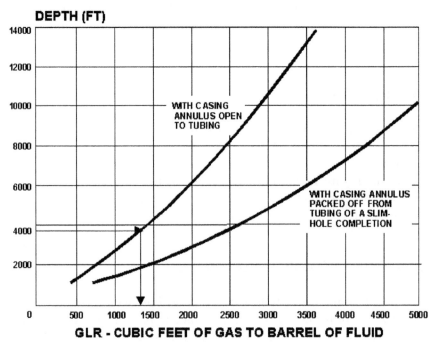

Figure 7-9: Gas Needed for Plunger Lift with or without a Packer in the Well [3]

For example, a well having a GLR or 1400 scf/bbl is sufficient to operate a plunger to a depth of 3900 ft if the well has no packer. With a packer installed, however, the operable depth is reduced significantly to 2000 ft.

Some industry rules simplify this chart saying 1000–2000 scf/bbl-1000' to operate plunger with a packer.

7.4.5 Plunger Lift Nodal Analysis

Reference [4] calculates the average bottomhole pressure for all portions of the cycle for one production rate. The average pressure includes the rise, the flow period, the fall period, and the build-up period. This is compared to various sizes of tubing and what pressure is required to flow up the tubing at various rates. Then the plunger lift performance can be compared to flowing up various sizes of tubing. The results of this type of analysis are shown in Figure 7-10.

Plunger vs Velocity String Performance

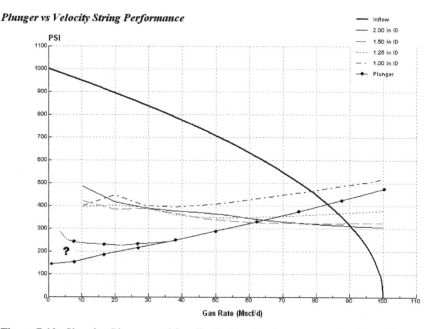

Figure 7-10: Showing Plunger and Smaller Tubing Performance on the Same Down-hole Nodal Plot [4]

Figure 7-10 shows the well inflow performance or IPR plot and the tubing performance of several sizes of tubing. For this example, none of the tubing performance curves are predicted to flow as they do not intersect the well inflow performance curve. However, the plunger performance [4] shows that for the low gas rates, using plunger gives a lower required flowing bottomhole pressure. As the well IPR declines to lower and lower pressure, only the plunger performance curve can intersect the inflow curve and achieve a flow rate. This program is being improved but the trends shown here are typical.

7.5 PLUNGER SYSTEM LINE-OUT PROCEDURE

The following section outlines hints and suggestions to incorporate into the procedures used to bring a plunger lift system online. The section covers procedures covering all aspects of plunger lift from the initial start up, considerations before and during the first kickoff of the plunger, methods to adjust the plunger cycle, and techniques to optimize the plunger cycle to maximize production. The following material on system operation and maintenance follows the Ferguson Beauregard Plunger Operation Handbook [5] with some updates and alterations. Although most of these functions may be done with computer control algorithms, the precautions are listed here so one can compare to computer control if control is not using manual set points.

7.5.1 Considerations before Kickoff

There are several parameters that must be considered before kicking off a plunger lift well. Most important is the casing pressure. As mentioned earlier, the casing annulus acts as energy storage, holding compressed produced gas that eventually is responsible for bringing the plunger and the liquids to the surface. It is this gas trapped in the casing that primarily determines the frequency of the cycles, and therefore the success of a plunger lift system.

Another key factor to consider is the liquid load or the amount of liquid accumulated in both the casing and the tubing. The rate of accumulation of liquids also plays an important role in determining the plunger cycle time. If the liquid volume is allowed to become too high, it is less likely that the plunger will be able to bring the liquids to the surface with the gas pressure available.

A third major factor to be considered is back-pressure. This includes back-pressure from all likely sources, whether it is from high line pressure, small chokes, or compressors that will not handle the initial surge of gas. Back-pressure is the pressure the well sees on the downstream of the tubing valve when it is opened.

Load Factor

It is extremely important to properly prepare the well before you open it to flow. First, it should be as clean, or as free of liquid, as possible. This may mean swabbing the well until it is ready to flow or it may mean leaving it shut in for several days to allow the well pressure to build high and to push liquids back into the formation.

The Load Factor can be used to see if the well is ready to be opened. This may be automated or could be used in manual operation. The definition is:

$$\text{Load Factor} = 100 \times \frac{\text{Shut-in Casing Pressure} - \text{Shut-in Tubing Pressure}}{\text{Shut-in Casing Pressure} - \text{Line Pressure}} \%$$

A good rule of thumb is to ensure that the load factor does not exceed 40 to 50 percent before opening the well to let the plunger and liquids rise.

Example 7-4

A given well has been shut in until the following conditions prevail. Determine whether the conditions are sufficient to start the plunger cycle.

Casing pressure:	600 psi
Tubing pressure:	500 psi
Sales Line Pressure:	100 psi

$$\text{Load Factor} = 100 \times \frac{600 - 500}{600 - 100} \% = 20\%$$

Since the Load Factor is less than the maximum limit of 50 percent, then the plunger and liquid slug are predicted to rise when the well is opened. Conditions are predicted to be acceptable to start the plunger cycle.

It pays to be patient while waiting for the well conditions to meet the initial load factor requirements. Should the well be opened too soon without sufficient casing build-up pressure, the plunger may not make it to the surface and the well will further load with liquids. It is important that the well be allowed to build an abundance of pressure, more in fact than actually is needed, prior to opening the well to production. Time permitting, the initial shut-in might be allowed to proceed until the pressures are static, just to insure that this vital first cycle can be accomplished.

A common mistake is to allow the well to flow too long following the production of the initial slug after swabbing. Once the well's production becomes gaseous and the casing pressure begins to drop, the well should be shut in and allowed to build pressure. The produced gas pressure is a vital component required to bring the plunger to surface and should be conserved, especially when just starting an installation.

In many cases, it is desirable to vent the gas above the liquid in the tubing on the initial cycle to a lower pressure. This creates more differential pressure across the slug and plunger, pushing the slug to surface. Regardless every effort should be made to remove as many restrictions in the flow line as possible. If a flow line choke is required, the largest possible choke for the system should be used. It's also good practice to put large trims in the dump valves of the separator. A slug traveling at 1000 ft/min. corresponds to a producing rate of 5760 bpd in 2-3/8 tubing. Frequently a larger orifice plate in the sales meter is used to measure the peak flow of the head gas.

7.5.2 Kickoff

Once adequate casing and tubing pressures have been reached, the well is ready to bring the plunger to the surface. The casing and tubing pressures required to kick off the well are obtained from the methods just outlined.

It is imperative that the motor valve open as rapidly as possible so that the tubing pressure is bled off quickly. If done, this quickly establishes the maximum pressure difference across the plunger and the liquid slug to move them to the surface.

Record the time required for the plunger to reach the surface. The current thinking is that the plunger should travel at between 500 and 1000 ft/min for optimum efficiency with a mid point of ~750 fpm being best unless plunger-specific rules are developed. Experience has shown that plunger speeds in excess of 1000 ft/min tend to excessively wear the equipment and waste energy, and plunger speeds lower than 500 ft/min will allow gas to slip past the plunger and liquid slug lowering the system efficiency. The plunger travel speed is controlled by the casing build up pressure and the size of the liquid slug that is produced with the plunger. Note that a plunger could be run slower if it had a very good sealing mechanism.

When the motor valve is opened, a surge of high-pressure gas from the annulus will be produced into the tubing to lift the plunger and liquid. As the gas rate at the surface bleeds down, a slug of liquid will be produced followed by the plunger. Often some liquid will follow the plunger. In most cases when just starting the plunger cycles, it is best not to let the well flow more than a couple of minutes after the plunger surfaces. If the well is allowed to flow for too long, the casing pressure will decrease below the recommended limit and allow too much liquid to accumulate in the annulus before the next cycle. If the volume of liquids becomes excessive, the well will not be capable of completing the next cycle. Rules such as seeing the casing pressure and tubing pressure begin to spread apart, indicating liquids accumulating downhole, are used. Or flow can be measured or inferred with a delta pressure cell, and when it drops near critical (+/−) flow can be ceased.

With the plunger at the surface initially, close the motor valve and allow the plunger to fall. Gas begins to pressurize the casing and tubing for the next cycle. The plunger must also be allowed to reach the bumper spring. New data from Echometer can help indicate when the plunger hits bottom, or you can use the Echometer system or the PCS (Denver) smart plunger to measure when the plunger hits bottom. Once the casing pressure has regained its initial value, the cycle can be set for automatic operation if a few of these manual cycles are used to start the plunger operation.

Some controllers will do the starting procedure without manual intervention.

7.5.3 Cycle Adjustment

Liquid loading can occur not only in the tubing but also in the reservoir immediately surrounding the wellbore. Liquid accumulation in the reservoir near the wellbore can reduce the reservoir's permeability. To partially compensate for this, it is recommended to run the plunger on a conservative cycle for the first several days. A conservative cycle implies that only small liquid slugs are allowed to accumulate in the wellbore and that the cycle is operated with high casing operating pressures.

When setting the operating cycle for a plunger lift installation, one proven method is to use a casing pressure sensor in combination with a plunger arrival sensor to shut the well in. This method provides consistent shut-in casing pressures for each cycle, insofar as the well is shut in immediately after the plunger arrives at the surface. In so doing, the method essentially minimizes the time required for the next cycle. If casing pressure above the line pressure is used as a control guide, it will prevent trying to bring the system on when line pressures have drastically increased from one cycle to another.

In summary, the kickoff procedure is outlined here:

1. Check (and record) both the casing and tubing pressures. Apply the rule of thumb demonstrated in Example 7.4.
2. Open the well and allow the head gas to bleed off quickly. Record the time required for the plunger to surface (plunger travel time).
3. Once the plunger surfaces and production turns gassy shut the well in and let the plunger fall back to the bottom.
4. Leave the well shut in until the casing pressure recovers to a satisfactory operating value in excess of 1.5 times line pressure. Better is to return to the casing pressure in excess of line pressure.
5. Open the well and bring the plunger back to surface and again record the plunger travel time. Shut the well in.
6. If this cycle has been operated manually, then set the timer and sensors to the recorded travel time and pressures.
7. If you have no casing pressure sensor or magnetic shut-off switch, then it is necessary to use time alone for the cycle control. Allow enough time for adequate casing build up and enough flow time to get the plunger to the surface. A two-pen pressure recorder can be a valuable asset under these conditions. By monitoring the charts, you can quickly compare the recovery time of the casing and adjust the cycle accordingly.
8. Whichever approach you use, once you see the cycle is operating consistently, leave it alone and allow the well to clean up for one or two days until the liquids in the reservoir wellbore area have been somewhat cleared.

Although many new controllers will take care of these steps, they are listed to show what physically should be considered to start a plunger installation, and also for when newer controllers are not being employed.

7.5.4 Stabilization Period

Because the formation adjacent to the well tends to load with liquids whereas the wellbore itself loads up, it generally takes some time for the well to clean up. Depending on the reservoir pressure and permeability, this cleanup could be accomplished in a day or it may take several weeks. Optimization procedures are easier to implement after the well has stabilized.

During the cleanup period, the plunger cycles should remain conservative. This implies longer shut-in cycles and shorter flow times than will be used after the well has had time to clean up. As the well produces liquids and stabilizes, the build-up casing pressures should rise and the rate of liquid production should decline. It is important to continue to keep plunger velocity about 750 ft/min. As the well stabilizes, the plunger travel time will initially decrease and then become stable, indicating that the well is sufficiently clean to begin optimization. Note that although the buildup pressures are changing such that they *can* become larger, production optimization dictates that the cycle times be adjusted for shorter cycles such that smaller operating casing pressures can be used. See the next discussion on optimization.

7.5.5 Optimization

Once the well has stabilized, the plunger cycle is ready to be optimized. The optimization procedure varies somewhat depending on whether the well is a gas well or an oil well. The first step in either case is to determine the operating casing pressure. This is done by incrementally dropping the surface casing pressure, required just before each cycle, by 15 to 30 psi then allowing the plunger to cycle four or five times before dropping the pressure again. At each incremental casing pressure, record plunger travel time to ensure that the plunger speed stays close (+/−) to an average speed of about 750 ft/min.

If the plunger speed drops below 750 ft/min then slightly increase the casing operating pressure and record the plunger travel time for several more cycles until the plunger speed stabilizes at a value slightly above the minimum. If, on the other hand, the plunger speed is above 1000 ft/min, allow the well to flow longer after the plunger surfaces to allow more liquid to feed into the wellbore each cycle. Eventually, the swings between the high and low casing pressures will stabilize with the plunger travel times within the desired operating parameters indicating that the well is again stable at the new casing operating pressure.

Actually, if you measure the speed of arrival and it's too slow, you can (1) increase the buildup time and/or (2) decrease the flow time (reduce slug size) for adjustment. If it rises too fast, then you can (1) decrease the buildup time (shut-in time) and/or (2) increase the flow time (increase the slug size).

This discussion assumes that many of the adjustments are made manually to clarify how the well can be controlled. However, many of these operations are now taken care of by the newer computerized controllers.

The next step is to adjust the time for the well to flow with the plunger surfaced. In this case, an oil well is generally much simpler to set than a gas well. Oil wells generally have much lower gas liquid ratios (GLRs, scf/bbl) and therefore have much less gas available to push liquids to the surface.

Oil Well Optimization

To fully optimize the flow time for an *oil well*, it is necessary to install a magnetic shut-off switch in the lubricator at the surface to shut the well in upon plunger arrival. Any reliable arrival transducer would serve the purpose. The switch activates the motor valve shutting the well in immediately upon plunger arrival, which saves the needed tail gas for the next cycle. The plunger then starts its return to bottom with only a small hesitation at the surface, shortening the cycle time and increasing liquid production.

This prevents the well from depleting the vital gas supply stored in the casing. Depleting this stored gas would require longer shut-in periods to rebuild pressure and in most cases would lower the overall liquid production.

In the event that the casing pressure remains too high after plunger arrival, rather than allowing the well to flow gas after the plunger has surfaced, the recommended practice is to lower the casing operating pressure. Lowering the operating pressure generally prompts an increase in production since the pressure against the formation is reduced. This is the type of cycle described in the Foss and Gaul [6] paper. The authors have witnessed oil wells on plunger making as much as 300 bpd from about 4000 ft; that does seems high, but it may or may not be exceptional.

Gas Well Optimization

Optimizing the flow time for a *gas well* requires more effort, if done manually, since the time that the gas is allowed to flow after plunger arrival is considerably longer than that of an oil well.

An older method of optimization is as follows: The flow time for a gas well is optimized by continually adding small increments to the amount of time allotted to gas flow while recording the plunger travel time. These small increments should be added over the period of several days to allow the well to regain stability after each change. As the flow time is increased the plunger travel time will decrease. Once the plunger travel time drops to approximately 750 ft/min the flow time used to be considered optimized. However, now the velocities mentioned are achieved, but attention is given to the average pressure on the formation during the cycle, and this is minimized by allowing only small liquid slugs to accumulate during the cycle.

Optimizing Cycle Time

The previously mentioned methods to examine rise velocity only did work to establish cycles but do not optimize production. For instance, a large slug can be brought into the well during the flow period, and then a large casing build-up pressure will allow the plunger and liquid to be lifted to the surface at 750 fpm. This would exert a high average pressure on the formation and as such the production would be reduced.

It would be better if a small slug of liquid is accumulated in the tubing during a brief flow period, and then only a small casing build-up pressure would be required to lift the slug, at an average rise velocity of about 750 fpm. This would result in a smaller average pressure on the formation and the production would be higher (Figure 7-11).

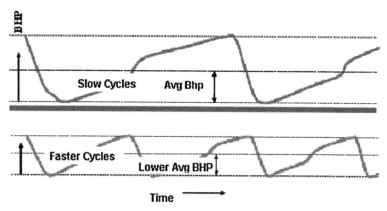

Figure 7-11: Faster Cycles with a Smaller Liquid Slug of Liquid Result in a Lower Average Flowing Well Pressure

7.5.6 Monitoring

Note that any changes to the conditions at the surface will have an impact on the operation of the plunger lift cycle. If, for example, the sales line pressure were to decrease due to a lower percentage of liquid in the flow, the optimum flow time would increase. On the other hand, if the sales line pressure were to increase, the flow time must be shortened. Similarly, if the orifice plate size or choke settings are changed, then the appropriate changes to the flow time must be made.

Once the well is reasonably optimized and the plunger lift system is in stable operation, it remains necessary to monitor the well for best performance. Well and reservoir conditions continually change, thus altering the performance of the plunger lift system requiring adjustments. Most controllers do this work and the operator does not have to check on the performance. Controllers will also maintain the plunger rise velocity near 750 fpm or some input average rise velocity. It is a good idea to physically check the plunger for damage and wear monthly as plunger wear will also impact the rise time for a given set of well conditions.

7.5.7 Modern Controller Algorithms

This section by Bill Hearn, Weatherford

Bill Hearn is the Plunger Lift Systems Business Unit Manager for Weatherford. Bill is responsible for supporting sales and marketing globally as well as new product development. He joined Weatherford in 2001 as a manager for Weatherford's Artificial Lift Location in Rock Springs, Wyoming specializing in Gas Optimization. Prior to Weatherford, Bill served as a Northern Area Manager for Integrated Production. In his 10-year career he has worked in the service industry of optimization in a variety of sales, service, and technical roles with most of this time spent on Gas Well Optimization. He currently resides in Houston, Texas.

The control system on the plunger application significantly affects your optimization opportunities. Control systems range from simple on/off timers to complete automated systems with pressure optimization and automatic adjustments.

Most control systems will go through a series of repeating states. The control system will begin with an initial On-Time, which is also called minimum on or A-Valve on-time. During this state, the control valve opens up and the plunger arrival timer begins. This timer is set for the maximum amount of time the well will produce in a cycle unless the plunger arrives. During this timer, the controller waits for the plunger to arrive. The next state depends on whether or not the plunger arrives. If the plunger does not arrive most control systems will either allow the well to vent, relieving the wellhead pressure that assists in differential to lift it, or it will go into an extended shut-in timer, building additional pressure to cycle again. In the case that the plunger does arrive, it immediately goes into sales or after flow timer. This is the time for the well to flow freely, accumulating fluid for the next cycle. In auto-adjust and pressure control systems this time varies depending on conditions. The effect of auto-adjust and pressure control is discussed later. Once this state finishes, the well goes into a plunger fall timer, or minimum off-time. The timer is based on the amount of time it takes the plunger to reach the bottom, or in the case of continuous flow, the time the plunger needs to fall from the wellhead. The plunger fall timer is established by either chasing it to the bottom with wireline or tracking it using acoustic sounding. Once the control finishes this state it transitions into an off-time or shut-in time. This transition allows the well the opportunity to build the pressure required for arrival. As with the sales or after flow time, specifically auto-adjust or pressure control, the time will vary depending on conditions.

Cost per system and benefits will depend on both the vendor and the level of automation required. Typically, costs increase as control goes from simple to complete automation, beginning at a low around $1,000 up to systems that may cost $25,000.

Simple On/Off Controllers

The simplest controllers will have only an on-time and off-time. During the on-time the well makes its arrival and produces its after flow. During the off-time, the plunger should fall to the bottom and shut in long enough to build the pressure necessary for the next arrival. These controllers are the most difficult to optimize because they operate independently of the plungers' movement.

Basic Plunger Control

These types of controllers are very common on plunger wells and will cycle through all the states listed in the first paragraph. They require the operator to establish the well and leave the timers set to optimally produce the well and ensure that it operates with the proper safety margin. This procedure also ensures that the well does not load up and begin to miss cycles. In many cases it is the control of choice due to its simplicity. However, with varying line pressure or changing well conditions it will consistently require attention to ensure optimal production.

Basic Auto-Adjust Control

This type of controller makes its changes to the sales and off-times based on the previous arrival time and where the plunger arrival occurs compared to an ideal arrival time or window depending on controllers. Ferguson Beauregard produces the Auto-Cycle™, which is a well-known version of this type of control, yet many other vendors also produce similar operating controls.

Fast plunger arrivals are defined as arrivals that are earlier than the ideal arrival time or window. Typically the controller will extend the sales or after-flow time and decrease the off or shut-in time. By making these changes the load size should increase and the energy at the beginning of the cycle should decrease, therefore decreasing velocity.

Slow Plunger arrivals are defined as arrivals that are later than the ideal arrival time or window. Typically the controller will decrease the sales or after flow time and increase the off or shut-in time. By making these changes the load size should decrease and the energy at the beginning of the cycle should increase, therefore increasing velocity.

Basic Pressure Control

This type of controller will usually use the current condition of the well to dictate the amount of after-flow time and the amount of off- time. Although the other parts of the cycle remain similar to basic plunger control, the changes to the after-flow time and off-time are based on the well conditions.

Once the well is in after-flow time the controller will monitor the flowing rate or flowing pressure conditions and send the well into plunger fall once conditions are met. This type of control ensures that the well flows below critical or that it has begun loading before shutting in. This ensures that a plunger cycle is necessary, then ends the after–flow timer once the conditions are met and sends the controller to plunger fall time or minimum off time.

Once the well completes its plunger fall or minimum off time the controller begins to monitor the shut-in conditions and waits for the pressure build necessary to ensure an arrival. In many cases this may be a single pressure reading, a comparison of pressure readings, or an algorithmic value that is the result of three pressure readings. Once the condition to cycle is met, the controller sends the well in to its on-timer.

Auto-Adjust Pressure Controls

This type of control combines the use of pressure control with the additional benefit of adjusting the set pressures based on arrival times.

Early arrivals suggest that the well needs less energy (pressure) to start the cycle and that the flow rate before shut-in could be decreased. This permits more fluid into the tubular before the next cycle. Late arrivals will indicate that the well needs more energy (pressure) to start the cycle and that the flow rate before shut-in should be higher, resulting in a smaller load for the next cycle.

Automated Systems

Automated systems will generally allow the application of some or all of these control theories to a well. These systems also add the ability to remotely make changes allowing for operation by exception and multiple well optimization.

7.6 PROBLEM ANALYSIS

The following section outlines solutions to some of the more common problems encountered with plunger lift systems. These items are grouped with respect to the system components and particular malfunctions.

Table 7-1 can be used as a quick reference for some general points. Many of the table entries are field specific but the user might develop a similar field specific table for a particular operation. The following material revised from Ferguson and Beauregard [5] contains more detail of troubleshooting procedures.

Table 7-1
Various Problems That Can Occur with Plunger Lift (Phillips and Listiak [7])

Problem	Check/Change Plunger	Optimize Program Settings	More Off Time	More Afterflow	Less Off Time	Less Afterflow	Check Well TBG(Restriction/Hole)	Check Wellhead (Design)	Clean Sensor/Check Wiring	Check Module/Wiring	Change(+) Lead Fuse Link	Power Down & Restart Module	Set Sensitivity of Sensor	Change Supply Gas Filter	Adjust Supply Gas Pressure (20–30 psi)	Clean Control Bleed Ports	Change O-Rings Under Latch Valve	Check/Change Battery	Check Solar Panel	Repair Motor Valve Trim	Eliminate Flow Restrictions	Check Catcher	Change Module	Change Latch Valve	Check Special Settings	Check Motor Valve Diaphragm	Inspect Plunger
No Plunger Arrival	6	3	2			1			5	7			4							9	10		8				
Slow Plunger Arrival	4	3	2			1	8	7												5	6						
Fast Plunger Arrival	3	3		1	2		6															4					5
Fast Plunger Arrival @ all Settings or plunger won't fall	1	1					4						2									3					
Slow Plunger Arrival @ all Settings or plunger won't come to surface	4	3	2			1	7	6													5						

Short Lubricator Spring Life		4		2	3		5															1					
Short plunger Life		3		1	2		5	4																			
Sensor Error									3	4			2									1	5				
Plunger error	6	3	2			1	12	11	5	7			4						9	9	10		8				
Good Trip, No Count (Plug-in Sensor)		1							3	4	5		2										6				
Good Trip, No Count (Strap-on Sensor)		1								1	4		2														
Fatal Error Code @LED		1							5	10		3													2		
LED Control Screen Blank									5			4						2	3				5				
Sales Valve Won't Open/Close		1												4	3	6	7	2	8	9			11	12		13	
Tank Valve Won't Open/Close		1												4	3	6	7	2	8	9			10	11		12	
Latch Valve Won't Switch														4	3	5	6	1	2					7			
Motor Valves Won't Close or Close Slowly														4	3	1	2	5	6	8			9	7			
Short Battery Life										1								2	3				4				1
Won't go to Afterflow		2								3													4		1		

7.6.1 Motor Valve

Valve Leaks

When motor valves leak, there are two possible sources. Under normal conditions, a valve will have from 20 to 30 psi on the diaphragm section of the valve, and much higher pressures on the body of the valve. External leaks are most commonly found at the packing section located between the diaphragm and the body of the valve. This occurs when the packing around the stem wears and leaks due to the high pressure from the body of the valve. All valves have some type of packing around the stem of the valve. In some cases it is possible to tighten a packing nut and stop the leak, but generally, it is necessary to replace the packing to eliminate the leak. Contact the valve manufacturer or the plunger lift company to help with the repair and/or parts.

The diaphragm portion of the valve can leak at one of two places. Either the valve will leak around the flange where the two portions of the diaphragm assembly are connected or at the breather hole (located on the opposite side from where the supply gas enters the diaphragm). In the latter case, the leak indicates a ruptured diaphragm. It is possible that a leak occurring at the flange can be the result of loose bolts, so it may be corrected by simply tightening the bolts and nuts, eliminating the need for replacing the diaphragm.

Internal Leaks

The most common leaks encountered in motor valves are internal leaks. Often ball and seat configurations normally are used as the sealing element. Because of the extreme pressure differential and high flow rates, the seat area is subject to fluid cut or erosion, which can be aggravated by abrasive materials. If the valve has an insert seat it will have an O-ring seal, which is also susceptible to cutting or deterioration due to gas composition.

If a valve is suspected to be leaking, the leak can be isolated by simply putting pressure on the upstream side with the valve closed and checking to see if there is any flow through the valve. If flow is identified, the leak is likely across the seat and can be corrected in the following ways:

- Check the valve adjustment. Depending on the size of the seat, the size of the diaphragm, and the flow path, there is a maximum pressure that a particular valve will hold. Manufacturers have charts for determining this differential pressure.

- If the valve seat is subject to a higher pressure difference, it is possible that the diaphragm and spring cannot contain the pressure. If the valve is equipped with an adjustment bolt on top of the diaphragm, tightening down on this bolt will put more pressure on the ball and seat to seal against the higher pressure. Be careful not to screw the bolt all the way in, as it will restrict the valve from fully stroking open.

- Also consider using a smaller seat. It is the differential pressure across the area of the seat that prevents the seat from holding. A smaller seat can dramatically reduce the force against the diaphragm spring. If a smaller seat is objectionable, consider larger diaphragm housing. The larger housing will have a larger spring and can hold a higher differential pressure.

- The valve may be turned around in the flow. This will put the higher pressure on top of the seat and that pressure will act to help hold the valve closed. Caution should be exercised, however, because if the pressure is in fact too high for the particular seat, then it will prevent the valve from opening. This is a last resort before new equipment is installed, as this idea will make the valve chatter.

- Another cause of a leaking ball and seat can be the formation of hydrates (an "ice" formed of hydrocarbons and water) in the seat area. An extreme pressure drop, across the ball and seat, in some service will prompt the formation of hydrates. Correcting the leak under these conditions is a matter of dissolving the ice or hydrate at the valve. With the hydrates removed, the valve should hold.

The prevention of ice or hydrate formation presents a somewhat more complex problem. The formation of hydrates might be prevented by either reducing the pressure differential across the valve or increasing the temperature. Simply using a larger trim in the valve will not reduce the pressure drop. The best solution is to lower the operating pressure of the entire system. This is not always possible, however, since operating pressure directly affects plunger system efficiency.

A common, but expensive, method to solve hydrate problems is to inject methanol just upstream of the freezing point. Alternately, a choke (larger than the valve seat) can be placed downstream of the valve. This will reduce the pressure drop across the valve seat and can reduce or eliminate the formation of hydrates by spreading out the pressure drop.

Valve Won't Open

Generally, there are four factors that play a part in the opening or closing of a motor valve:

- The size of the diaphragm
- The amount of pressure applied to the diaphragm
- The compression of the diaphragm spring
- The line pressure acting with or against the valve trim

A malfunction of any one or a combination of these components can prevent the valve from properly opening.

Earlier, it was pointed out that too much line pressure acting on top of the trim of the valve could hold the valve closed. In this situation it is possible to increase the supply gas pressure to the diaphragm to assist in opening the valve. Do not exceed 30 psi diaphragm pressure when attempting this procedure. If the valve still won't open and the adjusting screw has been backed out, then change to the next smaller seat or use a larger diaphragm. Exceeding the 30 psi limit placed on the diaphragm gas pressure can cause the valve to bang open, which can cause damage or rupture the diaphragm.

Another reason for a motor valve not opening is the adjustment of the compression bolt. The compression bolt puts tension on a closing spring that is connected to the trim by a short stem. If the compression bolt has been over-tightened, the valve will not fully open. When flowing over the seat, the tension should be at a minimum.

Finally, if these items have been checked and the valve still will not open, then the valve may have severe mechanical problems, such as a bent stem or a clogged valve. A bent stem or a frozen or clogged valve, although not common, is not out of the question.

Valve Won't Close

Many of the steps just mentioned are appropriate for troubleshooting a valve that will not close. In addition:

- Line pressure that is out of the operating range for the diaphragm size can prevent closure.
- The top adjusting bolt unscrewed too far could also prevent the valve from closing.

- Under certain conditions, it is not uncommon for ice to form in the trim, preventing the ball and seat from making a complete seal, thus keeping the valve open.
- Sand, paraffin, welding slag, or other foreign objects can get lodged between the ball and seat preventing valve closure.
- If the controller is not allowing the supply gas to bleed, the valve will not close. If this problem is suspected, the compression nut on the copper tubing link to the motor valve should be loosened while operating the controller open and closed. This should free the controller to bleed the supply gas.

7.6.2 Controller

The most complex part of the plunger lift system is the controller. There are many commercial controllers, and description and analysis of each is beyond the scope of this text. The following discussion covers only those basic components that might apply to the majority of controllers.

Basically, all controllers have similar operational characteristics. Generally, most controllers use a 20 to 30 psi pneumatic source, usually gas, which is utilized to open and close a motor valve. The motor valve is opened by directing supply gas through the controller to the valve diaphragm to force the valve open. The motor valve is closed when the controller blocks the supply gas and bleeds the gas from the diaphragm that opened the valve, thus allowing the valve to close. The discussion of controller troubleshooting will be covered in the next two sections, Electronics and Pneumatics.

Electronics

When the controller doesn't appear to be working properly and faulty electronic equipment is suspected, the first thing to check is the LCD (or LED) display. In addition to showing the time, most controllers are designed so that the display will indicate the mode of the controller (whether it is on or off), if it has power, or whether there are any outside switch contacts. No display may simply mean no power so check batteries for charge and proper contact.

In general there are so many controllers at this time, one must consult operating procedures or the manufacturer for troubleshooting.

Pneumatics

All controllers use some type of interface valve to control the pneumatic signal. The two most common are the latching valve and the slide valve. Each operates differently but essentially performs the same function. The latching valve is made up of an electromagnet and a small poppet valve. The valve operates when an electric "on" pulse from the electronics module activates the magnet and pulls the poppet off its seat then latches it back, directing supply gas to the motor valve. The off pulse from the electronics reverses the polarity of the electromagnet, releasing the poppet, and a spring moves it to the closed position. In the closed position, the poppet valve blocks the supply gas to the diaphragm and vents the gas, closing the motor valve.

The slide valve consists of housing and a small piston that slides through a cylinder in the housing. The travel of the piston is limited by end plates. The piston is fitted with three O-rings, one at each end for power and one in the middle. The position of the piston, either at one end of the cylinder or the other, directs the gas or determines whether the valve is in the open or closed position. A solenoid is fixed to each end plate of the housing. When either solenoid receives an electronic signal from the controller it directs a shot of gas to the power end of the shift piston, pushing it to the opposite end of the cylinder, thus opening or closing the valve. When the valve slides to the open position, supply gas is directed to the diaphragm of the motor valve; when it slides to the closed position, the supply gas is blocked and the diaphragm is bled.

Troubleshooting and maintenance of these valves is performed in the same manner. If a pneumatic problem is suspected, the gauges on the bottom of the controller should first be analyzed. With supply gas being fed to the controller, when the controller is pulsed to "on" both gauges should read the same pressure. Then if the controller is pulsed to "off" the pressure on the right-hand gauge should drop to zero. If this is not the case then the likely problem is a faulty valve (shifter) in the controller.

The fact that the shifter (latching valve or slide valve) is not working does not necessarily mean that it is damaged. The shifter does require voltage. Once you have determined the shifter is not operating, the next step is to check its supply voltage. Check the wiring to ensure there are no loose connections or broken wires. Next, with a voltmeter, check to see that there is power being supplied to the shifter from the electronics module. There should be no power supplied to the shifter until the controller is pulsed on or off when only a brief pulse is issued. If no pulse is evident then the electronic module must be replaced. If the pulse is being fed to the shifter and it is not operating then the problem is with the shifter.

The most common problem encountered with the shifters is fouling from contaminated supply gas. Fortunately, shifters are easily disassembled and cleaned. After a thorough cleaning, the slide valve must be lubricated with a thin coat of lightweight grease (such as Parker O-ring Lube). It is not recommended to disassemble the solenoid valves, located at either end plate of the sliding valve, for cleaning. The solenoid valves rarely malfunction, but when they do, they must be replaced.

To ensure smooth operation of either type shifter, it is recommended that a filter be installed in the supply gas line to keep impurities from entering the shifter mechanism. The supply gas should also be maintained as dry as possible. If casing head gas is to be used, it is good practice to install a drip pot upstream of the controller and keep it blown dry.

New controllers may contain features for which this discussion does not apply.

7.6.3 Arrival Transducer

The arrival transducer is a device that plays a very important role in most plunger lift installations. The function of the switch is to detect the arrival of the plunger in the lubricator. This then typically signals the controller either to shut-in the well (oil well), or to switch valves or just to register the cycle in a plunger counter (gas well). Most commercially available switches use a magnet to close a set of contacts on an electric switch. This switch closure completes a circuit that sends a signal to the controller. These switches are normally trouble free, but mechanical malfunctions are possible. There are some switches based on vibration.

To isolate a malfunction, the first step should be to determine if the switch is even capable of operation. This is done by removing the switch from the housing on the catcher nipple, and touching it to the lubricator. If making contact causes the controller to turn off or record arrival, then it can be concluded that the wiring is functional, and that the switch is at least capable of operation. Some styles cannot be unplugged. This type must be shorted manually by placing a small piece of metal across the switch (inside the catcher nipple). Again, if shorting the switch causes the controller to turn off or signal arrival, then the switch is capable of operation.

A closed magnetic switch and the entire off-time may be displayed if this is a controller function. If this occurs, first determine whether the plunger is up in the lubricator, which would indicate normal operation. If the plunger is not in the lubricator, then remove the switch from the housing and see if normal operation resumes. If the controller doesn't immediately start counting down, then disconnect the wiring from the controller. The countdown should resume unless the problem is within the controller itself. If the controller restarts the countdown and the plunger is not in the lubricator, then the problem is either in the switch or the wiring.

If these procedures have been followed and all components appear to be functional, further investigation is required to isolate the apparent malfunction with respect to the operation of the entire system. It might be possible for a plunger to travel at speeds too fast for the switch to detect. Most switches are sensitive enough to detect a plunger traveling at speeds in excess of 1000 ft/min. There are controllers on the market, however, that are not capable of detecting these high reaction rates. In such cases, it is possible for the arrival of the plunger to go undetected. Slowing down the plunger travel speed is the best way to determine if this is the cause of the apparent magnetic arrival transducer malfunction.

Another possible system malfunction that could mistakenly be attributed to a magnetic sensor problem is when the plunger does not surface or does not travel high enough in the lubricator for detection. In the former case, the best method to determine whether the plunger is truly arriving at the surface is by physical inspection. Although it is possible to receive an indication of the plunger surfacing on a chart or recorder, these indications are not totally reliable. The well response can indicate the plunger surfacing on the chart without the plunger actually making it to the surface.

When the plunger is surfacing, but not going far enough into the lubricator to trip the switch, adjustments must be made to the system to allow the plunger to pass further into the lubricator. To insure the plunger travels far enough into the lubricator to make contact with the magnetic sensor switch it is recommended that the upper flow outlet be open to allow flow to go past the sensor, carrying the plunger past the magnetic switch and allowing the sensor to signal arrival of the plunger. Some plunger wellheads try to use only one outlet, but the dual outlet used as described is a better setup for the arrival transducer.

7.6.4 Wellhead Leaks

Wellhead leaks must be repaired to maintain a safe and clean environment at the well site. On most wellhead hook-ups, leaks generally are due to faulty threads. Leaking around the wellhead bolts typically is caused by improper torque on the bolts, improperly repaired wellhead, or damaged bolts.

Other than the bolt connections, the most common place for wellhead leaks is at the catcher assembly or where the lubricator screws into the flow collar. The catchers usually are attached through some type of packing gland (not unlike those found on many valves). Leaks that occur at the catcher can normally be fixed by tightening the packing nut. If not, it may be necessary to replace the catcher assembly.

The lubricator upper section has a quick connect with an O-ring seal. These can leak and need to be replaced periodically. In most cases, tightening will not stop a leaking O-ring.

Wellhead connections may be screwed on or flanged. Flanged wellheads are thought to be safer if the plunger arrives dry.

7.6.5 Catcher not Functioning

For plunger inspection but not general operation, the catcher should be able to hold the tool in the lubricator to accommodate its removal. Plunger catchers catch or trap the tool and hold it, in one of two ways.

Some catchers use a spring-loaded cam-type device. To activate the catcher and catch the plunger, either a thumbscrew is unscrewed (which activates the catcher) or a catcher handle is released. In both cases, a cam is extended into the path of travel of the plunger. As the plunger moves past, the cam is pushed back, allowing the plunger to move past it. Once the plunger has moved past the catcher, the spring-loaded cam flips out beneath the plunger, preventing it from falling back downhole.

The other type of catcher, commonly found on older installations, uses a friction catch to hold the plunger at the surface. The friction-type catcher consists of a ball extending into the sidewall of the catcher, pushed by a coil spring. As the plunger moves past the ball, the compression of the spring on the ball causes friction against the side of the tool, preventing it from falling.

Before troubleshooting catcher problems, it is important first to verify that the plunger is arriving at the surface and then to make sure that upon arrival, the plunger is traveling far enough into the lubricator for the catcher to engage. One way to assist the plunger to go further up into the lubricator is to open the flow outlet above the catcher. Also, closing the lower outlet will direct all the flow through the upper outlet, driving the tool higher into the lubricator. If under these conditions the catcher still fails to capture the plunger, further inspection of the catcher itself is required.

The first step in troubleshooting the catcher is through visual inspection to determine if ice or paraffin or other produced solids have clogged the catcher. The removal of foreign material should restore catcher operation.

Next inspect the catcher nipple while manually engaging and disengaging the catcher. The nipple should move all the way out of sight and stay there. In the run position when the catcher is activated, the cam (or ball) should extend into the path of the plunger. If it is not extending or retracting back into the housing, then it requires repair or replacement.

Never operate with the plunger surfacing and the wellhead open at the surface.

7.6.6 Pressure Sensor not Functioning

A common method for starting many plunger cycles is with a casing pressure activated switch-gauge. The switch-gauge is a pressure gauge with two adjustable contacts on the face and a pressure indicator (needle), all of which are connected to electric wires. Changes in casing pressure, up or down, cause the needle to move toward one contact or the other. When the needle touches either the high or low contact it completes a circuit that signals the controller to open or close the motor valve. Switch-gauges seldom have malfunctions but problems do occur.

As a first step, before examining the switch-gauge itself, check the controller as outlined earlier. Often a properly functioning sensor is blamed for a controller malfunction. If the controller is functioning properly, check the contacts on the gauge. The contacts on either side of the gauge can become fouled and unable to complete the circuit to the controller. To check the contacts, try to operate the controller manually by moving first the high and then the low set points (on the face of the gauge) so that they make contact with the needle. If the circuit is intact, the controller should function. If there is no response, clean the contacts and try again. If there is no response, check the condition of the wires between the switch and the controller. Shorted or crushed wires may break the circuit.

Examine the gauge to ensure that all pressure lines are properly attached. Determine that all pressure valves leading to the gauge are open. If using the switch gauge as a flow line sensor, check and see if all valves downstream of the gauge are open. Bleeding the line connecting to the gauge should cause an appropriate response by the needle. If no response is registered make certain that the casing pressure is changing. This can require the temporary installation of a second pressure gauge. If the casing pressure is changing normally but not registering on the gauge, the gauge must be replaced. If only small casing pressure changes (or no changes at all) are being recorded then the plunger system is not operating normally and must be reoptimized.

This discussion may not apply to some controllers.

7.6.7 Control Gas to Stay on Measurement Chart

Controllers may be used to throttle motor valves open or closed while maintaining a set-sensed pressure. When used in conjunction with a plunger lift system, its purpose is to restrict the initial surge of head gas within the pressure limits of the system, in order to prevent the produced gas from going off the sales chart. These controllers most commonly are used on compressors and production units but are finding application with plunger systems. However it would be better to have an electronic sensor that will record the bursts of gas, because throttling back the surge of head gas can only serve to have some effect in *reducing the production*.

The unit works by sensing a pressure and then converting that signal to a proportional pneumatic pressure to the diaphragm of a motor valve, causing the motor valve to throttle. The sensed pressure pushes on a high-pressure flexible element, which in turn operates a pilot valve. This throttling of the pilot valve varies the pressure supplied to the motor valve, causing the motor valve to respond in a manner directly proportional to the sensed pressure signal. By throttling the motor valve, the unit attempts to maintain a constant sensed pressure. If the system (well) cannot supply enough pressure to meet the throttling range preset, however, the motor valve will remain wide open. On the other hand, if the sensed pressure exceeds the preset design pressure maximum, the motor valve will close completely.

This system is known to have two weaknesses. First, the supply gas entering the controller is metered through a small choke or orifice. When the gas supply leading to the output signal is slow to respond (build), this orifice should be examined. The choke is very small and is prone to get clogged with debris from dirty supply gas. It can usually be cleared with a small wire. It is good practice to place a filter in the supply gas line upstream of the choke to help prevent this clogging.

If the controller does not respond to sensed pressure, the sensing element should be inspected.

7.6.8 Plunger Operations

Plunger Won't Fall

Plungers are free-traveling pistons that depend solely on gravity to get back to the bottom of the well. If the plunger remains in the wellhead after the shut-in period or if it is back at the surface very quickly after opening the well, there is likely an obstruction either in the lubricator or down-hole keeping the plunger from falling to bottom.

In the event that the plunger returns to surface too quickly, first make sure that the plunger has been given ample time to reach bottom. Ideally, a plunger should travel up the hole between 500 ft/min and 1000 ft/min. On the other hand, plunger fall rates can be considerably slower. Plungers without a bypass, to allow gas to easily flow through the plunger on the down cycle, can fall at rates of only 150–500 ft/min or greater. Plungers equipped with a bypass or collapsible seal may fall at rates between 500–2000 ft/min. Fast fall is recommended to optimize a system for high production. If liquids have accumulated in the well during the last bit of after flow, then for maximum production the well should be opened as soon as the plunger lands on the bumper spring.

Echometer Company has devised a system [8] that tracks the plunger both during the rise and fall portion of the cycle. The measurements have been made both by acoustically recording the plunger depth using acoustic pulses and pressure change that occurs as the plunger travels past the tubing collar recesses. Figure 7-12 is a schematic of the Echometer setup to record plunger travel with time. Figure 7-13 is an example of the pressure and acoustic trace of a plunger cycle.

If the plunger ran smoothly during the initial installation, then it is unlikely that tubing is either crimped or mashed. If damaged tubing is

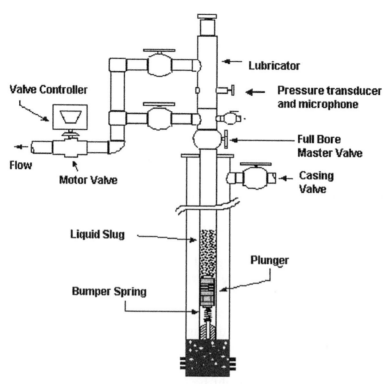

Figure 7-12: Echometer Well Configuration for Plunger Lift Analysis [8]

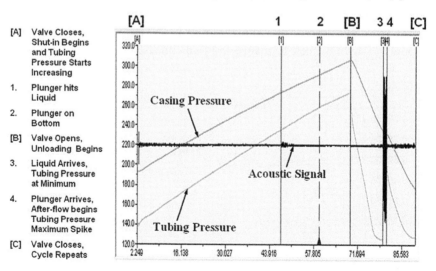

Figure 7-13: Sample Acoustic and Pressure Signals Recorded by Echometer [8] to Monitor Plunger Travel

suspected, then a wireline gauge ring should be run in the tubing having an OD corresponding to the tubing's manufactured drift diameter. It is also a good practice to run a gauge ring with a gauge length at least the length of the plunger. Care should be exercised while running the gauge ring, however, to prevent the ring from becoming stuck in the event that there is foreign debris in the tubing.

If it can be assumed that the tubing is of good quality, the two most common ailments that prevent the plunger from reaching bottom are ice (hydrates) or wax (paraffin) deposits. Typically, plungers will scrape the tubing clean of paraffin when cycling frequently. Severe paraffin build up generally requires that is be cut out of the tubing with a wire-line cutter.

Hydrate formation often occurs in particular gas wells at a depth where the gas is expanding rapidly. This occurs perhaps at depths less than 3000 ft. If the well is plagued with severe hydrate problems, a methanol injection system may be required to bring the well back to normal operation.

If the plunger will not drop out of the lubricator at the surface, the most likely cause is a faulty or damaged catcher. Review the preceding section on catchers.

Finally, if the well was recently worked over or other malfunctions have occurred, then there might be foreign debris (catcher parts, old swab cups, sand plugs, etc.) lodged in the tubing preventing the free travel of the plunger. Run a wireline gauge ring. In addition to foreign material in the tubing, the plunger may have damaged or bent parts impeding travel. Check to insure the pads on the plunger move freely.

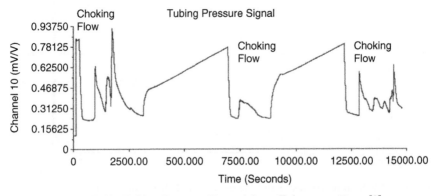

Figure 7-14: Tubing Pressure Record from Echometer Tests [8]

Sand behind the pads makes the tool stiff and difficult to fall and a brush plunger could be required for sandy production.

The Echometer [8] system can be used for on-site analysis and can reveal many of the potential problems just discussed or more. The "Smart Plunger," an instrumented plunger from PCS, Denver, can also analyze plunger travel and perform troubleshooting.

Plunger Won't Surface

Plunger lift operations require the tool to travel the full distance between the bottomhole spring and the lubricator each cycle. If the tool is not getting to the surface some or all of the liquid load will remain in the well. Isolating the source of the problem preventing the plunger from surfacing can be difficult. There are both mechanical and operational considerations.

The ideal travel time for a surfacing plunger is in the 500 to 1000 ft per minute range. This, however, is the ideal rate and many installations operate at much slower speeds. It is important, therefore, that ample time be given for the plunger to travel to the surface. It the plunger has been given sufficient time to surface (corresponding perhaps to an equivalent 100–200 ft/min rise time), then other problems must be investigated.

First inspect the system for mechanical malfunctions. Most of the mechanical problems that would prevent a plunger from falling to bottom would also prevent it from rising to the surface. Debris in the tubing, tubing quality, and plunger damage can all prevent the plunger from reaching the surface. In addition to restrictions, however, conditions that prevent the plunger from sealing in the tubing can prevent its reaching the surface. These would include ballooned tubing, mixed tubing strings with changes in the ID, tubing leaks, and gas lift mandrels installed in the tubing, among others. Typically when the plunger encounters enlargements and loses its seal, it will stop traveling at that point. It is vital that well completion records be checked closely before installing a plunger lift system.

Finally, the plunger itself may have been damaged, preventing it from surfacing. Plungers equipped with bypasses may develop leaks, preventing an adequate pressure seal across the plunger. The plunger should be checked regularly for wear and loose parts. Although uncommon, plungers can come apart in the hole. In some cases where the plunger will not surface under normal conditions, it may be possible to bring the plunger up by venting the head gas to a low-pressure separator. This provides extra pressure differential across the plunger that may be sufficient to bring the plunger to the surface. If this fails, the plunger must be wire-lined out of the well.

Operational problems that would prevent the plunger from surfacing all have been discussed in previous sections. It may be necessary to go through the initial kickoff procedure again to ensure that the well is ready to begin normal plunger operations. It is important to make sure that the casing is allowed to reach the required operating pressure. It might be necessary to allow the casing to come to equilibrium before attempting another plunger cycle. If the plunger has been idle for a time, it may be necessary to swab the well and produce most of the liquids or shut the well in for a period to drive liquids in the formation, before attempting to start the plunger cycle.

Plunger Travel Too Slow

The speed with which the plunger travels to the surface can greatly affect the performance of the plunger lift system. Plunger travel speeds that fall below the suggested 750 ft/min can significantly reduce the efficiency of transporting the liquids. For high rate gas wells this may not be a critical problem since these generally have ample gas production to replace that lost in inefficiencies. On weak or marginal gas wells where gas production is low and all the available casing gas is needed to surface the plunger, this can be a very important issue.

Bear in mind, the plunger and liquid slug rise with the aid of the gas stored up in the casing annulus with some help from formation production as well. If there is not a large volume of casing gas available or if it takes long shut-in times to rebuild casing pressure, the maximum possible number of plunger cycles per day is less. Experience has shown that the slower the plunger travels, the less efficient it becomes and the more gas it takes to move it to surface as gas leaks past the plunger. The seal between the plunger and the tubing is such that some gas always slips past the plunger, reducing its effectiveness since the pressure below the plunger is larger than above the plunger. When the plunger is traveling within the optimal speed range (750 ft/min–1000 ft/min), this gas slippage is presumed minimal. As the travel speed falls below the optimum, however, the amount of gas slippage is increased dramatically. This means that more of the casing gas is used each cycle, and so the shut-in (or build-up) time is longer. Ultimately this results in fewer cycles per day, which generally amounts to less liquid production per day. It is important to maintain plunger speeds near the optimum so as not to waste valuable casing gas, particularly on low rate wells. Note that many of these guidelines are from experience and may not have been tested extensively, so questioning and testing standard procedures for your particular wells is not a bad idea.

There are a number of ways to increase the plunger rise speed while conserving casing gas and maintaining adequate liquid production. The plunger travel speed is a function of the size of the liquid load and the amount of net casing pressure (less sales line pressure), plus the rate that the head-gas is removed at the surface. As mentioned earlier, the plunger speed also has a major effect on the efficiency of the plunger seals. It is important to first analyze the well conditions and determine whether smaller slug size or a higher casing operating pressure is warranted.

By raising the casing operating pressure, more effective pressure is exerted against the formation, thus lowering the liquid influx and reducing the slug size.

Reducing the size of the liquid load then allows higher plunger rise speeds. Similarly, lengthening the shut-in time again raises the net pressure on the formation while increasing the amount of gas in storage in the casing annulus. With roughly the same liquid load (or possibly less) but more compressed gas in the annulus at a higher pressure (more energy), the speed of the plunger again is increased.

On the other hand, reducing the sales line pressure has the same effect as increasing the casing pressure on the plunger travel time (more differential pressure across the plunger) without the adverse effects of longer shut-in periods and more pressure against the formation. Plungers with more efficient seals can also operate with reduced plunger travel times, by reducing the amount of gas slippage. Often just replacing plungers with worn seals will have a dramatic impact on performance.

Finally, plunger performance can be improved by rapid evacuation of the head gas (the reverse of trying to choke back gas surges to keep them on a recording chart) above the liquid slug. This could require replacing an existing orifice plate with a larger size, or opening up a choke or enlarging the dump-valve trim to allow greater use of the gas that is available.

Velocity controllers control the flow time and the build up time to maintain the correct rise average velocity but do not necessarily trend to low operating pressures and shorter cycles needed for production optimization.

Plunger Travel Too Fast

A plunger traveling up the well too rapidly could have bad consequences. Although the efficiency of the plunger sealing mechanism is not dramatically affected by higher speeds, well safety and equipment longevity dictate that the plunger rise speed be maintained below the 1000 ft/min maximum. The plunger and lubricator undergo fairly severe punishment under normal operating conditions. As the plunger speed increases the impact force imposed on the lubricator by the plunger increases roughly by the square of the speed. Although the plunger and lubricator are designed to withstand plunger impacts under normal speeds, higher speeds can quickly wear out and destroy both. In general, the economic benefits brought about through longer equipment life far outweigh those of shorter plunger travel times. A large plunger coming up dry such as for 2-7/8 inch or 3½ inch tubing can cause the most damage.

From an operational standpoint, either decreasing the casing build-up pressure or increasing the size of the liquid slug can reduce plunger travel speed. This can be accomplished by allowing the well to flow for longer periods of time after plunger arrival at the surface. Another way to accomplish this is to reduce the shut-in period. Choking the well, however, to slow the plunger is not the recommended although it will sometimes accomplish the objective. Choking or operating with too large of a liquid slug reduces production.

One reason that a plunger is coming up too fast is that even though there was liquid in the tubing when the plunger fell, the liquid can be displaced from over the plunger to the casing during the shut-in period. This could be due to bubbles entering the tubing during shut-in, or perhaps the casing liquid is dropped below the tubing end that might accelerate the loss of liquid from over to under the plunger.

One method to control this is to run a standing valve below the bumper spring. However, a standing valve would trap any random slug that might be too large to lift, and then you could not raise the tubing pressure to push the slug below the plunger to start the cycle again. The standing valve would hold the large slug over the plunger regardless of pressure changes.

One common method used to attack this problem is to use a standing valve, but notch the seat of the valve so it will leak. It will then give some resistance to liquids leaking back to below the plunger during the build-up cycle, but liquids can still be forced below the plunger through the leak if the slug should be too large to continue the cycles.

In general low pressure, low liquid rate wells should probably all be equipped with a standing valve unless sand or scale or such dictate otherwise.

Another method is to use a device: a new spring-loaded seat on the standing valve (Figure 7-15). The standing valve holds liquids over the plunger during the off cycle, but if the need arises to add tubing pressure to pressure liquids back below the plunger, then enough pressure can be applied to force the seat down and allow the liquid to leak back from over to under the plunger.

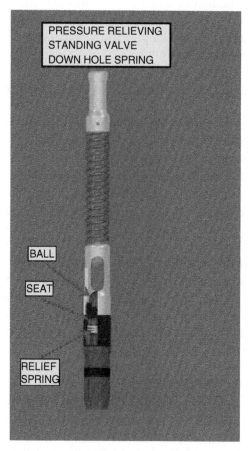

Figure 7-15: Schematic of a Spring-Loaded Seat for a Standing Valve to Be Placed below the Bumper Spring (Ferguson Beauregard, Tyler, Texas)

7.6.9 Head Gas Bleeding Off Too Slowly

Bleeding the head gas off too slowly can reduce the differential pressure needed to surface the plunger. The slower the bleed, the less the differential pressure across the tool and the less the chance of the plunger surfacing. The faster the head gas is allowed to bleed, the better the plunger performance.

Small chokes and high flow line pressures act as large barriers for the head gas to overcome, keeping the system from performing at optimum efficiency. Getting rid of the head gas as quickly as possible is critical. If it is necessary to choke the well, the choke should be as large as possible. To accomplish this it may be necessary to modify the surface facilities, but the benefit of doing so far outweighs the cost. If the sales line pressure is too high, then efforts should be directed toward reducing that pressure although this is a potentially expensive process that may require compression.

7.6.10 Head Gas Creating Surface Equipment Problems

A common complaint about intermittent operations is that they create problems with the surface equipment and gas measurement. Plunger lift falls within this category.

When a well fitted with plunger lift is first opened, generally a surge of high-pressure gas is forced at high rates through the surface equipment. Often the surface equipment was designed for an average flow rate and cannot handle the short duration surge that ends up going off the charts. One common, but not recommended, way of handling this problem is to install a positive choke in the flow line. Although the choke will restrict the initial gas surge to manageable levels, it will also restrict the flow of the remainder of the gas and the liquid slug. In particular, when a liquid slug passes through a gas choke the flow is drastically reduced, presenting a wall of liquid to the plunger that has a similar effect as closing a valve. The consequence of this is almost always a loss of production.

Fortunately, the problem often is negated once the well has been optimized. If this is not the case, however, other methods can be employed. One of the most effective ways to correct the problem is to install a valve with a throttling controller (discussed earlier) to limit downstream pressure while allowing the motor valve to be opened slowly to minimize production loss. This type of controller can be optimized to adjust to the pressure capabilities of the surface system and can therefore eliminate problems like selling off the chart, overpressuring separators and surging compressors. Finally, in installations where several wells are on plunger lift the surge effects can be negated by producing a number of wells through a manifold. In this manner, the surges produced by various wells can be timed to occur at different intervals and any single surge will make up a smaller percentage of the total flow and therefore be less likely to peg the sales meter.

7.6.11 Low Production

Optimizing or fine-tuning a plunger lift well can make a difference in the production. Consider testing short flow times to bring in small slugs of liquid. Then short build-up times required to build smaller casing pressures are required to lift smaller liquid slugs. The result is a lower average flowing bottomhole pressure and more production. Limits are that a too-short flow period could result in no liquid slug and a too-short shut in period would not allow the plunger to reach bottomhole.

In general whatever can be done to lower the average casing pressure per cycle will add to gas production.

7.6.12 Well Loads Up Frequently

Many wells are found to be very temperamental, where any small change in the operation can greatly affect their performance. Marginal wells tend to be particularly sensitive and are often easily loaded up. Liquid loading on a plunger lift well is usually a result of too long of a flow time or too little casing pressure during the shut-in period. Also trying to run plunger lift in small tubing can aggravate this problem.

Generally, imposing a more conservative plunger cycle can alleviate liquid loading of a plunger lift well. This means, as stated earlier, higher casing operating pressures and longer shut-in periods. Once the cycle has been changed, the well should be allowed to stabilize, which might take several days. Then continue with the optimization procedures outlined earlier, making only small incremental changes to the system times and pressures and then allowing the well to achieve stability between each change. It is possible to eventually adjust the well back to the original cycle settings once the well has had a chance to clean itself up.

If a well is completely loaded with liquids then it must be brought through the kickoff procedures from the beginning. First shut the well in and allow it to build pressure. With the well loaded, it may be necessary to swab the well to clean it up before starting the kickoff. Remember to work slowly, making small incremental changes to the system and then allowing the system to become stable before continuing to the next step. Many new controllers now adjust cycle times and pressures to follow optimization algorithms.

Operation with Weak Wells

Two methods are mentioned here using plunger lift for weak wells. One is use of the casing plunger and the other is using a side string for gas injection.

7.7 TWO-PIECE PLUNGER: TYPE OF CONTINUOUS FLOW PLUNGER

A new two-piece plunger (MGM Well Service, Corpus Christi, TX), shown in Figure 7-16, is designed to trip to bottom while the well is producing at considerable rate. In some wells, the plunger falls to the bottom while the well is producing at 1,000 Mscf/D or more. Both pieces

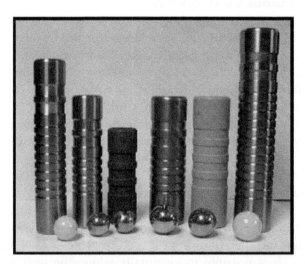

Figure 7-16: New Two-Piece Plunger Concept with Plunger Hardware (Pacemaker Plunger, a Division of MGM Well Service, Corpus Christi, Texas, Now Available through IPS)

Continuous Flow Plunger Cycle

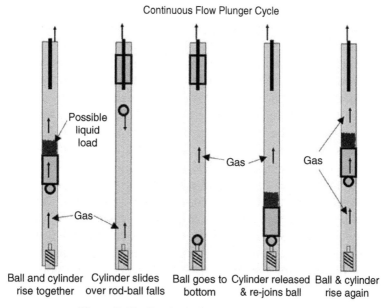

| Ball and cylinder rise together | Cylinder slides over rod-ball falls | Ball goes to bottom | Cylinder released & re-joins ball | Ball & cylinder rise again |

Figure 7-17: Continuous Flow Plunger Cycle

of the two-piece plunger have considerable bypass area when they are falling independently in the well, allowing the well to produce around the bottom piece (the ball) and through the top piece (the piston). They join at the bottom and are held together by the flow from the zones below as it pushes the plunger (now one unit) and any liquid in the tubing to the surface as a conventional plunger system would. The surfacing plunger strikes a shifting rod and a gas powered catch cylinder. The shifting rod separates the two pieces and the piston is held at the surface by the catch cylinder or in some cases by just the flow around the cylinder. The ball falls back to the bottom to await the arrival of the piston. When released from the surface the piston arrives at the bottom of the well and joins with the ball, beginning the process again. The cylinder can be released by a short shut-in time so that pressure and fluid drag will cease holding the plunger at the surface. If the arrangement is not such that pressure and drag are holding the plunger cylinder at the surface, then a mechanical catch system may be employed.

The plunger can trip to the bottom at speeds in excess of 1,000 ft/min or faster, while the well is flowing at a considerable rate. The high round trip speed allows the plunger to lift smaller amounts of liquid with each trip so it can make more liquid per day with less average bottomhole pressure than conventional plunger lift systems.

Another advantage of the two-piece plunger system is that it seems to perform very well without using the casing/tubing annular volume for pressure storage. The plunger is stated to rely more on volume than trapped pressure to move the plunger to the surface. The two-piece plunger works in 2-7/8-inch slim hole or wells with a packer and no communication with the annulus. Wells with on-site compressors usually are adapted to the plunger, because the shut-in time of only seconds has almost no effect on the suction pressure of the compressor. The smaller liquid loads have less effect on suction pressure, and the compressor may not need a recirculating valve. A shut-in time of only seconds does not create high spikes in wellhead pressures or volumes. The effect of the two-piece plunger is similar to a normal well flowing head, so it should not be necessary to oversize the compressor to accommodate the volume spikes common with the conventional plunger systems.

Since much of the operational practice and some of the feasibility charts in this chapter for conventional plunger systems, consider the energy stored in the casing before opening the well to allow the plunger to rise, then these practices and charts should not be applied to the use of the two-piece plunger.

A general rule of when the two-piece plunger will work is when the flow rate is still above 80 percent of critical. Rules of when it no longer applies include when a shut-in time of over 20 minutes is needed to continue operation of a two-piece plunger then a conventional plunger, perhaps with a better seal, should be used instead.

So with the many plunges available, how do you select one for your application? See the next section.

7.8 SELECTION OF PLUNGER

This section is by Bill Hearn, Weatherford; see his biography, earlier.

A traditional plunger lift system is defined as the use of one plunger with a turbulent seal to travel the length of the tubular to lift the fluid from the bottom of the wellbore to the surface. A plunger lift system is categorized into two major divisions: continuous flow and conventional plunger lift. Continuous flow refers to a well that does not require down- (or off-) time to build pressure in order to cycle the plunger. In this division, the plunger can fall against the natural flow to the bottom of the well, reset, and then return to surface using only the velocity of the flowing gas. Excellent examples of continuous flow include the Pacemaker™ (1) (Figure 7-18), spiral RapidFlo™ (2) (Figure 7-18), padded RapidFlo™ (2) (Figure 7-19), and the FreeCycle™ (Figure 7-20). Conversely, the conventional plunger lift system refers to a well that does require down- or off-time to build pressure in order to cycle the plunger. The two divisions include both solid plungers with and without bypass, and in some cases collapsible plungers (Figure 7-20). Both divisions are designed for some minimal fall time to reach the bottom in order to reset. The evaluation of both divisions includes one complete process for each different scenario.

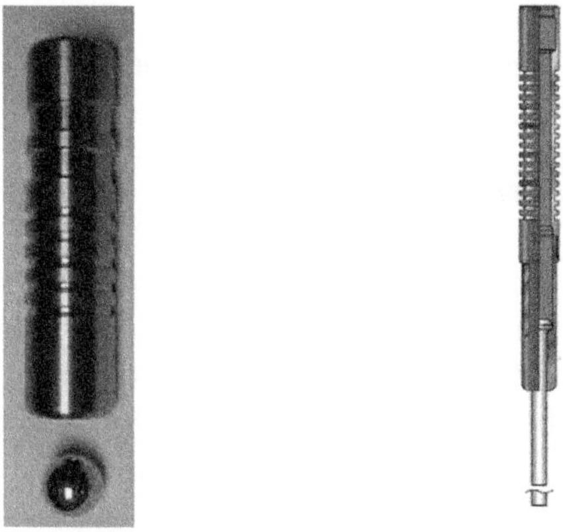

High Speed Continuous Flow Plungers

Pacemaker™   RapidFlo™

Figure 7-18: High Speed Continuous Flow Plungers

Padded Continuous Flow Plungers

MacLean™ and Padded RapidFlo™

Figure 7-19: Padded Continuous Flow Plungers

Pictures of Conventional Plungers

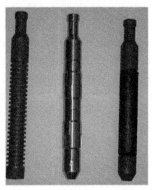

Figure 7-20: Conventional Plungers

7.8.1 Continuous Flow Plunger Lift

Continuous flow plunger lift references the concept that if gas can maintain a minimal velocity necessary to make a turbulent seal, without shut in-time to build pressure, then a plunger should travel to the surface once it has closed its bypass method. For example, some plungers require that a valve be reset upon arrival on a solid contact, usually a bumper spring, whereas others will be reset once the ball and sleeve make contact and seal. The development of the turbulent seal will occur only if the gas moves with enough velocity to make this seal.

7.8.2 Types of Continuous Flow Plungers

Continuous flow plungers are divided into solid ring type: contact padded or brush plungers, each with their own application. Depending on both tubing and plunger conditions, it is usually necessary for over 15 ft/s velocity to ensure a solid ring arrival and a minimal of 10 ft/s velocity for a padded plunger or new brush. The lower velocity requirement is a result of the better metal-on-metal or metal-on-brush seal that is obtained in a brush or padded plunger.

Solid Ring (High-Speed Continuous Flow)

In general, high-speed continuous flow plungers are a ring type of plunger lift with some form of valve on the bottom. Examples include the Pacemaker™ and the solid ring RapidFlo™. Advantages to these plungers include reduced down-time, minimal amounts of moving parts, relatively low cost, and overall operating simplicity. The greatest advantage to the operators is the number of trips in a day. With fewer moving parts, the plungers are able to travel with minimal fluid loads above the spring without causing damage. The plungers are especially effective in low line pressure/high velocity applications, or in single well compression where down-time is critical to compressor operation. High-speed continuous flow plungers also have great application in winter conditions where continually moving plungers can reduce hydrate problems. Disadvantages to solid ring plungers are the necessity of the 15 ft/s gas velocity and the requirement of some moving parts to operate the internal valves.

Solid Contact (Padded and Brush)

Continuous flow plungers are generally a brush, pad, or any combination of the two and have some form of valve system. Some are mounted on the bottom, similar to the padded RapidFlo™ and FreeCycle™, whereas others contain an internal valve system. The advantages to these plungers are similar to those of the solid ring plungers (earlier), including the reduction of down-time. However, depending on the seal, these plungers have the potential to travel with as little as 10 ft/s velocity. The primary disadvantage of solid contact plungers is the possibility of damage due to added moving parts.

7.8.3 Conventional Plunger Lift

Conventional plunger lift requires a build-up pressure in order to cycle a plunger and is categorized into padded, solid ring, brush, or any combination of the three. In some instances with the same concept, it is necessary to have a valve bypass for quicker fall times. The plunger falls to the bottom and once enough build-up pressure is obtained the plunger can cycle to the surface. Padded plungers generally provide the best seal for extended periods of time, but require maintenance to avoid breakage. Solid ring plungers provide the least efficient seal, but require very little maintenance. Brush plungers, primarily used in sand in-flow wells, provide the best initial seal, but tend to wear out very quickly. Conventional plungers can also come in combinations with a variety of sealing capabilities.

7.8.4 Evaluation Process

In order to properly select your plunger, the evaluation process is extremely critical for a successful plunger lift program. The first step in the evaluation process is to gather data in order to decide whether the well is an appropriate plunger lift candidate.

Data Gathering

There are several factors to consider during the data gathering process:

- The more data collected allows for accurate estimates of well potential using decline curves.
- More recent data provides the most current possible picture of the well's potential.
- Producing pressure data can help indicate other potential problems and conditions.
- Downhole and surface details help to identify potential problems for a plunger candidate.

Data is divided into four categories:

- Past production data, such as flowing and static pressure data, and producing pressure data (casing, tubing, line pressures)
- Current production data and pressure, such as flowing and static pressure data, and producing pressure data (casing, tubing, line pressures)
- Downhole details such as tubing detail and perforation depths
- Surface information including wellhead, facility, and gathering system information

Past Production Data usage:

1. Establishes a decline curve. This may be difficult due to the lack of information, but is important if available.
2. Establishes what the current flowing rate and pressures might be if the well were following this decline.
3. Establishes potential economics: uses production data versus the actual current production.

4. Attempts to identify the point at which the well acquired liquid loading issues and compares this to the wells' predicted critical velocity.
5. Establishes the potential fluid rate of the well.
6. Establishes the velocity of gas throughout the tubing string. In order to properly calculate the velocity, it is very important that the rates used are the unloaded state conditions. Graphs of flow versus pressure estimate into which velocity range the well falls. These graphs will change when all well data is entered including all reservoir properties (Figures 7-21–7-23).

Current Production Data and Pressure

Current flowing data indicates the magnitude of the liquid loading problem.

1. The casing-to-tubing differential, when open-ended, indicates the amount of liquid currently in the tubing and identifies additional backpressure.

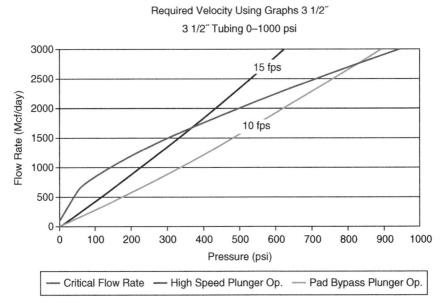

Figure 7-21: Required Velocity Using Graphs 3½ Inch Tubing

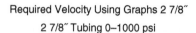

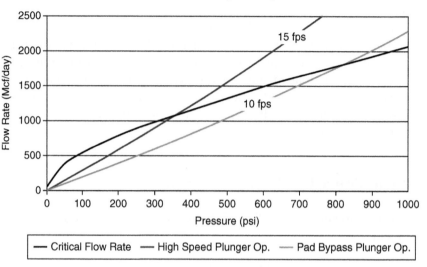

Figure 7-22: Required Velocity Using Graphs 2-7/8 Inch Tubing

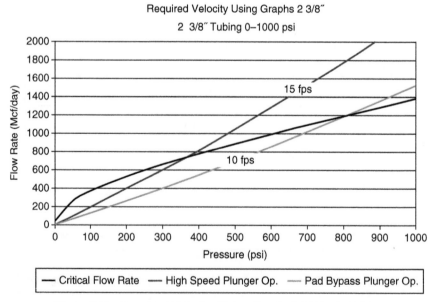

Figure 7-23: Required Velocity Using Graphs 2-3/8 Inch Tubing

2. Flowing gradient highlights fluid presence as well as Flowing Bottom Hole Pressure (FBHP) in packer completions.
3. Varying Liquid Gas Ratio (LGR) can indicate either erratic fluid entry or liquid loading.
4. If available, minute-by-minute flowing data points to sweeping recognized on flow charts as a "painting" effect.
5. Operator input shows what is being done to keep the well producing. Operators can also provide insight to potential losses (e.g., losses of blowing the well to atmosphere or shutting in for build).

Down-Hole Details

1. Tubing Depth
   a. Ideally, tubing depth is far enough into the perforations so that the most prolific part is completely unloaded. It is possible to unload perforations that are below the End of Tubing (EOT) using dead-strings or critical velocity reduction tools.
   b. A consistent Internal Diameter (ID) of tubing from top to bottom without tight spots that can't be broached.
   c. One nipple close to the bottom of the tubing.
2. Packer/packer-less completion
   a. A packer in the hole makes conventional plunger lift more difficult because it depends on the casing energy.
   b. Packers will not necessarily result in problems if the well is a potential continuous flow candidate.
3. Perforations
   a. Clear (no fill).
   b. No skin damage or blockage.

Surface Information

1. Can the surface facility handle the fluid as it arrives in a slug?
   a. Can the gathering system handle the high and low flow rate manner in which the gas is delivered in conventional application?
2. Is the wellhead sizing correct, do the IDs match the tubing?

Process of Elimination

Once the data is gathered, a step-by-step process of elimination determines whether the well is a good plunger lift candidate.

1. Consider the downhole and surface data gathered earlier. This information is important to project cost if there are major problems. In some cases rig work is required to change downhole configurations. Other instances may require major piping changes to modify surface equipment. These factors are important to keep in mind as the evaluation continues.

2. If possible, establish a decline curve, and the current potential flowing and actual rates. This will be important for establishing economics as well as calculating velocity (see next step). If no decline is established due to the well always flowing below critical velocity, an In-Flow Performance Relationship (IPR) is appropriate if current static and flowing bottomhole pressure with rates are available. This information estimates the potential due to the decrease in flowing bottomhole pressure that should occur with a plunger lift installation.

3. Use the potential rates and the current surface pressures to establish a velocity of the gas. Also, estimate a bottom hole velocity using an unloaded condition. This is achieved by using either a velocity equation or a graph of velocity versus depth for the tubular. The other option is to use a graph of flow rate versus pressure with plotted velocity that also estimates which bucket a well matches (Figures 7-21–7-23).

   a. +15 ft/s wells can operate as continuous flow with a high-speed sold ring plunger. The amount over the 15 ft/s will determine how much tolerance the well has to run in this type of system, as well as how much the wear and tear on the plunger will affect the well production. Wells that barely meet this criterion are better candidates for a padded continuous flow plunger.

   b. 10–15 ft/s wells can function as continuous flow; however, these wells will usually require a better seal than the solid ring plunger can provide. In this case, consider a padded continuous flow plunger. For wells that barely meet these criteria, consider a conventional plunger.

   c. Consider <10 ft/s wells for conventional plunger lift because they are going to require a build-up pressure to establish the necessary velocity to bring a plunger upward.

7.8.5 Deciding Which Type of Plunger to Use

The preceding process and information contributes to the decision as to which type of plunger is best suited for the well. Using this information, once the bucket is established, reevaluate fluid rates and the amount of fluid per load to ensure proper running conditions.

Solid Ring Continuous Flow Plungers

Solid ring continuous flow plungers are used under the following circumstances: first, when flowing conditions require the maximum number of cycles due to fluid intake, and second, when the velocity is readily available. Although calculating the maximum number of cycles can be difficult, it is possible to make estimates. Using the velocity of the gas, the rise time can be accurately estimated. Depending on the bypass area, the fall time can also be calculated; however, fluid gradients, plunger, tubing, and changing flowing conditions make it difficult to match exactly. A common rule of thumb is that the fastest the plunger will fall with flow velocity over 15 ft/s is the same as half the rise time. Meanwhile, the slowest a continuous plunger will fall will be around three times slower than the speed of its rise time. Any rate slower than three times the rise time indicates that the plunger dropped against flow rates and above critical velocity. If the well flows above critical velocity, there is no benefit to the plunger falling as the well is able to unload on its own. For example, a 9,900 ft well with 15 ft/s velocity would be 5 minutes fall and 10 minutes rise, leading to 15-minute cycles, four per hour, 96 per day maximum.

Ideally, an attempt should be made to maintain a minimum of 100 ft cushion of fluid. For example, in 2-3/8 in tubing this would be around one quarter of a barrel per run, or a minimum of 24 barrels per day of fluid if the cycle is continuous. Based on the well's flowing capabilities, it may be able to produce more. If the well makes less fluid, some afterflow is recommended between cycles to acquire a fluid load.

Padded Continuous Flow Plungers

Padded continuous flow plungers are used when fewer cycles are required and 15 ft/s velocity is not available. A padded continuous flow plunger will provide a better seal necessary to travel at lower velocity. This type of a plunger usually has a method, some form of orifice assembly for slowing down the fall to the bottom in order to avoid damage to the bottom hole equipment. This will usually result in less cycles being possible.

The rise time is estimated by velocity, but the restrictions should be changed in order to keep the fall time below 10 ft/section. For example, in an application in a 12,000 ft well containing 1,000 ft of fluid and gas moving with a velocity of 10 ft/s, the rise time is approximately 20 minutes and the fall time is less than 20 minutes, which makes the maximum number of cycles per day 36. With one-quarter barrel per run, a minimum of nine barrels a day is needed to run a continuous flow plunger. However, if that is not the amount produced, either after-flow is needed or a smaller orifice to slow the fall time further.

Conventional Systems (added use for each type of plunger conventionally)

When establishing how much fluid a conventional system can move, there are many good methods from which to choose. The classic Foss and Gaul method developed in the 1960s provides a very mechanical approach to plunger lift [1]. This method can generally be used to establish if a plunger can operate mechanically. If an operator uses the Foss and Gaul method, it allows for the use of the decline to show potential rates.

Slightly more complex is the Hacksma method, which includes a reference to the IPR and with enough data provides more information on daily rates. Hacksma uses the Foss and Gaul method to calculate the average casing pressure then considers the gradient and uses this on the IPR to calculate potential uplift [2].

If an understanding of the well dynamics in addition to the reservoir characteristics is substantial enough, something more complex like the Dynamic method of Jim Lea is applicable. This method is substantially more complete and includes instantaneous velocity, acceleration, and inflow capability [3]. Ultimately, the Dynamic method should produce the most accurate results, but any model must compare to the actual results in the field [4].

Once a method is chosen, the next step is to apply the actual physical results, including accurate plunger fall times. Typical estimates of fall times for padded plungers may be as slow as 150 ft/min in gas and below 100 ft/min in fluid where bypass plungers may fall as fast as 1,800 ft/min in gas and slow only to 1,000 ft/min in fluid if the well is shut-in. The fall velocity is a key component in establishing the number of cycles available in a day. Thus, an Echometer plunger tracking program and hardware provide a solid basis for field trials. These field trials establish fall velocity as well as a good basis during the initial clean-up and line-out phases. By checking the fall time with an acoustic tracking device, they assure that the plunger reaches the bottom and the length of time that it takes to do so. Depending on the plunger selected, one that falls in a 9,000 foot well may take as long as 90 minutes or as quick as five minutes to reach the bottom. The key aspects to consider in a conventional plunger system are the speed at which the well builds pressure to cycle, and how much fluid the plunger falls into in order to avoid damage. These key aspects are vital in the evaluation process to help estimate uplift.

Depending on the results of your evaluation process, the selection of the plunger will help to assure the lowest flowing bottomhole pressure and maximum in-flow.

7.8.6 Progressive/Staged Plunger Systems

In wells where conventional plunger systems struggle due to low gas to liquid ratios, high line pressure, packer, or slimhole completions without sufficient GLR, or low reservoir pressure, there is an application for Progressive™(4) or Multi-Staged™(5) plunger lift. A staged plunger system essentially operates as two separate, yet intricately linked systems within the same well (Figure 7-24). In a conventional system, the gas above the plunger and fluid level is produced rapidly

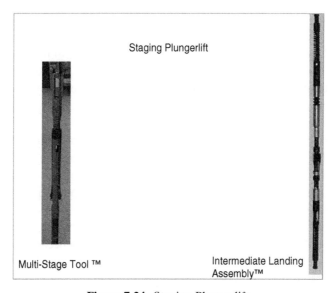

Figure 7-24: Staging Plungerlift

down the line in order to create a differential. This differential allows the casing or near wellbore gas to expand into the tubing, forcing the plunger and its fluid to surface by attempting to drive gas past the plunger at a high velocity that forces a turbulent seal. In staging plunger lift the gas above the top system produces the same as in a conventional system. However, the gas that sits below the top stage and above the fluid in the bottom system is used as the energy source for the top plunger. This allows the bottom plunger to require only the energy to travel one-quarter to halfway up the well depending on the design. This makes the plunger system significantly more efficient. It also reduces the necessary GLR to drive the plunger system and decreases the amount of pressure necessary. For this reason, staging a plunger well significantly increases the operating window and decreases the necessary pressure to produce the plunger that extends the well's life.

From the bottom or near the end of tubing to surface, the composite of a staged plunger system usually requires the following: a bottomhole bumperspring with hold down, a conventional plunger (usually a spiral but dependent on conditions), an Intermediate Landing Assembly™(6) or Multi-Stage Tool™, another conventional or quick-trip plunger, and a typical surface lubricator with control system.

Typically a staged plunger system begins with a conventional plunger system evaluation due to low GLRs or pressure build that require significant shut-in time that the original system does not optimally lift the well. At this point, the well is evaluated using two or more systems. The results of the evaluation process will typically mean a decrease in necessary GLR between 1 and 3 Mcf/Barrel. The decrease in necessary pressure may be significant in a typical installation. Thus, the necessary casing pressure to drive the plunger system decreases between 50 to, in some extreme cases, 500 psi or greater. This results in lower average flowing bottomhole pressures and increased inflow. In most cases, the staged plunger system will normally mean two to three times more trips with smaller controlled load sizes due to the decreased gas usage per cycle. Usually the necessary casing pressure will decrease, which provides significant value in high line pressure applications where the initial casing pressure necessary is quite high.

Although staging plunger lift is a significant advantage to wells performing beyond conventional applications, it does not necessarily improve production on typical conventional candidates. This is due to the bottom plunger system functioning as a choke during after-flow, causing it to run dry if the well cycled at a high frequency. The staged plunger system, unlike the conventional, also requires slickline work in order to inspect the bottom plunger due to the fact that it does not travel to surface.

7.9 CASING PLUNGER FOR WEAK WELLS

The casing plunger travels in the casing only and there is no tubing in the well. The plunger senses when a head of liquid appears above the plunger and then the internal bypass valve closes and the well gas production lifts the plunger and slug of liquid to the surface. Then at the surface, the plunger internal valve will open and the plunger will drop. The casing plunger rises and falls slowly. It has rubber cups that fit the casing. The cups will not last long enough for satisfactory service if the casing is very rough. Figure 7-25 is a picture of a casing plunger:

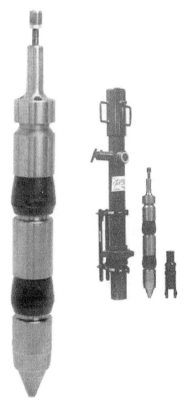

Figure 7-25: Casing Plunger Showing the Rubber Cups That Seal against the Casing ID and the Lubricator and Downhole Casing Stand (Multi Products Co., Millerburg, OH)

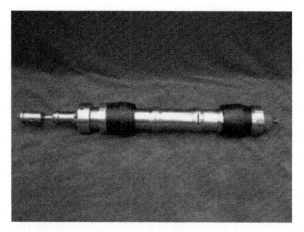

Figure 7-26: OptiFlow Casing Plunger (www.optiflow.ca)

In summary, the casing pressure can be used for shallow low pressure wells with relatively good casing condition to extend the life of the well to very low pressures. The casing plunger can also be made for 5½ inch casing. Multi-Systems, IPS, and others can supply a casing plunger.

The OptiFlow casing plunger pump (Figure 7-26) has a different operating principle.

- Between cycles, the pump rests on the catch mechanism of the lubricator at surface.
- As the cycle begins, the pump is released from the surface lubricator with the internal valve open. The pump freefalls in the casing allowing gas and liquids to pass through the open valve. Gas production is uninterrupted.
- The pump's internal valve closes upon contact with the down hole stop. The formation liquids, which have accumulated in the casing aboe the perforations, are effectively trapped and isolated from the gas in the reservoir by the positive sealing swab cups.
- As formation gas enters the casing, the pump and column of trapped fluid ascend to the surface due to the differential pressure produced by the reservoir.
- The pump reaches the surface, enters the surface lubricator and is captured by the catch mechanism. The internal valve opens which

allows continuous uninterrupted gas production to the sales line until the pump is released to start the next cycle.

7.10 PLUNGER WITH SIDE STRING: LOW PRESSURE WELL PRODUCTION

Plunger lift with a side string can be used to produce gas or oil wells with low bottomhole pressures where a source of higher pressure make up gas is available at the well head.

A plunger lift system in combination with a side string for injecting make up gas and pressure for lift is used for this system.

The plunger lift system with side string injection requires that the tubing be removed from the well. As the tubing is run back in the hole, ½- or ¾-inch coiled stainless steel tubing is banded to the production tubing. On the bottom is a standing valve with a side port injection mandrel above it. Above the injection mandrel is a bottomhole spring assembly and a plunger.

Makeup gas is injected from the surface down the side string directly into the production tubing. As the gas enters the tubing, it is prevented from entering the wellbore by the standing valve. The gas is forced to U-Tube up the production tubing, driving the plunger ahead of it, which in turn removes liquid from the tubing.

The injection gas is injected for only a short period, just long enough to cause the plunger to surface. Once the plunger surfaces, the well is allowed to bleed down to sales line pressure. As this occurs, liquid enters the production tubing from the wellbore and the plunger drops back to bottom on its own weight.

As the plunger continues to remove liquid from the wellbore, the liquid level in the casing drops. As the liquid in the casing drops, the perforation zone is relieved of hydrostatic pressure, and formation gas enters the casing. The formation gas is produced out the casing.

PLSI (Midland, TX) developed this technique for low bottomhole pressure wells in 1992. Initial installations occurred in the Antrium gas zones of northern Michigan. The technology has been economic in this area and to date over 500 installations of this system are in place.

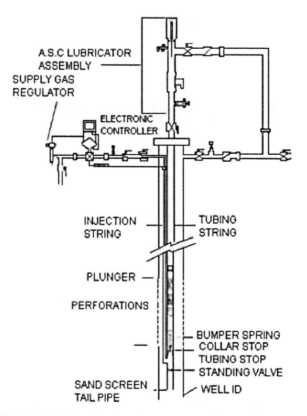

Figure 7-27: Side String Gas Supply for Plunger Lift (PLSI, Midland, TX)

Side- String Summary

If you have a compressor or a source of higher pressure gas, you can use this concept with the side-string to lift liquids from low pressure gas wells where it is not feasible to run a conventional plunger system.

7.11 PLUNGER SUMMARY

Plunger systems work well for gas wells with liquid loading problems as long as the well has sufficient GLR and pressure to lift the plunger and liquid slugs.

Plunger lift works well with larger tubing so there is no need to downsize the tubing. Conventional plunger lift works much better if there is no packer; this can be a problem if the old packer should be removed.

Plunger lift can take the well to depletion although the recoverable production may not be quite as much as using a more expensive beam pump system, for example, to pump liquids out of the well in the latter stages of depletion.

The two-piece plunger concept is discussed, which requires little or no shut-in period and also possibly being able to operate better with a packer present. Application may be when the rate is at least 80 percent of critical or see Weatherford criteria. Other bypass plungers can run on the continuous flow cycle (few seconds shut-in).

7.12 REFERENCES

1. Lea, J. F. "Dynamic Analysis of Plunger Lift Operations," Tech. Paper SPE 10253, Nov. 1982, 2617–2629.

2. Beeson, C. M., Knox, D. G, and Stoddard, J. H. "Part 1: The plunger Lift Method of Oil Production," "Part 2: Constructing Nomographs to Simplify Calculations," "Part 3: How to User Nomographs to Estimate Performance," "Part 4: Examples Demonstrate Use of Nomographs," and "Part 5: Well Selection and Applications," *Petroleum Engineer*, 1957.

3. Otis Plunger Lift Technical Manual, 1991.

4. Lea, J. F. "Plunger Lift Versus Velocity Strings," *Journal of Energy Resources Technology*, Vol. 121, December 1999, 234–240.

5. Ferguson Beauregard Plunger Operation Handbook, 1998.

6. Foss, D. L. and Gaul, R. B. "Plunger Lift Performance Criteria with Operating Experience–Ventura Field," *Drilling and Production Practice*, API 124–140, 1965.

7. Phillips, D. and Listiak, S. "How to Optimize Production from Plunger Lifted Systems, Pt. 2," *World Oil*, May, 1991.

8. McCoy, J., Rowlan, L., and Podio, A. L. "Plunger Lift Optimization by Monitoring and Analyzing Well High Frequency Acoustic Signals, Tubing Pressure and Casing Pressure," SPE 71083, presented at the SPE Rocky Mountain Petroleum Technology Conference in Keystone, CO, May 21–32, 2001.

Additional plunger lift references:

Hacksma, J. D. "Users Guide to Predict Plunger Lift Performance," Presented at Southwestern Petroleum Short Course, Lubbock, TX, 1972.

White, G. W. "Combining the Technologies of Plunger Lift and Intermittent Gas Lift," Presented at the Annual American Institute Pacific Coast Joint Chapter Meeting Costa Mesa, CA, October 22, 1981.

Rosina, L. "A Study of Plunger Lift Dynamics," Master Thesis, U. of Tulsa, Petroleum Engineering, 1983.

Ferguson, P. L. and Beauregard, E. "How to Tell If Plunger Lift Will Work in Your Well," *World Oil*, August 1, 1985, 33–37.

Wiggins, M. and Gasbarri, S. "A Dynamic Plunger Lift Model for Gas Wells," SPE 37422, Presented at the Oklahoma City Production Operations Symposium, 1997.

Hacksma, J.D. "Predicting Plunger Lift Performance," presented at the South West Petroleum Short Course, Lubbock, TX. October 5–7, 1981.

Avery, D.J. and Evans, R.D. University of Oklahoma, "Design Optimization of Plunger."

"Lift Systems," Paper SPE 17585, presented at the SPE international meeting on Petroleum Engineering held in Jinjin, China, November 1–4, 1988.

USE OF FOAM TO DELIQUIFY GAS WELLS

8.1 INTRODUCTION

Foams have several applications in oil field operations. They are used as a circulation medium for drilling wells, well cleanouts, and as fracturing fluids. These applications differ slightly from the application of foam as a means of removing liquid from producing gas wells. The former applications involve generating the foam at the surface with controlled mixing and using only water. In gas well liquid removal applications, the liquid-gas-surfactant mixing must be accomplished downhole and often in the presence of both water and liquid hydrocarbons.

The principal benefit of foam as a gas well dewatering method is that liquid is held in the bubble film and exposed to more surface area resulting in less gas slippage and a low-density mixture. The foam is effective in transporting the liquid to the surface in wells with very low gas rates when liquid holdup would otherwise result in sizable liquid accumulation and/or high multiphase flow pressure loses.

Foam is a particular type of gas and liquid emulsion. Gas bubbles are separated from each other in foam by a liquid film. Surface active agents (surfactants) generally are employed to reduce the surface tension of the liquid to enable more gas-liquid dispersion. The liquid film between bubbles has two surfactant layers back to back with liquid contained between them. This method of tying the liquid and gas together can be effective in removing liquid from low volume gas wells.

Campbell *et al.* [1] describe the foam effect on production of liquids using the critical velocity. Equation 3-3 is repeated:

$$V_t = \frac{1.593\sigma^{1/4}(\rho_l - \rho_g)^{1/4}}{\rho_g^{1/2}} \; ft/\text{sec} \tag{8-1}$$

where:

σ = the surface tension between the liquid and gas, dynes/cm
ρ = density, lbm/ft^3
The subscripts l and g indicate liquid and gas. Vt is the terminal velocity or the gas critical velocity.

Campbell *et al.* [1] discuss that foam will reduce the surface tension and therefore reduce the required critical velocity. The surface tension should be measured under dynamic conditions. They also discuss that foam will reduce the density of the liquid droplets to a complex structure containing formed water and/or condensate and gas. Weatherford has shown that a rule of thumb is that foaming water will reduce the critical velocity by about two-thirds in a gas well by reducing the surface tension and the liquid density in Equation 8-1 simultaneously.

Water and liquid hydrocarbons react differently to surfactants. Liquid hydrocarbons do not foam well. This is particularly true for light condensate hydrocarbons. The gas-condensate bubble dispersion can be accomplished but the resulting foam is not stable and will readily separate. Light hydrocarbon liquids must be continuously agitated to maintain foaming.

One reason why hydrocarbons do not foam well is because hydrocarbon molecules are nonpolar and therefore have less molecular attraction forces between molecules. On the other hand, water molecules are polar and can build relatively high film strengths with surfactants. When both water and liquid hydrocarbons are present in the wellbore, foam is created mainly within the water phase and the water foam assists in carrying along the liquid hydrocarbons. Laboratory observations indicate that when both water and light hydrocarbon liquids are present, the liquid hydrocarbons tend to emulsify and the foam is generated in the external water phase. The percent gas in the foam mixture at operating pressure and temperature is termed foam quality; for example, foam that is 80 percent gas is called 80 quality foam.

If this foam is caused to flow, a certain minimum stress will be required to overcome the interlocking of the bubble structures. This minimum stress is called a yield point. Thus, foams have an apparent viscosity that is dependent upon the shear rate operating within the moving stream.

The application of foam to unloading low rate gas wells generally is governed by two operating limitations: economics and the success of foam surfactants in reducing bottomhole pressure. Both limits are defined by comparison to other methods of unloading wells.

Low rate gas wells with producing GLRs between 1000 and 8000 cu ft/ bbl are among the better candidates for foaming although there is no real upper GLR limit. For high GLR wells, plunger lift may give better performance; that is, produce with less bottomhole pressure than foam. Downhole pumps may be better suited for the lower GLR ranges. The producing gradients expected with foam surfactants ultimately are controlled by the producing rates and well conditions and by the performance of specific surfactants in the well. Multiphase flow programs generally predict performance in gas wells where the foam is treated as the liquid (although it is both liquid and gas) in a two-phase system.

Table 8-1 lists some advantages and disadvantages for foam lifting of liquids that should be considered or evaluated before selecting this method for unloading gas wells.

Table 8-1
Well Diagnostics, Surfactant Selection, and Application
of Foam Assisted Lift

Advantages	Disadvantages
1. Foam is very simple and inexpensive method for low rate wells. Chemical costs are proportional to the liquid water rate.	1. Surfactant used may result in foam carryover or liquid emulsion problems.
2. No downhole equipment is required. (However, a capillary injection system may be very beneficial to low rate wells tending to produce in slugs.)	2. The foaming tendency for various systems depends on the amount and type of well fluids and on surfactant effectiveness. Well producing substantial condensate (say greater than 50% condensate) may not foam.
3. Method is applicable to wells with low gas rate where gas velocities may be on the order of 100 to 1000 fpm in the production string. The value is about 1000 fpm for critical velocity in un-foamed wells.	

8.2 FOAM ASSISTED LIFT (FAL)

By Butch Gothard and Brian Price

Butch Gothard is Director of Technical Services and Gas Enhancement Groups for Multi-Chem Production Chemicals. Mr. Gothard has 25 years of experience in the oilfield production chemical industry with the last 10 years focused primarily on the application of foaming agents to increase production in liquid loaded gas wells. He has developed gas enhancement programs and products for application across the United States, Canada, Europe, and the Far East. (E-mail: butch_gothard@multi-chemgroup.com)

Brian Price is Director of Technical Marketing and Development for Multi-Chem Production Chemicals. Mr. Price has worked in the oilfield production chemicals sector of the oil and gas industry for the last 20 years. His varied background includes laboratory, technical, field application, training, and management experience. He has had the good fortune to be involved with production chemical applications globally in many oil and gas production areas. (E-mail: brian_price@multi-chemgroup.com)

8.2.1 Introduction

Virtually all producing wells produce a liquid of some type—either hydrocarbon or water. For the production process to occur, liquid must be removed from the wellbore. Accumulated liquid can create a hydrostatic head pressure that impedes removal of fluid from the wellbore. As reservoir pressure declines, production rates are reduced. The production of a well will always result in a reduced reservoir pressure over the life of the well. Loading occurs on all wells. The question that must be answered for each well is, "When will this well load?"

Wells with low bottomhole pressure require a means of artificial lift to transfer fluid from the wellbore to surface treating equipment. Conventional methods of artificial lift are pumps (rod, hydraulic, ESP, etc.), gas lift, and plunger lift. Conventional methods of unloading wells include flaring, "blowing down," nitrogen stimulation, or "rocking." In recent years, foaming agents have been applied broadly with success as a means of artificial lift and for unloading loaded wells. Foam assisted lift (FAL) has become an integral part of extended production plans for many wells considered to be marginal producers.

Foaming agents have been used globally for removal of water from loaded wellbores for many years. Most producers think back to the days

of using soap sticks to unload wells. In some areas, liquid and/or pow-
dered detergents also have been utilized to create foam for liquid
removal from a wellbore. Often, the use of soap was a part of a last ditch
effort to get a well flowing—this was one of the tricks of the trade in
the oil and gas industry.

Recent applications indicate that FAL can not only be an integral
part of mature production, it can be utilized on adolescent producers to
improve production rates by returning wells to their unimpeded production potential. With gas consumption becoming an increasingly more
important part of global energy supply, producers are focusing on efforts
to support the demand by increasing production from all wells.

This section will address issues related to FAL that must be considered for long term success. Three steps for FAL success are (1) proper
well diagnostic, (2) proper foaming agent selection, and (3) proper
application and assessment.

8.2.2 Well Diagnostics

Evaluation of a well for FAL starts with properly diagnosing the well.
This process involves a bit of education, a bit of experience, and a bit
of speculation. Data collected for this effort is as follows:

- Well name
- Bottomhole temperature, static
- Flowing wellhead temperature
- Bottomhole pressure, flowing
- Bottomhole pressure, static
- Flowing wellhead pressure
- System pressure at each component
- Tubing inside diameter
- Casing inside diameter
- Packer depth (if applicable)
- Depth to end of tubing
- Depth to top perforation
- Total casing depth
- Capillary/velocity string/coil (if yes, information on capillary/velocity
  string/coil: size, material, depth)
- Complete water analysis
- Oil/condensate gravity
- Gas analysis

- Static fluid level
- Flowing fluid level
- Gas production rate
- Water production rate
- Oil/condensate production rate
- Well deviation
- Horizontal completion (if yes, diameter and length of each lateral)

A completion diagram, production history, and operator notations are also very useful for the evaluation.

The following information can be used to evaluate the well and characterize it into one of several categories.

Flowing Well—Partially Loaded—Producing Water

This is a well that flows below its potential. It maintains a measurable fluid level in the wellbore under flowing conditions. Measurable water typically is associated with condensed water from the vapor phase.

Flowing Well—Partially Loaded—Not Producing Water

This is a well that flows below its potential. It maintains a measurable fluid level in the wellbore under flowing conditions. It does not produce water, as condensation of water vapor occurs in the tubing and condensed water falls back into wellbore.

Flowing Well—Transient Loading

This is a well that flows at its potential some of the time and below its potential some of the time. Loading likely occurs due to transient production events. Water may cycle from the wellbore in slugs.

Loaded Well

This is a well that will not flow due to a standing fluid column exerting hydrostatic pressure. This may be due to production circumstance that allowed loading to occur. Typically occurs as reservoir pressure declines.

Well with Inflow Issue(s)

This is a well that has restricted flow into the wellbore. Water cannot migrate from the reservoir to the wellbore due to scale, iron sulfide, paraffin, sand, salt, and so on. Remediation must be performed to return this well to production.

Depleted Well

This well may be in a depleted reservoir. In some instances, the completion package for the well may not meet the reservoir's capability (basically an oversized completion). A determination should be made to distinguish reservoir depletion from "completion depletion."

Characterization of a well is key to returning the well to its full FAL production potential. Each of the well types mentioned exhibit unique challenges that must be addressed. Unlocking a well's potential can be expedited by correctly diagnosing the issue causing reduced production. It is important to note that most wells that producers initially select for FAL evaluation are (1) Loaded Well, (2) Depleted Well, or (3) Well with Inflow Issue(s). Many producers do not contemplate the application of FAL to flowing wells, yet this is where the most potential lies for increasing production and revenue. Each well requires relatively the same amount of time for evaluation. With most companies operating with limited resources, we strongly recommend that flowing wells be evaluated prior to nonflowing wells if a producer's goal is to increase revenue from a portfolio of wells.

The well can be profiled using various assessment programs as illustrated in Figures 8-1 through 8-4.

The assessment program should allow the user to determine important information about the well. Correlation can be made between hydrostatic head pressure and fluid volumes. The location of the fluid can be addressed to determine the best application method for removal. Figures 8-3 and 8-4 illustrate the fluid residence in a wellbore and the corresponding placement of the fluid above the perforations. Velocities can also be evaluated to determine if the well can flow above critical rate as illustrated in Figures 8-1 and 8-2.

The use of a computer to crunch data can yield results for an educated consideration of the issues at hand. It should be noted, however, that

OPTION 1	Enter U.S.		Metric Conversion		
Mist Foam Analysis					If terminal
Brine Density	8.500	lb/gal	'	1.019	g/mL	velocity with
Gas Water Content	15.000	lb/mcf	'	0.240	kg/m³	foam is less
Gas Rate	1,500,000	cfd	'	42.48	km³	than well
Gas Temperature	120	°F	'	49	°C	velocity, the
Pipe Diameter	4.000	inches	'	10.160	cm	well is a good
Pressure	150.000	psi	'	10.34	Bars	candidate for a mist foam application.
Terminal Velocity without Foam	40.507	ft/sec		12.347	m/sec	
Well Velocity	18.427	ft/sec		5.617	m/sec	
Terminal Velocity with Foam	11.990	ft/sec		3.655	m/sec	
Stable Foam Analysis						If bottom hole
Depth of Well	10,000	ft	'	3,048	m	pressure is less
Foam Density	2.125	lbs/gal	'	0.255	g/mL	than head pressure look
Bottom Hole Pressure	3,500	psi	'	241.32	Bars	for other alternatives.
Head Pressure with Foam	1,105	psi	'	76.19	Bars	

Figure 8-1: Simplified Foamer Candidate Modeling Program

Foam Analysis				
Brine Density (lb/gal)	Foam Density (lb/gal)	Gas Dew Point (lb/mcf)	Gas Flow Rate (MCFD)	Gas Temperature (°F)
8.500	2.125	15.0	1500	120

Tubing Inside Diameter (inches)	Flowing Pressure (psi)	Terminal Velocity without Flam (ft/s)	Terminal Velocity with Foam (ft/s)	Actual Well Velocity (ft/s)
4.000	150	40.507	11.990	18.43
THIS WELL MAY FLOW WITH FOAMER				

Batch Treatment Analysis			
Distance to Surface (feet)	Brine Density (lbs/gal)	Foam Density (lbs/gal)	Bottomhole Pressure (psi)
10000	8.500	2.125	3500

Head Pressure without Foam (psi)	Head Pressure without Foam (psi)		
4420	1105		
THIS WELL MAY UNLOAD WITH FOAMER			

Figure 8-2: Foam Analysis Modeling Program for Continuous Application (Top); Foam Analysis Modeling Program for Batch Application (Bottom)

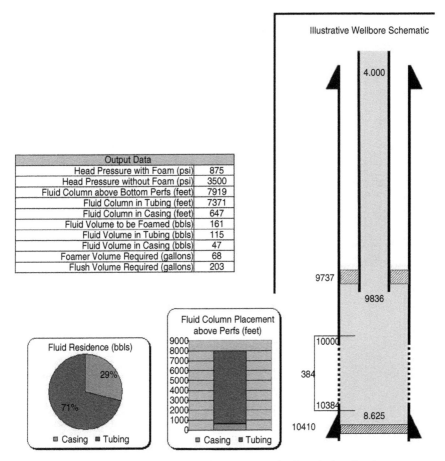

Figure 8-3: Simplified Modeling Program for Batch Application

reliance on computer-generated data can lead to erroneous assumptions and conclusions. The data derived from such exercises is only as valid as the data entered into the process. Many times, static data does not reflect the reality of a dynamic production system. Additionally, analyzed data may be erroneously applied to incomparable field observations. Caution and experience should be utilized when applying processed data to the well assessment. We recommend avoiding broad extrapolations of processed data.

After an assessment is conducted, it may be advisable to consider further tests to assess fluid levels, bottomhole pressures (flowing and static), inflow capabilities, impact of surface pressure reductions, and others.

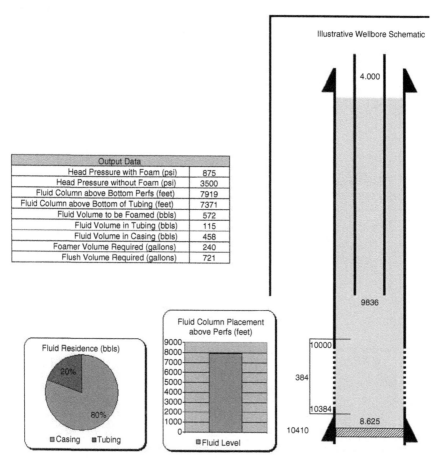

Figure 8-4: Simplified Modeling Program for Batch Application

Any well that is stifled by fluid production (related to fluid accumulation and increased hydrostatic pressure) and would benefit from the reduced density of a gas/liquid mixture is a good candidate for FAL. Once it is determined that a well is a suitable candidate for FAL, the factors that must be addressed are related to the foaming agent and identification of a suitable application method.

8.2.3 Foaming Agent Selection

An effective FAL program must have a foaming agent suitable for completion of the task. Testing must be conducted on fluid from each well in order for a program to achieve a sustainable and/or substantial increase in production. Foaming agents are not all the same—they will perform differently on fluid of varying compositions. A product that works on a given well may or may not be effective on wells within a reasonable proximity. You should always apply a product that has been specifically tested on fluid from the applicable well.

Experience has demonstrated that selection of a foaming agent cannot be completed in a lab under all circumstances. Field testing and selection of foaming agents is necessary to obtain accurate results. There are several well-documented factors for this phenomenon:

- Components of the water oxidize as the sample ages. This leads to solids present in the water that can accumulate at the gas/liquid interface and impact foam quality. Oxidized water will also have a different interfacial surface tension, thus impacting the performance of surfactants used in foaming agents.
- Dissolved gases dissipate, leading to pH changes in the fluid. Changes in pH impact the performance of the products.
- Oil/condensate will oxidize as the sample ages. Volatile components of the sample will be lost to flashing. Naturally occurring surfactants that may be present in the hydrocarbons will likely be altered by the oxidation process and can have an impact on product selection.

In all cases, we recommend that fresh field fluids be utilized for selection of the foaming agent. Synthetic fluids do not accurately reflect the organic content of field samples and likely will not be fully representative of the complete inorganic content. Testing on synthetic fluids will likely lead to erroneous results and the increased likelihood of application failure.

There have been several methods of testing recommended by various companies involved in FAL programs. These methods include blender testing, malt mixer testing, sparge column testing, or well simulators. All the various methods have merits and the methodology for each method can be verified for given circumstances. It has been our experience that malt mixer testing works best with our selection and application methodology. This method has proven to be reliable and delivers results that can be correlated to application performance.

Beyond the selection of product based on performance as a foamer, it is also important to consider other product related concerns prior to applying the foaming agent. Items that should be addressed are:

- Compatibility with produced fluids (solubility/dispersibility)
- Incorporation of other treating chemicals
- Product compatibility with metals and elastomers in the injection system and production components
- Temperature stability of the product in relation to temperatures encountered in the application
- Residence time of the product in the injection system and related production components

In many cases, foaming agents are formulated for specific fluids and/or specific applications. Having the right product in place will increase the likelihood that favorable results will be achieved from the FAL program.

8.2.4 Application and Assessment

Application of the foaming agent may be oversimplified in many instances. The idea that the product can merely be pumped down the tubing or annulus can lead to failure of the treatment. Many wells do not respond to simple treatments of a foaming agent. The initial application(s) of a foaming agent to a well should be viewed as part of the diagnostic process for the well. Application and assessment of foaming agents should be broken down into three segments:

1. Diagnostic Treatment(s)
2. Routine Application
3. Application Administration

A diagnostic treatment with a foaming agent should allow the producer to gauge the well's response to the effort. Depending on the complexity of the loading condition, various treatments may be required to fully assess and remediate the loading condition. These types of treatments may also incorporate mechanical methods to apply product and manipulate fluid or pressures on the well. Correlation or repudiation of information derived from the well diagnostic phase can often be determined during the diagnostic treatment process.

After a well has been successfully unloaded, routine applications can be established to maintain production at the well's full FAL potential.

The optimum application method will vary from well to well—this goes back to the initial characterization of the well discussed previously in this work.

The diagnostic treatment results, well characterization, and the completion design are the primary factors to consider when establishing the routine application method.

Common methods for routine application of foaming agents are:

- Manual batch treatment—tubing
- Manual batch treatment—annulus
- Automated (cyclical) batch treatment—tubing
- Automated (cyclical) batch treatment—annulus
- Continuous application—annulus
- Continuous application—capillary string
- Continuous application—velocity string
- Continuous application—coil tubing
- Continuous application—gas lift
- Intermittent application—plunger lift

The method applicable to a given well is determined by the parameters associated with the well. Some wells may require a trial of different methods to determine the best mode of application to obtain full FAL potential. The chosen method should consider other production challenges such as surface equipment constraints, other production treating concerns (corrosion, scale, salting, bacteria, dehydration, H_2S, etc.), and reliability of injection equipment.

Application administration will allow for extended success of a FAL program. Personnel responsible for maintaining production of the well should have an understanding of the FAL process as it relates to the well. Transient conditions will likely occur that will cause the well to load. Unloading of the well and returning the well to the routine application should be understood. Inevitably, characteristics of the well will change. These changes will have an impact on the FAL program and the viability of the application. The initial program established for a well utilizing FAL will likely change as the well ages. Periodic well reviews should incorporate the three steps for FAL success as outlined in this work: (1) well diagnostic, (2) foaming agent selection, and (3) application and assessment.

Case Study 1

This strategy was applied to a package of wells in a large field of approximately 700 gas wells. The wells were approximately 8000 foot deep completions with no packer.

The producer had been using liquid foamer from supplier B. The product was mixed with methanol and continuously applied down the annulus to 154 wells in the field. The application resulted in an average gain in production of 22 Mscf/D per well. Emulsion problems related to the foamer created operational cost increases that detracted from the benefits of gained production. Production increase with this application was 3,388 Mscf/D.

The treated wells were reevaluated using the steps outlined in this work. A detailed FAL process was developed and implemented. Foaming agent from supplier M was chosen to replace the previously applied product. Wells were unloaded via batch treatments and placed on continuous injection down the annulus. The revised application resulted in an average gain in production of 114 MSCF/D per well. Emulsion problems were eliminated. Production increase with this application was 17,556 Mscf/D.

The increase in production related to the change in application resulted in a net production gain of 14,168 Mscf/D, or a 418 percent increase.

At a gas sales price of $5.00 per MCF, this change resulted in a revenue increase of $70,840 per day ($25,856,600 per year).

Case Study 2

A single well producing 600 Mscf/D would load after flowing for two week intervals. The well would remain shut-in for three weeks for build-up before it would flow again.

An evaluation indicated that the well was subject to transient conditions that caused loading to occur. A batch treatment program was initiated that allowed the well to unload within 12 hours of loading. The reduction in well downtime allowed for a significant revenue increase. Prior to FAL, the well flowed 40 percent of the time. The FAL application increased flowing time to 98 percent.

At a gas sales price of $5.00 per MCF, the FAL program increased revenue by $635,100 annually.

Case Study 3

A well that was thought to be depleted had been purchased from the initial owner. The well was not producing at the time of the sale. Prior to shutting in, the well had been producing 600 Mscf/D and 400 BWPD. Initial production from the well was 12 MMscf/D with no water. The well was a horizontal completion under packer with ~14,000 foot of 3½ inch tubing. A capillary was installed and foamer from supplier X had been applied. Nitrogen stimulation also was applied, but the well would not flow.

The treated well was reevaluated using the steps outlined in this work. A detailed FAL process was developed and implemented. Foaming agent from supplier M was chosen to replace the previously applied product. A continuous application of foaming agent was initiated via the capillary. The response to this application was unloading of the well and consistent production of 1.6 MMscf/D with 650 BWPD.

At a gas sales price of $5.00 per MCF, the properly applied FAL program increased revenue by $2,920,000 annually.

Case Study 4

A mature offshore field was evaluated to determine if FAL would be applicable. Five wells were reviewed. Of the five wells, two were flowing and three were not flowing. Review and diagnostic treatments indicated that four wells were experiencing loading and one well had inflow and/ or depletion issues.

The four FAL candidate wells were batch treated for diagnosis and observation. The treatments resulted in a production gain of 9.5 MMscf/ D. Work is ongoing to determine the best long-term application strategy related to FAL.

At a gas sales price of $5.00 per MCF, the properly applied FAL program has the potential to increase revenue by $17,337,500 annually.

8.2.5 Conclusions

Foam assisted lift has been utilized with great success as a method for restoring production to loaded wells. This technology can be broadly applied to a variety of adolescent and mature wells to boost production.

Restoring production to liquid loaded wells has allowed producers to further exploit the potential of booked reserves. Extending the life of a wellbore by implementing a FAL program can allow for substantial revenue increases with minimal capital expenditures.

The keys to successful utilization of this technology are:

- Proper well diagnostic
- Proper foaming agent selection
- Proper application and assessment

The three key areas must all be addressed in detail to realize the full benefit of a FAL program. Each of the three areas requires emphasis for full realization of a well's FAL potential. Lack of attention to detail can lead to failure in the FAL process or under realization of capability.

Every well loads over the course of its existence. Increased understanding of FAL technology can only increase the odds that the technology will be appropriately utilized in applicable situations.

The global energy supply can be significantly augmented by improving production rates from existing wellbores. The net result for producers is a substantial increase in delivery of reserves to the market. The net result for consumers is an extended supply of natural gas for improved quality of life.

8.3 METHODS OF APPLICATION OF SURFACTANTS

There are three primary methods of surfactant introduction to a well. The methods are:

- Dropping soap sticks down the tubing
- Batch treating down the annulus (with no packer of course)
- Lubricating a capillary string down the tubing for injection of surfactants

Although soap sticks can be launched down the tubing in a variety of ways, Figure 8-5 shows one type of an automated soap stick launcher.

If no packer is present, batching down the annulus is a very acceptable way of surfactant injection (see Figure 8-6).

There is evidence that sometimes soap sticks do not find their way to the bottom of the well. New weighted and shaped sticks may help, but

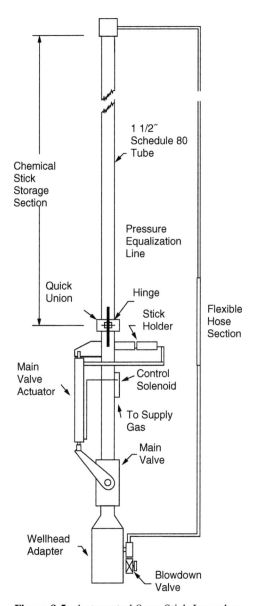

Figure 8-5: Automated Soap Stick Launcher

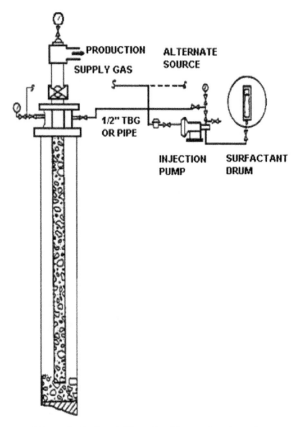

Figure 8-6: Batch Treating Down the Annulus

use of the capillary string to inject chemicals to the bottom of the tubing is a sure way of getting the chemicals to the pay zones.

8.4 CAPILLARY LIFT TECHNOLOGY

By Steve Turk, Weatherford International Ltd.

Steve Turk, BSPE, Marietta College, 1984, is currently Weatherford International's Director of Gas Well Deliquification Technologies. His responsibilities are to develop and globally market artificial lift technology applications that deliquify gas wells. In February 2003, he started up their Capillary Technologies Business Unit. In April 2004, he assumed additional responsibilities of USA Sales Manager—Artificial Lift Product

Lines and held that role until he assumed his current responsibilities in February 2007.

Prior to joining Weatherford, Mr. Turk worked in operational, sales, and management capacities for Mitchell Energy, Halliburton Energy Services, Bargo Energy, and Mission Resources. He has extensive experience in reservoir, production, and subsurface engineering as well as broad industry perspective from both an operating and service company point of view.

Adding surfactants to gas wells to enhance the production of liquids is an increasingly popular method used to deliquify gas wells [7,8]. A common method to deploy surfactants in gas wells is to use capillary strings, small diameter tubing placed either inside or on the outside of the production tubing.

Capillary systems provide a means to precisely convey lift enhancing surfactants in foam lift applications in gas wells. In essence, the capillary string installation is a "microtubing" system that is hung in the well mechanically in a similar fashion to a regular oilfield tubing. Capillary tubing systems commonly are installed using one of two basic techniques: (1) conventional systems hung inside the production tubing string and (2) nonconventional installations that are banded to the outside of the production tubing.

8.4.1 Conventional Capillary System Installations

Conventional capillary system installations are snubbed into and hung off in the well directly inside the production tubing string. The tubing typically is installed using a Capillary Coiled Tubing Unit (CCTU). All the system components can be lubricated and snubbed into the well under "live" flowing conditions in similar fashion to conventional coiled tubing operation (see Figures 8-7 and 8-8), avoiding the usual costs of a conventional work over unit. CCTUs can generally run or pull capillary tubing at speeds up to 130 feet per minute, therefore installation of a 10,000 ft conventional capillary system can often be done in 2 to 3 hours with a CCTU and a two-man crew. This speed, lack of equipment spread cost, and manpower efficiency deliver much lower service charges when compared to conventional work over operations both during initial installation and on eventual service intervention applications.

Ease of installation, versatility and low overall system costs are the primary drivers for installing conventional capillary injection systems.

Figure 8-7: Capillary String Installation System—CCTU

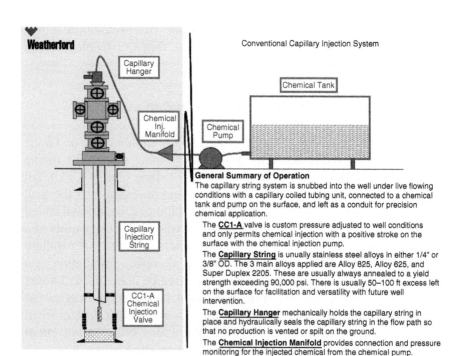

Weatherford

Conventional Capillary Injection System

Capillary Hanger

Chemical Inj. Manifold

Chemical Pump

Chemical Tank

Capillary Injection String

CC1-A Chemical Injection Valve

General Summary of Operation

The capillary string system is snubbed into the well under live flowing conditions with a capillary coiled tubing unit, connected to a chemical tank and pump on the surface, and left as a conduit for precision chemical application.

The **CC1-A** valve is custom pressure adjusted to well conditions and only permits chemical injection with a positive stroke on the surface with the chemical injection pump.

The **Capillary String** is unually stainless steel alloys in either 1/4" or 3/8" OD. The 3 main alloys applied are Alloy 825, Alloy 625, and Super Duplex 2205. These are usually always annealed to a yield strength exceeding 90,000 psi. There is usually 50–100 ft excess left on the surface for facilitation and versatility with future well intervention.

The **Capillary Hanger** mechanically holds the capillary string in place and hydraulically seals the capillary string in the flow path so that no production is vented or spilt on the ground.

The **Chemical Injection Manifold** provides connection and pressure monitoring for the injected chemical from the chemical pump.

Figure 8-8: Conventional Capillary String Components

Capillary systems can generally be installed at roughly one-fifth the cost of a conventional 1-1/2 inch coiled tubing installation (see Figure 8-9). However, in conventional systems the capillary tubing is exposed to the production fluids, are relatively fragile and unsupported, and must be pulled to swab/kick the well off to initiate flow.

Conventional capillary systems are comprised of four major components, all of which connect to a surface chemical conveyance system (pump and tank) (see Figure 8-2). These include:

- A chemical injection valve or foot valve, which is pressure adjusted to well conditions and only permits chemical injection with a positive stroke on the surface with the chemical injection pump.
- The capillary string that is usually made from stainless steel alloys having either 1/4 inch or 3/8 inch OD. The three main alloys applied are Alloy 825, Alloy 625, and Super Duplex 2205. These are generally annealed to a yield strength exceeding 90,000 psi. In a typical installation, there is 50–100 ft excess left on the surface to facilitate future well intervention.
- The capillary hanger that mechanically holds the capillary string in place and hydraulically seals the capillary string in the flow path to prevent venting or spilling of production fluids.

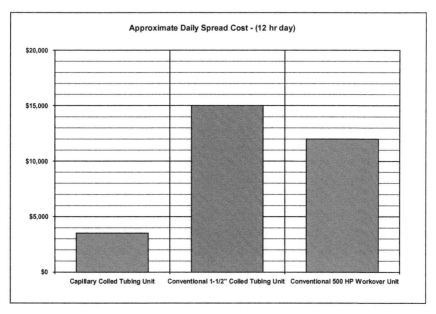

Figure 8-9: Conventional Capillary String Spread Cost Comparison

- The chemical injection manifold that provides connection and pressure monitoring for the injected chemical from the chemical pump.

8.4.2 Foot Valves

The foot valve is located at the bottom of the string and is designed to prevent backflow up into the capillary tubing. Proper foot valve selection is critical to the success of a capillary string installation. The foot valve serves two main purposes:

- Prevent well production entry to the inside of the capillary injection tubing string, which can cause corrosion, plugging, and other operational maladies.
- Provide down-hole regulation for chemical injection by limiting the rate at which the chemical is introduced to the flow stream.

Chemicals for artificial lift application generally are applied using one of two different techniques: controlled siphoning or controlled positive injection. The application method dictates the type of foot valve used in the capillary system.

8.4.3 Controlled Siphoning Applications

A controlled siphoning application is one where the lift chemicals are injected continuously. The main function of the foot valve in this application is to prevent produced fluids from entering the capillary tubing. The foot valve used for controlled siphoning applications typically consists of a spring closed check valve having a set pressure adjusted so that the check valve opens at a set differential pressure acting against the fixed spring in the seat of the check (see Figure 8-10). Generally, these valves are used only on systems where large volumes of chemicals are intended to flow continuously into a well in high production applications, for example high volume combination surfactant/salt inhibitor injections. These valves are appropriate for continuous operations only if the surface chemical flow ceases; the foot valve will continue to flow chemicals into the production stream until the preset spring force in the check valve can overcome the differential forces and close the valve. Valve failure can occur when production pressures exceed the hydrostatic pressure of the chemical in the capillary tubing. When this occurs, the combined hydrostatic and production forces prevent valve closure and the production stream enters the tubing, either plugging or corroding it.

.750" O.D. 316L Stainless Steel Check Valve for Capillary Strings

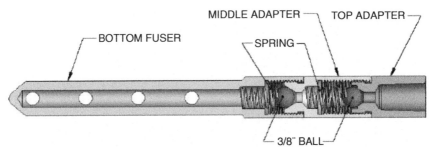

Figure 8-10: Capillary Check Valve

Figure 8-11: Differential Pressure Valve

8.4.4 Controlled Positive Injection Applications

A controlled positive injection application is one where chemical injection occurs only when the positive hydrostatic pressure of the chemical in the capillary tubing plus the applied pressure from the chemical pump overcome the seat pressure force of the valve (see Figure 8-11). Foot valves used in controlled positive injection applications are typically fully adjustable differential pressure controlled check valves that open when the differential pressure across the seat of the check exceeds the valve's spring set pressure. The valve is preset to allow fluid passage only when a positive pressure is applied by the chemical pump. Accurate knowledge of the static and flowing bottom hole pressures is essential to properly design controlled positive injection systems. Underestimating the flowing or static bottom hole pressures can render the surface chemical injection pump incapable of overcoming the preset valve seat pressure to open the valve and permit downhole injection of the chemicals. The valve set pressure is computed using the following equation (see Figure 8-12):

$$P valve = P pump - P bhp + P hydro$$

where:

$P valve$ = valve setting pressure (psi)
$P pump$ = pump injection pressure (psi)
$P bhp$ = bottom hole pressure (psi)
$P hydro$ = hydrostatic pressure inside tubing (psi)

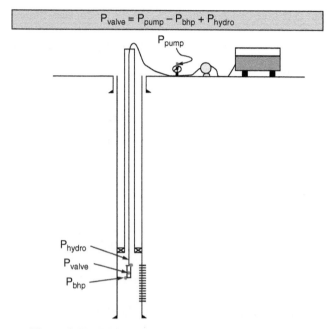

Figure 8-12: Differential Pressure Valve Calculations

Example 8-1: Calculation of Foot Valve Set Pressure Based on 500 psi Pump Operating Pressure

Given the following data, compute the differential pressure setting of the controlled positive injection foot valve.

Depth to valve = 10,000 ft
Chemical Wt = 8.45 ppg
Flowing BHP = 1500 psi
Pump Operating Pressure = 500 psi

The hydrostatic pressure of the chemical in the capillary tubing is calculated as:

Phydro = 0.052 × 10,000 ft × 8.45 ppg = 4394 (psi)

The valve set pressure is then computed as:

Pvalve = 500 − 1200 + 4394 = 3394 psi

Example 8-2: Calculation of Flowing BHP When Pump Is Actually Running at 1500 psi

Compute the actual flowing BHP if the pump in Example 8.1 was running at 1500 psi.

In this case, the foot valve has been set to 3394 psi. At the surface, the chemical pump is set to operate at 1500 psi, rather than the designed 500 psi. The actual flowing bottomhole pressure can be calculated by:

Pbhp = Phydro + Ppump – Pvalve
= 4394 + 1500 – 3394
= **2500 psi**

The valves used for controlled positive injection applications provide the best barriers to backflow into the capillary tubing. Properly set, the valves open only with a positive stroke of the chemical injection pump on the surface. When the pump pressure is applied, the valve snaps open, instantaneously injecting chemicals into the production stream, thus preventing backflow into the capillary tubing. Due to their complexity relative to a spring check valve, differential pressure valves are generally more expensive than spring-check type foot valves. The additional cost, however, is usually insignificant when compared to the cost associated with potential remedial operations resulting from an improperly functioning foot valve.

8.4.5 Capillary Tubing Strings

The purpose of the capillary tubing string is to provide a conduit to precisely apply chemicals down hole. Hanging in gas wells, capillary tubing strings are subject to the well environmental conditions and must be designed accordingly. When selecting capillary tubing strings, many of the same considerations used to select conventional production tubing must be considered with, of course, economics being foremost. Considerations like the anticipated volume of chemical to be ported through the tubing, the corrosiveness of the well environment, and what actions the tubing must withstand during remedial operations must all come into play. The cost of capillary tubing is controlled largely by the tubing length, the cross-section or size of the tubing, and the metallurgy. The length of the tubing is dictated by the depth of the desired point of injection, leaving the size and metallurgy to be considered when selecting capillary tubing strings. The economics must consider both the original cost of the system and any costs associated with the anticipated system maintenance. Capillary tubing string selection typically considers the following:

- Size
- Chemical Injection Rate
- Wall Thickness of the Tubing
- Metallurgy
- Bottomhole Temperature
- Produced Fluids Chlorides
- Produced Fluids pH
- Partial Pressures of H_2S and CO_2
- Yield Strength of Material

Size

Generally, capillary tubing used for foam lift application is run in one of two basic configurations: ¼ inch OD × 0.035 inch wall thickness (WT) or 3/8 inch OD × 049 inch WT. Selection is based primarily on the necessary chemical injection rate; however, consideration also should be given to provision for extra wall thickness that could extend string life by delaying potential hole development or breakthrough communication in the string.

The ¼ inch OD × 0.035 inch WT tubing is capable of delivering approximately 125 gallons of chemical per day through 10,000 feet of tube with a 500 psi pressure differential, assuming a friction coefficient for the chemical equal to that of fresh water (see Figure 8-13). This is by far the most common size used for capillary foam lift installations, especially in controlled positive injection applications.

When larger chemical volumes are necessary, particularly high volume controlled siphoning applications, the larger 3/8 inch OD × 0.049 inch WT tubing is required. The larger 3/8 inch OD × 0.049 inch WT tubing is capable of delivering up to 450 gallons per day with the same 500 psi differential pressure (see Figure 8-14). In some applications, the heavier wall thickness of the 3/8 inch tubing is preferred to prolong tubing life in extreme corrosion environments. It important to note, however, that the 40 percent increase in wall thickness achieved using the larger tubing does not necessarily produce an equal percentage increase in tubing life, and the 3/8 inch tubing is significantly more costly than the ¼ inch tubing.

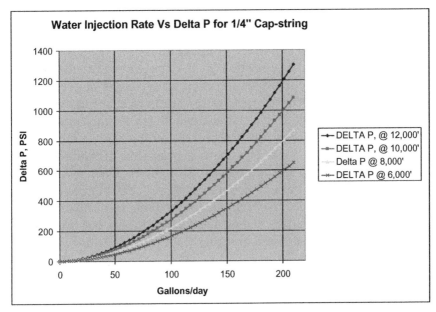

Figure 8-13: Flow Volume Delivery—1/4 inch Capillary Tubing

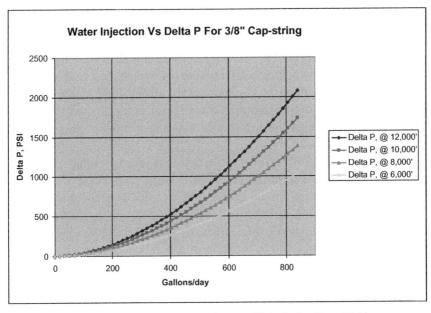

Figure 8-14: Flow Volume Delivery—3/8 inch Capillary Tubing

Metallurgy

Probably the most important consideration in a capillary injection system design is the metallurgy of the components that make up the system [9–11]. Improper selection often results in predictable failure and calamity. Considerable effort has been spent to develop a reliable process for selecting a suitable metallurgy for the components in capillary string systems. Such a process must consider the effects of temperature, CO_2 and H_2S partial pressures, produced water chlorides, system pH, and mechanical properties of the tubing such as applied stresses (weight loads), cycling (cold-working in the injector head), and welding technique, and their respective interactions. Most often capillary tubing mechanically fails due to one of two effects: (1) stress-assisted cracking (SAC) or (2) stress corrosion cracking (SCC).

Stress-Assisted Cracking (SAC)

Stress assisted cracking (SAC) failures are much less common in capillary string systems than stress corrosion cracking (SCC). On the other hand, SAC failures are more predictable and therefore more preventable than SCC failures. SAC occurs when a material subject to intergranular corrosion (e.g., an austenitic stainless steel suffering from weld decay) prematurely breaks when a mechanical load is applied to the area weakened by corrosion. Examples of external mechanical load stresses are loads applied by the capillary coiled tubing injector head, the weight of the tubing hanging in a well, and thermal expansion and shrinking of the capillary string caused by temperature changes during installation.

SAC failure is usually preventable through good welding practices and a thorough understanding of how the welding process affects the welded joint. Typically an automated TIG welder is used to weld two sections of capillary tubing having the same metallurgy. For this case the TIG welder is preset and tested, through trial and error, to deliver a weld that is as strong or stronger than original material. Once the TIG is calibrated it delivers consistent, highly repeatable computer controlled welds.

It is also common practice to weld two dissimilar capillary tubing sections together in an effort to place more corrosion resistant alloys in the more environmentally harsh sections of the well. In such cases, care should be taken to ensure that the area surrounding the new weld has sufficient corrosion resistance and mechanical strength to support all anticipated loads over its life in the well. Bear in mind, however, that it is sometimes difficult to TIG weld two dissimilar materials and produce a welded joint having equal or greater strength and/or higher corrosion resistance than the original material. It is generally preferable to pilot test the TIG welded joint between two dissimilar capillary strings prior to installing the welded strings in the well.

Stress Corrosion Cracking (SCC)

Failure due to stress corrosion cracking is the most common form of capillary string failure. It is the result of microscopic cracks that develop on the surface of the capillary tubing as a result of the well environment. Common causes are system salinity, pH, temperature, and the partial pressures of gas impurities. Once SCC begins, the cracks continue to propagate, ultimately penetrating deep into the tubing wall until the effective cross-sectional area of the tubing can no longer support the imposed loads.

An alloy's resistance to SCC can be correlated directly to the amount of chrome, nickel, and molybdenum in the alloy. Nickel is effective in protecting capillary tubing against SCC caused by the presence of chlorides. The combination of nickel and molybdenum aids in protecting the tubing against sulfide stress cracking, and the combination of chromium and molybdenum protects against crevice and pitting corrosion. A simple empirical formula is used to rank alloys according to their resistance to pitting or corrosion is to calculate the Pitting Resistance Number (PRE). This is given by:

$$PRE_{Number} = \underline{\hspace{1cm}}\% \ Cr + 3.3 \times \underline{\hspace{1cm}}\% \ Mo + 16.0 \times \underline{\hspace{1cm}}\% \ N_2.$$

The higher the PRE number, the more resistant the Corrosion Resistant Alloy (CRA) is to SCC. Typically the constituents found in alloys are specified over ranges. For this reason, PRE numbers for specific alloys usually are specified as maximum and minimum values. Example 8.3 demonstrates PRE number calculations for typical alloys.

Example 8-3: PREnumber Calculation

Calculate the PREnumber for the following two alloys:

Inconel 625 (N06625)   Duplex 2205 (UNS 31803)
  % Cr = 20.0 – 23.0   % Cr = 21.0 – 23.0
  % Mo = 8.0 – 10.0   % Mo = 2.5 – 3.5
  % N_2 = 0.0   % N_2 = 0.0

The PREnumber for these alloys is computed as:

PREmin (Inconel 625) = 20.0 + (3.3 × 8.0) + (16.0 × 0.0) = **46.4**
PREmax (Inconel 625) = 23.0 + (3.3 × 10.0) + (16.0 × 0.0) = **56.0**
PREmin (Duplex 2205) = 21.0 + (3.3 × 2.5) + (16.0 × 0.0) = **29.3**
PREmax (Duplex 2205) = 23.0 + (3.3 × 3.5) + (16.0 × 0.0) = **34.6**

Thus Inconel 625 has roughly 60 percent more resistance to pitting than Duplex 2205.

Figure 8-15 shows PRE numbers for several common stainless steel alloys. Note that as the PRE number increases, the cost of the tubing usually does too.

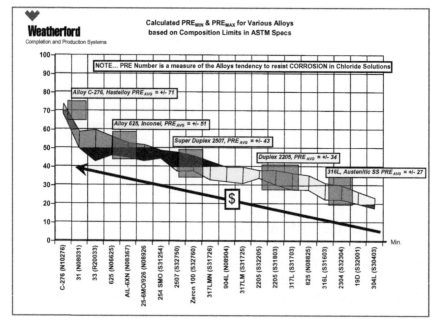

Figure 8-15: PREnumbers for Common Alloys [11]

SCC has been the topic of numerous studies over the past several decades. Schillmoller [11] developed a useful relation to select the metallurgy of capillary tubing in a well based on the partial pressures of CO_2 and H2S. Figure 8-16 (Figure 2 in his paper) presents the selection criteria for four common tubing alloys. The figure suggests that unless the tubing is to be exposed to extremely high partial pressures of H_2S (>10 psi), lower grade Duplex 2205 materials have sufficient SCC resistance, provided the well temperatures are lower than 150°C (300°F). If this was the only consideration used in the selection of corrosion resistant alloys (CRAs), Duplex 2205 metallurgy would suffice for the majority of all gas field applications since most gas wells fall within these limits of partial pressure and have maximum temperatures below 150°C.

Schillmoller [11], however, pointed out that the SCC resistance of CRAs is likely even more affected by their exposure to acid brine media, which are much more prevalent in gas fields than high H_2S partial pressures. Acid brines typically are generated when CO_2 and salt water, common gas well production streams, mix at high temperatures. In an effort to develop a more robust CRA selection guide, Schillmoller separated acid brine environments into three categories:

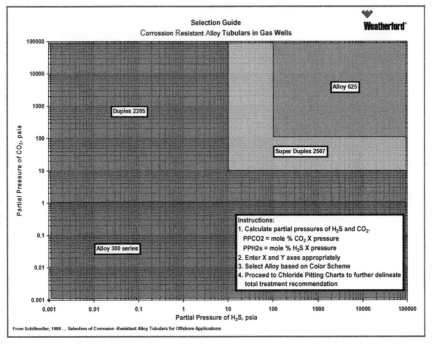

Figure 8-16: Selection Guide for CRA Materials in Sweet and Sour Wells [11]

1. Severe Corrosion—Depicts the presence of oxygen or free sulfur in the system, operating temperatures in the 175°C to 260°C (347°F–500°F) range, and high partial pressures of CO_2 and H_2S.
2. Moderate Corrosion—No oxygen or free sulfur in the system, operating temperatures in the 110°C to 200°C (230°F–392°F) range, and moderate to high partial pressures of CO_2 and H_2S.
3. Mild Corrosion—No oxygen or free sulfur in the system, operating temperatures below 175°C (230°F), and moderate to high partial pressures of CO_2 and low partial pressures of H_2S.

Most gas fields fall under the Moderate or Mild Corrosion environment classification. Figures 8-17 and 8-18 (adapted from Figure 3 in Schillmoller's paper [11]) have been adapted as follows by Weatherford.

The plots provide a means of selecting one of three CRAs based on bottomhole temperature, produced fluids pH, and produced water chlorides. Due to the conservative nature to which Weatherford has applied Schillmoller's data, these charts have proven to be very effective at minimizing catastrophic failure in installed chemical strings. Note that extreme care should be taken when the application falls near a transition line between two CRA selections.

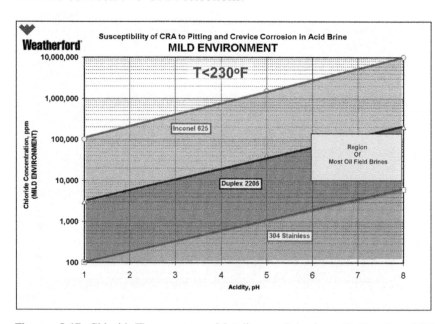

Figure 8-17: Chloride/Temperature Metallurgy Selection Guide for Mid Environments [11]

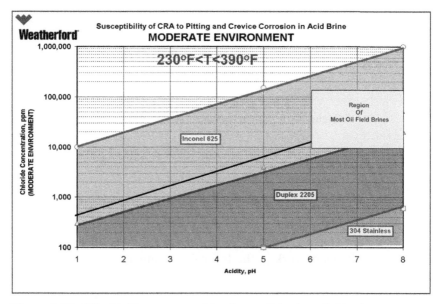

Figure 8-18: Chloride/Temperature Metallurgy Selection Guide for Moderate Environments [11]

8.4.6 Capillary Hanger Systems

Conventional capillary strings are suspended inside the production tubing of a gas well. These systems are connected to surface chemical equipment and left as permanent installations, thus, the downhole system must be "hung-off" in the well. Capillary hangers, like tubing hangers, serve the purpose of suspending and sealing the capillary string in the wellhead. There are several adaptations, but basically each hanger is comprised of a body that either screws or flanges onto the treecap receptacle with a mechanical slip hanger system and a hydraulic pack-off (see Figure 8-19).

The particular design and merit of each manufacturer's hanger system is beyond the scope of this text, but some basic considerations are worth mentioning. The primary considerations when selecting a hanger system for capillary tubing strings are:

- Metallurgy
- Hanger slip configuration
- Externally installed
- Internally contained

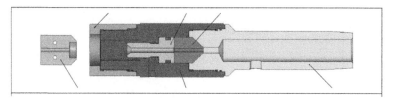

Figure 8-19: Circle Capillary Tubing Hanger

- Nonexistent
- Pack-off elastomer composition
- Sealing method

In general hangers work in the following manner:

1. Capillary tubing is stripped through the elastomer pack-off on a hanger until sufficient length has been run into a well to place the foot valve at the depth of desired operation.
2. Once on depth, the elastomer pack-off is hydraulically energized to squeeze it down on the exterior surface of the capillary tubing, thereby providing the necessary seal to prevent produced fluid escape through the tree.
3. This sealing force is captured and contained mechanically with a needle valve that captures the stored energy inside the elastomer pack-off, a mechanical cap screw, or some variation of both.
4. Once the capillary tubing is sealed, mechanical slips are placed that grab the external surface of the tubing, thereby preventing tubing slip into or out of the well.

Some operations/operators have special needs/requirements. For this reason, there are several hanger accessories that can also be included in the hanger system. Some examples of capillary hanger special accessories are:

- Y-Tube adaptors—Prevent loss of top master valve operation
- Multiple seal extended hangers—Provides multiple flow barrier seal protection
- Mechanical BOPs—Provides multiple flow barrier protection

8.4.7 Basic Operating Procedures for Installing and Removing Conventional Capillary Systems

Operating procedures for installing and removing capillary strings vary between the different service providers, but there are several common considerations that should be considered. The following general operational procedures supply general guidelines for capillary tubing installation, removal, and fishing operations.

Basic Procedure for Installing Capillary Tubing

It is highly recommended that prior to rigging up a Capillary Coiled Tubing Unit (CCTU) to install a capillary injection system on a well, the wellbore should be checked with wire line gauge ring (dummy run) to be reasonably sure that there are no obstructions in the well at least to the desired depth.

1. Arrive on location. Hold safety meeting and identify hazards. Eliminate hazards if possible. Determine the safest way to work with all hazards that cannot be eliminated.
2. MIRU CCTU. Close top master valve on tree.
3. Snub the end of the capillary tubing through the CCTU injector head, snubbing guide, BOPs, and strip on permanent capillary string hanger. Attach the BHA including the foot valve and necessary guidance accessories to the end of the cap string in preparation for installation in the well. (If using an adjustable differential pressure controlled foot valve, set valve to appropriate differential opening pressure prior to attaching to capillary string.)
4. Prepare the capillary tubing hanger to adapt to the wellhead and CCTU injector head. May require cross-over adaptors (typically available with the CCTU service rig). Remove the protective tree cap from the wellhead. Safely swing the rigged up capillary injection system assembly to well head using the CCTU crane. Connect hanger in tree cap internal threads and connect CCTU injector head and BOPs to capillary tubing hanger.
5. Slowly open top master valve and equalize pressure. Pack off snubbing guide and begin lubrication to stop any blow by. Set weight

indicator on electronic shut down on CCTU for 200# above line weight. Maintain 200# diff. while RIH.

6. HU chemical pump to tubing swivel. Begin loading capillary tubing with chemicals while RIH.

7. RIH slowly through injector head for at least 600 ft to ensure well is receiving capillary tubing appropriately. Continue increasing line speed to 140 ft max.

8. After reaching desired depth: RIH 5 ft below desired depth. Pack-off permanent capillary string hanger/pack-off. Bleed pressure off of BOP, snubbing guide, and BOPs. Disconnect injector head assembly from permanent hanger system. PU tubing 5 ft (carefully strip capillary tubing through the permanent hanger with CCTU head). Set permanent hanger system slips on tubing. Slack off on all weight on tubing and wait 5 min. to ensure all is held adequately with no movement.

9. Run 50 ft of tubing ("pigtail") through injector head and cut tubing on spool side of injector head.

10. Strip capillary hanger thread protector over "pigtail" and screw onto hanger system.

11. Lay down injector head and RD capillary unit.

12. Coil up 50 ft of capillary tube left on surface and connect end of tubing to chemical pump.

13. Set chemical pump to desired rate.

14. RDMU capillary unit.

Basic Procedure for Pulling (already installed) Capillary Tubing

Many times capillary intervention or service becomes necessary. The following basic steps provide operational guidelines for this practice:

1. Arrive on location. Hold safety meeting and identify hazards. Eliminate hazards if possible. Determine the safest way to work with all hazards that cannot be eliminated.

2. Move capillary unit into position and rig up unit.

3. Using blowdown valve at chemical injection manifold, bleed pressure from capillary tube.

4. Prepare the capillary tubing "pigtail"—the extra surface length—for load into the capillary injector head.

5. Remove the protective hanger cap from the tubing hanger and slide it off the capillary string.

6. Prepare the capillary tubing hanger to adapt to the CCTU injector head. May require cross-over adaptors (available with the CCTU service rig).
7. Feed the end of the capillary up through the BOP and into the injector head.
8. Raise the injector to a position above the wellhead and align it with the top of the wellhead/tubing hanger/pack-off.
9. With the injector, pull the excess tubing through the injector and on to the storage spool.
10. Feed the tubing through the side of the spool and attach the end to the tubing swivel.
11. Slowly spool tubing on to the spool and align each wrap for smooth spooling.
12. When all excess tubing is on the spool, lower the injector head/BOP assembly and rig up (connect) to the capillary tubing hanger. Take care to lower and spool the tubing at the same rate to be sure not to kink the tubing.
13. Equalize well pressure to the BOP by opening the master valve(s).
14. Release the pack-off pressure in the capillary tube hanger.
15. Note the well depth and calculate the string weight.
16. Set the proximity switch on the weight indicator to 200 pounds above the tubing weight.
17. Begin pulling tubing out of hole. Pull tubing slowly until the BHA is inside the production tubing.
18. Continue to pull tubing out of hole keeping the proximity switch setting 200 pounds above tubing weight. Do not exceed 145 feet per minute.
19. When the BHA is within 100 feet of the surface, slow the rate to 20 feet per minute until the BHA bumps up in the hanger.
20. Slowly close the master valve, being sure to count the rounds on the valve. (Most of the time, this will be 12-1/2 rounds.) If fewer rounds on the hand wheel are counted, insure that the BHA is above the valve gate.
21. Bleed the pressure off of the BOP and hanger assembly.
22. Disconnect the injector head/BOP assembly from the capillary tubing hanger. Slowly raise the injector/BOP assembly off of the Bowen nut. Watch the weight indicator to make sure there is enough slack in the tubing to raise the assembly. Take care to not to kink the tubing.

23. Back the hanger pack off out of the well head.
24. Swing the assembly to the side of the well head and lower to a level that gives safe and comfortable working conditions.
25. Install the tree cap on the tree and tighten.
26. Remove "Do Not Operate" tags and install the valve handles.
27. Put the well in service.
28. Remove the BHA and hanger pack-off.
29. Spool the remaining tubing through the injector and secure the tubing to the spool.
30. Move the injector head into the cradle and secure the injector to the cradle.
31. Secure crane into cradle and turn off power switch.
32. Secure all tools and ladders.
33. Hold a postjob prereturn trip safety meeting.
34. Move off of location.

Basic Procedure for Fishing (stuck or broken) Capillary Tubing

Fishing operations, by their nature, are unpredictable. Most often, the integrity and geometry of the down-hole fish is unknown. In fishing operations safety is of primary concern. When a capillary string breaks or is parted, it no longer makes connection through the capillary hanger/pack-off and there is a good chance that well production will be blowing through the hole in the top of the capillary tubing hanger. When this is the case, it is important to close the master valve immediately, being careful to count the turns on the valve to determine that the master valve is fully closed. There is always a possibility that the fish is hung in the valve and haphazard closing of the master valve could cut the capillary string, making fishing operations more difficult.

Once the surface conditions are stabilized (no more gas blowing, etc.), MIRU a SLU and a CCTU. Hold a JSA with all company personnel and service company representatives on location. Jointly discuss the detailed plan of operation. Point out that *every person* on the location has the right and responsibility to shut the job down if unsafe conditions arise during the operation. The following outlines a recommended fishing operation:

1. Nipple up a BOP to control the well.
2. MI and RU a wire line unit with tools to catch capillary tubing.

3. The fishing tool should be either:
   - Bull Dog grapple
   - Kilo overshot with cutrite on the flapper to catch stainless steel
   - An alligator grab
   - A barrel grab
4. Be sure to have enough wire line lubricator to pull up 4 feet or more of capillary tubing below the fishing tool.
5. RIH with fishing tools. Tag up lightly on the capillary tubing. Do not spud or run into the top of the tubing as this will bend the tubing over or make a bird nest at the top of the tubing.
6. Continue the preceding process until the capillary tubing is caught. Pick up on tubing noting the weight of the pick up. The ¼ in capillary string weighs 81 pounds per 1000 feet. The 3/8 in tubing weighs 181 pounds per 1000 feet.
7. When the capillary tubing is caught, POOH with tubing. Be careful not to exceed yield strength pull as this will stretch the capillary and permanently deform and ruin the capillary tubing to future use.
8. Pull the capillary tubing as far up into the wire line lubricator as possible.
9. Close the capillary BOP to hold the tubing.
10. Blow down the pressure in the wire line lubricator.
11. Disconnect the wire line BOP and lubricator. Strip the capillary out of the bottom of the wire line lubricator and release the capillary tube from the grapple.
12. RD the wire line unit and RU the CCTU injector head.
13. Cut a piece off of the end of the recovered capillary tubing to help determine the cause of failure.
14. Strip the capillary tube through the capillary unit BOP and into the capillary injector head.
15. Work the injector head and capillary unit BOP down on to the capillary BOP that is on the well head.
16. Equalize the pressure from the wellbore into the capillary unit BOP.
17. Open the BOP on the well head and control well pressure with the snubbing pack-off.
18. Pull enough capillary tube through the injector to reach the spooler.
19. Pull the capillary tube through the slot in the flange on the spooler and attach the capillary tubing to the pump swivel.
20. Open the bleeder valve on the swivel manifold and monitor the capillary tubing pressure.

21. If a blow is noted on the capillary tubing, fill the capillary with enough fluid to stop the blow. The volume in the ¼ in tubing is 1.31 gallons per 1000 feet. The volume in the 3/8 in tubing is 3.13 gallons per 1000 feet.
22. Pull up on tubing and note weight on the weight indicator.
23. Set the proximity switch on the weight indicator to 200 pounds over the tubing weight.
24. POOH slowly monitoring the tubing weight OOH.
25. Watch the weight indicator to notice the "bump up" of the bottom-hole assembly when it reaches the surface. Note: There may be more than one break. When the weight indicator reaches zero, the end of the tubing is near.
26. When the tubing "bumps up" in the bottom of the capillary snubbing guide, close the master valve on the tree. Be sure to count the rounds on closing the master valve. If the master valve becomes hard to turn with less than total turns required to close the valve, do not force the valve.
27. When the master valve can be closed, close the wing valve and blow down the pressure on the capillary unit BOP and snubbing pack-off.
28. Disconnect the capillary BOP from the well head BOP.
29. Slack off on the capillary tubing and work the injector head and capillary unit BOP off of the well head.
30. ND well head BOP and secure the well cap.
31. Open the master valve and slowly open the wing valve to put the well back online.
32. RD capillary unit and conduct postjob JSA to include postjob travel to home base or next job.
33. Examine capillary tubing and determine cause of the failure. Repair or replace the tubing.

8.4.8 Hooke's Law—Basic Capillary String Stretch Calculations

Capillary strings stretch and react to external forces similarly to conventional tubing strings. And similarly, Hooke's Law is applicable in calculating "tubing movements" due to external applied forces from:

- Tubing hang weight
- Internal chemical weight
- Chemical pump force
- External pull force
- Temperature change (shrinking/elongation) forces

Hooke's law is expressed by the equation:

$$S = \frac{F \times L \times 12}{A \times E}$$

Where

S = stretch, inches
F = force, pounds
L = length, feet
A = X-sectional area of tube metal
E = Modulus of Elasticity, psi (30×10^6 for metal)

The overall effect of tubing stretch is the cumulative sum of the individual components listed earlier. The following example demonstrates typical tubing stretch calculations.

Example 8-4: Hook's Law

Given:
¼"OD × 0.035"WT tbg set @ 12,000′
Surface Temperature = 80°F
Bottom Hole Temperature = 275°F
Tubing is filled with 8.33 lb/gal chemical
Foot valve is set for Controlled Positive Injection.
Chemical pump pressure is 500 psi.

Find:
(1) Elongation due to tbg weight
(2) Elongation due to chemical weight
(3) Elongation due to pump force
(4) Elongation due to temperature effect
(5) Total Elongation in conditions

1. Elongation due to tubing weight calculation:

$$S_{weight} = \frac{F \times L \times 12}{A \times E}$$

F = 80 lb/1,000 ft × 12 = 960 lb.
L = 12,000 ft × 12 = 144,000 in.
A = (3.1412 × (0.25/2)^2 in^2) − (3.1412 × ((0.25/2) − .035)^2 in^2)
A = 0.0491 in^2 × 0.0254 in^2 = .0237 in^2
E = 30,000,000

$$S_{weight} = \frac{960 \times 12,000 \times 12}{.0237 \times 30,000,000}$$

$$S_{weight} = 194 \text{ inches} = \underline{16.2 \text{ ft.}}$$

2. Elongation due to chemical weight calculation:

$$S_{chemical\ weight} = \frac{F \times L \times 12}{A \times E}$$

ID X_{area} = 3.1416 × ((0.25/2) – 0.0352)^2 = .0254 in^2
Chemical Force (F) on ID X_{area} = 0.052 × 12,000′ × 8.33 lb/gal/ = 132 lbs.
L = 12,000 ft × 12 = 144,000 in.
A = (3.1412 × (0.25/2)^2 in^2) – (3.1412 × ((0.25/2) – .035)^2 in^2)
A = 0.0491 in^2 – 0.0254 in^2 = 0.0237 in^2
E = 30,000,000

$$S_{chemical\ weight} = \frac{132 \times 12,000 \times 12}{0.0237 \times 30,000,000}$$

$$S_{chemical\ weight} = 26.7 \text{ inches} = \underline{2.2 \text{ ft.}}$$

3. Elongation due to chemical pump force calculation:

$$S_{chemical\ pump\ force} = \frac{F \times L \times 12}{A \times E}$$

ID X_{area} = 3.1416 × ((0.25/2) – 0.0352)^2= .0254 in^2
Chemical Pump Force (F) on ID X_{area} = 500 psi′ ×. 0254 in^2 = 12.7 lbs.
L = 12,000 ft × 12 = 144,000 in.
A = (3.1412 × (0.25/2)^2 in^2) × (3.1412 × ((0.25/^{2}) – .035)^2 in^2)
A = 0.0491 in^2 – 0.0254 in^2 = 0.0237 in^2
E = 30,000,000

$$S_{chemical\ pump\ force} = \frac{12.7 \times 12,000 \times 12}{0.0237 \times 30,000,000}$$

$$S_{chemical\ pump\ force} = 2.6 \text{ inches} = \underline{0.2 \text{ ft.}}$$

4. Elongation due to thermal expansion:

In this equation, the coefficient of linear thermal expansion for stainless tubing is assumed to be 7.2 microinches/inch-°F. This coefficient has been field proven to accurately reflect field applications.

$S_{thermal}$ = Expansion Coeff × L × 12 × (T2 − T1) (in.)
$S_{thermal}$ = 7.2 × 10^{-6} × 12,000 × 12 × (275 − 80)
$S_{thermal}$ = 202.2 in./12 = 16.8 ft.

5. Total elongation

Total Elongation is simply calculated by adding the components of elongation. Thus the total elongation is given by:

$S_{TOTAL} = S_{tubing\ weight} + S_{chemical\ weight} + S_{pump\ force} + S_{thermal}$
S_{TOTAL} = 16.2 ft + 2.2 ft + 0.2 ft + 16.8 ft. = 35.4 ft.

It is vitally important to make these stretch calculations to accurately determine the location of the end of the tubing and to ensure that the depth of injection is in fact the desired depth. In addition, the capillary tubing's yield strength and tensile strength are important properties to consider when installing capillary strings. These properties dictate the depth limits to which the tubing can be installed and the external forces that can be applied to the tubing when pulling or servicing the string. Figure 8-20 depicts the effects of yield and tensile strength of several tubing alloys under various conditions of well depth and well pressure. The table indicates when the weight of the tubing exceeds the tubing tensile strength (tubing parts), when the weight of the tubing exceeds 80 percent of the tensile strength, when the weight of the tubing exceeds the yield strength, and when the weight of the tubing is less than the yield strength.

8.4.9 Nonconventional (externally banded) Capillary System

The Patent Pending Weatherford Critical Velocity Reducing System (CVR™) consists of a combination of three generally accepted artificial lift methods usually used individually to dewater a liquid loading gas well. The Weatherford CVR™ System combines the attributes of plunger

	Physical Characteristics in the Annealed Conditon @ 100 degrees F													
	0.250″ OD, .035″ WT Capillary Tubing (5.250 psi WP)							0.375″ OD, .049″ WT Capillary Tubing (4.900 psi WP)						
	300 Series Stainless (Factor = 1.000) Weight (lbs/100 0 ft)	Alloy 400 (Factor = 1.100) Weight (lbs/100 0 ft)	Alloy 826 (Factor = 1.025) Weight (lbs/100 0 ft)	Alloy 625 (Factor = 1.052) Weight (lbs/1000 0 ft)	Alloy C-276 (Factor = 1.107) Weight (lbs/10 00 ft)	90k Duplex 2206 (Factor = 0.993) Weight (lbs/10 00 ft)	120 k Duplex 2205* (Factor = 0.993) Weight (lbs/10 00 ft)	300 Series Stainless (Factor = 1.000) Weight (lbs/1000 ft)	Alloy 400 (Factor =1.100) Weight (lbs/100 0 ft)	Alloy 825 (Factor =1.026) Weight (lbs/100 0 ft)	Alloy 626 (Factor =1.052) Weight (lbs/100 0 ft)	Alloy C-276 (Factor = 1.107) Weight (lbs/100 0 ft)	90k Duplex 2206 (Factor = 0.993) Weight (lbs/10 00 ft)	120 k Duplex 2206 (Factor = 0.993) Weight (lbs/10 00 ft)
Depth	81.2	89.3	83.3	85.4	89.9	80.6	80.6	172.2	189.4	176.7	181.1	190.6	171.0	171.0
0	0	0	0	0	0	0	1	0	0	0	0	0	0	1
500	41	45	42	43	45	40	40	86	95	88	91	95	98	98
1,000	81	89	83	86	90	81	81	172	189	177	181	191	171	71
1,500	122	134	125	128	135	121	121	258	284	265	272	286	257	257
2,000	162	179	167	171	180	161	161	344	379	363	362	381	342	342
2,500	203	223	208	214	225	202	202	431	474	442	463	477	428	428
3,000	244	268	250	256	270	242	242	517	568	530	543	572	513	513
3,500	284	313	292	299	315	282	282	603	863	618	634	687	599	599
4,000	325	357	333	342	360	322	322	689	758	707	724	762	684	684
4,500	365	402	376	384	405	363	363	776	882	795	816	858	770	770
5,000	406	447	417	427	450	403	403	861	947	884	906	953	855	855
5,500	447	491	458	470	494	443	443	947	1,042	972	986	1,048	941	941
6,000	487	536	500	512	539	484	484	1,033	1,136	1,060	1,087	1,144	1,026	1,026
6,500	528	581	541	566	584	524	524	1,119	1,231	1,148	1,177	1,239	1,112	1,112
7,000	568	625	583	598	629	564	564	1,206	1,326	1,237	1,258	1,334	1,197	1,197
7,500	609	670	626	641	674	605	605	1,292	1,421	1,325	1,358	1,430	1,283	1,283
8,000	650	715	666	683	719	645	645	1,378	1,516	1,414	1,449	1,526	1,368	1,368
8,500	690	759	700	726	764	685	685	1,464	1,510	1,502	1,539	1,620	1,454	1,454
9,000	731	804	750	769	809	725	725	1,550	1,705	1,590	1,630	1,715	1,539	1,539
9,500	771	849	791	811	864	768	768	1,638	1,799	1,679	1,720	1,811	1,625	1,625
10,000	812	893	833	864	899	806	806	1,722	1,894	1,767	1,811	1,906	1,710	1,710
10,500	853	938	876	897	944	846	846	1,808	1,989	1,865	1,902	2,001	1,796	1,796
11,000	893	983	916	939	989	887	887	1,894	2,083	1,944	1,992	2,097	1,881	1,881
11,500	934	1,027	950	982	1,034	927	927	1,980	2,178	2,032	2,083	2,192	1,967	1,967
12,000	974	1,072	1,000	1,026	1,079	967	967	2,088	2,273	2,120	2,173	2,287	2,052	2,052
12,500	1,015	1,117	1,041	1,068	1,124	1,008	1,008	2,153	2,368	2,209	2,264	2,383	2,138	2,138
13,000	1,066	1,161	1,082	1,110	1,168	1,048	1,048	2,220	2,462	2,297	2,354	2,478	2,223	2,223
13,500	1,096	1,206	1,126	1,153	1,214	1,088	1,088	2,326	2,567	2,385	2,446	2,573	2,309	2,309
14,000	1,137	1,250	1,166	1,196	1,259	1,128	1,128	2,411	2,662	2,474	2,536	2,668	2,394	2,394
14,500	1,177	1,296	1,208	1,238	1,304	1,169	1,169	2,497	2,748	2,562	2,626	2,764	2,400	2,400
15,000	1,218	1,340	1,250	1,281	1,349	1,209	1,209	2,583	2,841	2,661	2,717	2,859	2,565	2,565
15,500	1,269	1,384	1,291	1,324	1,393	1,249	1,249	2,659	2,938	2,739	2,807	2,954	2,661	2,661
16,000	1,299	1,429	1,333	1,366	1,438	1,290	1,290	2,755	3,030	2,827	2,898	3,050	2,736	2,796
16,500	1,340	1,474	1,374	1,409	1,483	1,330	1,330	2,841	3,125	2,916	2,988	3,146	2,822	2,822
17,000	1,380	1,518	1,416	1,452	1,528	1,370	1,370	2,927	3,220	3,004	3,079	3,240	2,907	2,907
17,500	1,421	1,583	1,458	1,495	1,573	1,411	1,411	3,014	3,315	3,092	3,189	3,336	2,993	2,993
18,000	1,462	1,608	1,499	1,537	1,618	1,451	1,451	3,100	3,409	3,181	3,280	3,431	3,078	3,078
18,500	1,502	1,652	1,541	1,590	1,683	1,491	1,491	3,186	3,504	3,269	3,350	3,626	3,164	3,164
19,000	1,543	1,807	1,583	1,623	1,708	1,531	1,531	3,272	3,599	3,367	3,441	3,621	3,249	3,249
19,500	1,593	1,745	1,624	1,665	1,753	1,572	1,572	3,358	3,683	3,446	3,531	3,717	3,336	3,336
20,000	1,624	1,741	1,666	1,708	1,798	1,612	1,612	3,444	3,789	3,534	3,622	3,812	3,420	3,420
20,500	1,665	1,831	1,708	1,751	1,843	1,652	1,652	3,530	3,883	3,622	3,713	3,907	3,506	3,506
21,000	1,705	1,876	1,749	1,793	1,898	1,683	1,683	3,616	3,977	3,711	3,803	4,003	3,591	3,591
21,500	1,746	1,920	1,791	1,836	1,933	1,733	1,733	3,702	4,072	3,799	3,894	4,098	3,677	3,677
22,000	1,786	1,965	1,833	1,878	1,978	1,773	1,773	3,789	4,167	3,887	3,984	4,193	3,762	3,762
22,500	1,827	2,010	1,874	1,922	2,023	1,814	1,814	3,875	4,262	3,976	4,075	4,289	3,848	3,848
23,000	1,868	2,054	1,916	1,964	2,068	1,854	1,854	3,961	4,356	4,064	4,166	4,384	3,933	3,933
23,500	1,908	2,099	1,958	2,007	2,113	1,894	1,894	4,047	4,451	4,152	4,258	4,479	4,019	4,019
24,000	1,949	2,144	1,999	2,060	2,158	1,934	1,934	4,133	4,546	4,241	4,346	4,574	4,104	4,104
24,500	1,990	2,188	2,041	2,092	2,203	1,976	1,976	4,219	4,640	4,329	4,437	4,670	4,190	4,190
25,000	2,030	2,233	2,083	2,136	2,248	2,015	2,015	4,306	4,735	4,418	4,528	4,766	4,276	4,276
25,500	2,071	2,278	2,124	2,178	2,292	2,056	2,056	4,391	4,830	4,505	4,618	4,850	4,361	4,361
26,000	2,111	2,322	2,166	2,220	2,337	2,096	2,096	4,477	4,924	4,594	4,709	4,956	4,446	4,446
26,500	2,152	2,367	2,207	2,263	2,382	2,136	2,136	4,563	5,019	4,683	4,789	5,051	4,532	4,532
27,000	2,192	2,412	2,249	2,306	2,427	2,176	2,176	4,649	5,114	4,771	4,890	5,146	4,617	4,617
27,500	2,233	2,456	2,291	2,349	2,472	2,217	2,217	4,736	5,209	4,860	4,980	5,242	4,703	4,703
28,000	2,274	2,501	2,332	2,391	2,517	2,257	2,257	4,822	5,304	4,948	5,071	5,337	4,788	4,788

WEIGHT OF TUBE EXCEEDS TENSLE STRENGTH OF TUBE (TUBE PARTS)

WEIGHT OF TUBE EXCEEDS (80%) TENSLE STRENGTH OF TUBE (CAUTION TUBE MIGHT PART)

WEIGHT OF TUBE EXCEEDS YELD STRENGTH OF TUBE (TUBE STRETCHES)

WEIGHT OF TUBE BELOW YELD STRENGTH OF TUBE (TUBE STATIC)

Assumes:
1. No buoyancy on outside of tube no hydrostatic on inside of tube
2. Isothermal conditions ... No temperature changes

Weatherford
Completion and Production Systems

Figure 8-20: Capillary Tubing Strength Table

lift, area reduction, and foam use to reduce or eliminate some of the limitations of the individual lift systems.

Candidate wells for potential CVR system installation would be any well with an extended perforation interval of 100 feet or more. To demonstrate the CVR system, consider a gas well having an operating field line pressure or producing wellhead pressure of 200 psi and with 2-3/8 in tubing inside 4-½ in casing with an all-water gradient. Using the 200 psi wellhead pressure, the critical velocity from the Coleman Equation required to unload 2-3/8 in tubing is 15.70 feet per second, giving an estimated gas volume flow rate of 406 Mscf/D. The real issue when attempting to unload gas wells having extended perforation intervals is the extremely high gas rates required to unload the casing below the end of the production tubing. For example, to unload the 4½ in casing with 200 psi of wellhead pressure, the Coleman equation calculates the same 15.70 feet per second as the tubing, but given the larger ID of the casing (casing area = 12.5683 in^2) the required flow rate becomes 1.63 Mscf/D. Under these conditions, many gas wells may not have sufficient production to unload the 4½ in casing section, even from initial production or after first completion.

By installing a 2-7/8 in dead string below the production tubing the flow area in the casing is reduced from 12.5683 in^2 to 6.0768 in^2. Based on the critical velocity computed earlier, the flow rate required to unload the 2-7/8 in dead string is reduced to 790 Mscf/D (from 1.63 Mscf/D required for the 4½ in casing). This reduced gas volume required to unload the 4½ in casing is likely more achievable by the well or will allow it to flow longer under its natural decline.

To further reduce the minimum critical flow rate provided by the reduced area of the 2-7/8 in dead string, the use of a chemical foamer or soap is added to the system. In order to get the foamer to the deepest point in the well, at the end of the dead string, it was necessary to externally band the capillary tubing to the outside of the production tubing. The capillary tubing is connected into the top of the dead string in a CM-1 conventional gas lift mandrel (see Figure 8-21). The gas lift mandrel is installed in the tubing string below the equalizing plug so that the foamer is injected into the top of dead string and not back up the production tubing. The CM-1 conventional gas lift mandrel allows for a chemical injection valve to be used that will support the chemical weight and prevent hydrostatically draining the capillary line. Once the foamer enters into the top of the dead string, it will fall through the inner diameter of the string and reach bottom, where it will mix

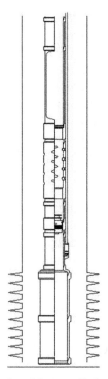

Figure 8-21: Nonconventional (externally banded) Capillary System

with the produced gas and produced water and begin to generate foam.

By injecting foamer at the bottom of the dead string, it is possible to reduce the required gas velocities even more by reducing the surface tension and density of the produced water. Generally the use of chemical foamer to reduce surface tension and density of water can reduce the required critical gas velocities by a factor of three. Therefore, by adding foamer to this installation, the required critical velocity inside the 2-3/8 in production tubing can feasibly be reduced from 15.90 feet per second (406 Mscf/D) to roughly 6.50 feet per second (168 Mscf/D). Similarly, inside the 2-7/8 in dead string the minimum critical flow rate to unload the well is reduced to 327 Mscf/D. Therefore, with the CVR system the required gas volumes needed to unload the 4½ in casing below the tubing has been reduced from 1.63 Mscf/D to 327 Mscf/D.

This represents a much more achievable rate for most gas wells, particularly after a period of normal decline.

The final advantage of the CVR System is the ability to install plunger lift inside the production tubing. An operator can install a normal 2-3/8 in bumper spring in 1.875 in X-LOK in the top of the Heavy Duty Flow Sub. The Heavy Duty Flow Sub is essentially the end of the tubing and it is generally installed at or near the top third of the perforation interval. This tubing placement at or near the top third of the perforation interval generally is accepted as the optimum placement for best plunger lift performance.

8.5 REFERENCES

1. Campbell, S., Ramachandran, S., and Bartrip, K. "Corrosion Inhibition/ Foamer Combination Treatment to Enhance Gas Production," SPE Paper 67325, presented at the SPE Production and Operations Symposium, Okla. City, OK., March 24–27, 2001.

2. Blauer, R. E., Mitchell, B. J., and Kohlhaas, C. A. "Determination of Laminar, Turbulent, and Transitional Foam Flow Losses in Pipes," SPE Paper 4888, presented at the 44th Annual California Regional Meeting, April 4–5, 1974.

3. Dunning, H. N., Eakin, J. L., Reinhardt, W. N., and Walker, C. J. "Foaming Agents for Removal of Liquids from Gas Wells," Bull. 06-59-1, Am. Gas Assoc., New York, NY, 14 pp.

4. Libson, T. N. and Henry, J. R. "Case Histories: Identification of and Remedial Action for Liquid Loading in Gas Wells . . . Intermediate Shelf Gas Play," SPE Paper 7467, presented at 53rd Annual Fall Meeting of SPE of AIME, Houston, TX, October 1–3, 1978.

5. Vosika, J. L. "Use of Foaming Agents to Alleviate Liquid Loading in Greater Green River TFG Wells," SPE/DOE 11644, presented at he 1983 SPE/DOE Symposium on Low Permeability, Denver, CO, March 14–16, 1983.

6. Letz, R. S. "Capillary Strings to Inject Surfactants," SWPSC School on De-Watering Gas Wells, April 24, 2001, Lubbock, TX.

7. Neves, T. R. and Brimhall, R. M. "Elimination of Liquid Loading in Low-Productivity Gas Wells," SPE–18833.

8. Saleh, S. and Al-Jamea'y, M. "Foam-Assisted Liquid Lifting in Low Pressure Gas Wells," SPE–37425, 1997.

9. Sullivan, III, Hinson, D. L., and Hendrix, D. E. "Benefits and Challenges Associated with CRA Injection Tubing in Corrosive Gas Wells," SPE 91692, 2004.

10. Craig, B. D. "Guidelines for Corrosion Resistant Alloys in the Oil and Gas Industry," The Nickel Development Institute.

11. Schilmoller, C. M. "Selection of Corrosion-Resistant Alloy Tubulars for Offshore Applications," NiDi Technical Series No. 10 035, 1989.

CHAPTER 9

HYDRAULIC PUMPING

By Toby Pugh, Weatherford

Toby Pugh is Business Segment Manager for Latin America for Weatherford Completion & Production Systems, a division of Weatherford International Ltd. Prior to joining Weatherford, he served as International Sales Manager for Halliburton. Most of his 30+ year career was with the Guiberson Division/Dresser, where he held positions from Manager of Research and Engineering to Regional Manager. He holds a BSME and an MS in Aerospace Engineering from UT Arlington. He is a registered PE, and has papers, patents, and presentations related primarily to hydraulic pumping systems.

9.1 INTRODUCTION

Since the 1930s, several thousand oil wells have been, and continue to be, produced with hydraulic pumps, and the number of new hydraulic installations is increasing yearly. As the volume, weight, depth, and well deviation of producing wells continues to increase, the application of hydraulics will also continue to increase.

The hydraulic pumping system takes liquid (water or oil) from a liquid reservoir on the surface, puts it through a reciprocating multiplex piston pump or horizontal electrical submersible pump to increase the pressure, and then injects the pressurized liquid (power fluid) down-hole through a tubing string. At the bottom of the injection tubing string, the power fluid is directed into the nozzle of a jet pump or to the hydraulic engine of a piston pump, both of which have been set well below the producing fluid level. The surface injection pressures normally range from approximately 2000 psi up to 4000 psi, with some going up to but rarely above 4500 psi. An electric motor, diesel engine, or gas engine is used to drive the multiplex pump.

The fundamental operating principle of subsurface hydraulic pumps is Pascal's Law, postulated by Blaise Pascal in 1653. This law states that:

> "Pressure applied at any point upon a contained liquid is transmitted with equal intensity to every portion of the liquid and to the walls of the containing vessel."

This principle makes it possible to transmit pressure from the surface by means of a liquid-filled tubing string to any given point below the surface. The energy of the power fluid is transmitted to the appropriate components of the downhole pump and/or the produced fluid(s) in order to bring the reservoir fluids to the surface. Figure 9-1 illustrates the surface and subsurface components of a typical hydraulic pumping system.

Artificial lift is needed if a well no longer flows or if it flows at a rate lower than desired. The walking-beam, sucker-rod method of pumping is the most common form of artificial lift in use today. It has been in use since at least 476 A.D. when the Egyptians used the principle for drawing water, and has been used in the petroleum industry since the days of Drake's discovery in Pennsylvania. In comparison, hydraulic pumping (reciprocating downhole pump or a jet pump) is much newer. Even compared to gas lift, which was first used to lift oil from some wells in Pennsylvania in 1846, hydraulic pumping is a relatively new method of artificial lift.

Hydraulic Pumping System
Schematic Flow

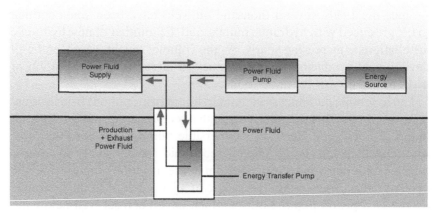

Figure 9-1: Flow Schematic for a Typical Hydraulic System

9.1.1 Applications to Dewatering Wells—Gas and Coal Bed Methane

The 1-1/4 in coiled tubing jet pump has become the hydraulic pump of choice for dewatering gas wells. This small jet pump has been developed so that it can be used either as a free style pump inside of 1-1/4 in tubing or attached to it. In both cases, the 1-1/4 in tubing, which can be either coil or standard, is run inside another string of tubing and the produced fluids plus spent power fluid returns to the surface through the tubing–tubing annulus. The gas is free to flow to the surface through the casing–tubing annulus (see Figure 9-4).

The jet pump is suited to this application because it is highly tolerant to sand and other particles and can typically be used in applications where the GLR through the pump is less than 1000. It is capable of producing up to 500 bpd of liquids, but it cannot create low formation pumping pressures near pump-off conditions before cavitation occurs in the pump.

The first printed references to jet pumps can be found as far back as 1852 in England. However, consistent mathematical formulas for jet pumps were not published until 1933, when J. E. Gosline and M. P. O'Brien of the University of California published their paper entitled, "The Water Jet Pump." That paper, and others, were used to develop currently used equations. Due to the iterative nature of these equations, the proper application of jet pumps had to wait until there was a widespread availability of computers. The following sections will discuss jet pumps in more detail.

A slow stroking, hydraulic piston pump is an alternate type of hydraulic pump that can be used in this application but only in selected wells. The size of the smallest piston pump available is for 1-1/4 in tubing, but the smallest available slow stroking pump is for 2-3/8 in tubing (see Figure 9-16). In order to use it in a dewatering application that allows gas to flow to the surface through the casing–tubing annulus, a parallel installation with two strings of tubing is required (see Figure 9-9). One string must be large enough to allow the pump to go from surface to its housing in the well (a BHA) and would also be a conduit through which the power fluid goes to the pump. The other string returns the spent power fluid and production fluids to the surface. This arrangement limits the casing size to 6-5/8 in–20 ppf or larger.

The 1-1/4 in pump can be used in either a parallel installation or concentric tubing installation (see Figure 9-8). The casing size for the parallel installation must be 5-1/2 in–26 ppf or larger, and the casing size for the concentric installation must be 3-1/2 in tubing or larger.

Solids and gas in the production fluids will create problems for piston pumps and all pumps that use moving parts with close tolerances.

The operation of hydraulic piston pumps is based on the principle that force equals pressure times area. Faucett first employed that principle for pumping oil in 1875. The Faucett bottomhole pump was a steam-operated device that required a very large diameter hole. Because of that diametrical requirement, the Faucett pump found no commercial application in the oil patch. By the 1920s, increasing well depths resulted in renewed interest in the use of hydraulic pumping, and the first serious hydraulic installation was set up on March 10, 1932, in Inglewood, California by C. J. Coberly. Piston pumps will be discussed more fully in the following sections.

The essential differences between the typical gas well application and a CBM well are that both the production volumes and the flowing bottomhole pressures usually are required to be less in CBM wells. A producing bottomhole pressure near zero is not an issue for a piston pump, but a jet pump will cavitate long before the pressure gets to that level. To help in identifying cavitation, a gauge can be installed on a jet pump, that will sense and record the producing pressure.

Another difference that has been reported between gas and CBM wells has to do with the coal fines as opposed to the flow back of frac sand. In those wells that have damaging fines, the areas that are typically the most damaged are the outside of the nozzle and the inlet of the throat. It has been found that using a throat made of silicon carbide, instead of the standard tungsten carbide throat, greatly improves the life of the pump. The only reported problems with frac flow back sand is when it comes back in slugs, which would be true for any form of artificial lift. In that case it has been known to plug the tubing standing valve but not the pump. It has not been found to cause problems in the pump.

9.1.2 Limitations of Other Forms of Lift

It is has always been recognized that the limiting factor in sucker rod-pumping systems is the sucker rod itself. The thousands of feet of rods needed to transmit the reciprocating motion from the surface to the bottomhole pump cannot be made strong enough to lift large loads from great depths. Even with the high strength Class D rods and tapered string designs, it would not be possible to have more than a 40,000 lb. peak load without overstressing the top rods and causing failures. The top rod must lift not only the well fluid on every stroke but also the weight of the submerged rods. The combined effects of the weight of the rods and the dynamics of cyclic loading along with rod/tubing wear in less-than-straight wells impose serious limitations on pumping depths and associated production volumes (see Figure 9-2a).

The use of high volume electric submersible pumping is increasingly limited with depth. Problems include the loss of power in the cable, the pressure limitations of the pump discharge housing, the large number of stages, and the horsepower of the motor (see Figure 9-2b).

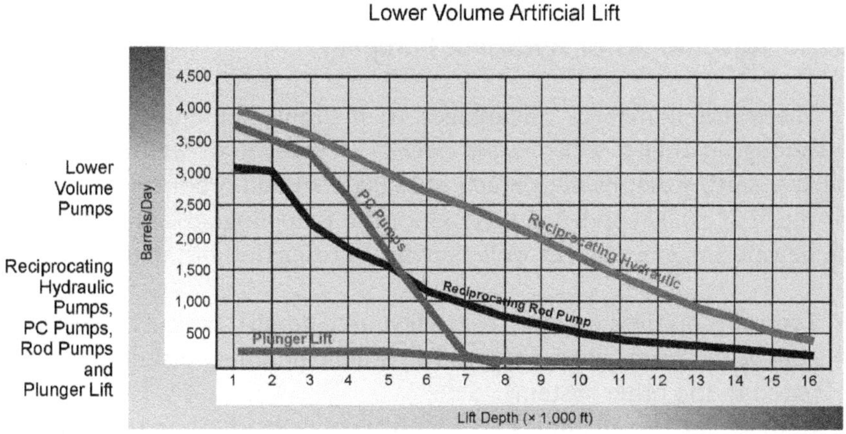

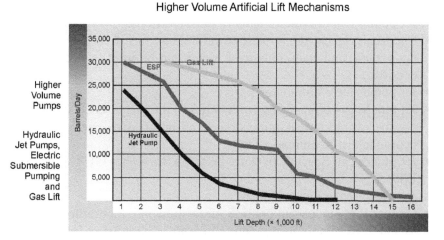

Figure 9-2a: Lower Volume Artificial Lift Mechanisms

Figure 9-2b: Higher Volume Artificial Lift Mechanisms

The use of gas lift is also restricted due to producing bottomhole pressure requirements. As a rule, it is not possible to obtain as much drawdown of the reservoir with gas lift as with pumps, provided gas interference is not a problem with the pumps. In addition, deep wells may require high compressor pressures, which can adversely affect the casing. Gas lift can still be advantageous, however, in gassy or sandy wells or wells that are very expensive to service due to pulling the tubing through the use of gas lift valves that are installed/retrieved with wireline. For low liquid rates, such as dewatering a gas well, lower formation producing pressures may be expected.

9.1.3 Advantages of Hydraulic Pumping

There are numerous advantages to hydraulic pumping. A major advantage is that it will operate over a wide range of well conditions such as setting depths of as much as 18,000 feet and production rates of as much as 50,000 bpd. Virtually all of the following advantages apply to dewatering gas wells as well as typical production installations.

- Typically, no rig is required to retrieve free pumps. In many cases, this may be the primary advantage of hydraulic pumping systems as compared to the other systems.
- Both jet and piston pumps are highly flexible in adjusting to changing production rates.
- Both jet and piston pumps are able to produce at higher rates from greater depth than a rod pump, ESP, or gas lift.
- Jet pumps can operate reliably in deviated wells.
- Chemicals can be added to the power fluid to control corrosion, paraffin, scaling, and such, plus fresh water can be used to dissolve salt deposits.
- Jet pumps have no moving parts.
- Jet pumps can typically perform better in the higher GLR wells than positive displacement pumps, such as progressive cavity, rod, or hydraulic piston pumps.
- Jet pumps can typically perform better in the higher GLR wells than ESPs.
- Jet pumps have long run lives.
- Standard jet pumps can operate successfully in temperatures as high as 400 °F by simply using high temperature elastomers for their O-rings and seal rings.

- Jet pumps have low maintenance costs.
- Jet pumps are field repairable.
- Jet pumps can be installed in sliding sleeves, wireline nipples, and across gas lift mandrels as well as their own bottomhole assemblies.
- Jet pumps have a high tolerance of solids in the production fluids.
- Jet pumps have a high tolerance to corrosive fluids through the use of CRA materials and/or inhibitors entrained in the power fluid.
- Jet pumps can produce high volumes (see Figure 9-2b).
- Jet pumps can be circulated through a subsea flow line loop with a radius of no more than 5 feet (TFL).
- A hydraulic piston pump has better efficiency at depth than a rod pump because there is no rod stretch and no rod/tubing wear.
- Multiwell installations can be operated from a single power source.
- The power fluid serves as a diluent when producing viscous crudes.
- The power fluid can be heated (usually water) to produce heavy crudes or crudes with high pour points.

9.1.4 Disadvantages of Hydraulic Pumping

The disadvantages connected with hydraulic pumping include:

- It is often misapplied (a common problem for all forms of lift).
- There is a lack of knowledge about the system.
- It requires knowledge by operating personnel (this is a common problem for all forms of artificial lift).
- The complexity of manufacturing hydraulic piston pumps (this is a problem for other forms of artificial lift such as ESPs).
- Surface pressures of as much as 5000 psi can be a safety hazard.
- Conditioning of the power fluid is required. Sand or other particles in the power fluid must be removed as they can damage the surface power fluid pump, the nozzle in a jet pump, and the engine piston/barrel in a piston pump.
- A jet pump cannot "pump-off" a well. It requires a minimum flowing bottomhole pressure in order to avoid power fluid cavitation. That minimum pressure can be as much as 10 to 30 percent of the hydrostatic based on TVD depending on the makeup of produced fluids.
- Piston pumps, rod or hydraulic, have a limited ability to tolerate solids in the production fluids.

- The use of hydraulic pumps offshore typically has been limited to those platforms where a water injection system is already in place, as the deck space requirements can be large.
- Casing pressure capability can be a limitation for reverse flow installations (only jet pumps can be used in reverse flow).
- Jet pumps have low operating power efficiencies requiring more horsepower.
- Power fluid rates for jet pumps will vary from 1 to 4 times the production rate.
- The back pressure on a jet pump has a strong influence on the power fluid injection pressure and can increase the injection pressure by 1.5 to 4 psi for each psi the back pressure is increased. This is determined by the area of the nozzle divided by the area of the throat (called the area ratio).

9.1.5 Types of Operating Systems

There are two basic types of hydraulic pump systems: the open power fluid system and the closed power fluid system. In an open power fluid system (OPF), the operating power fluid mixes with the produced fluid while down hole, and then both fluids are returned to the surface in a commingled state. In a closed power fluid system (CPF), the production and operating power fluids are never allowed to mix. Because jet pumps commingle the production and power fluid, the CPF system is limited to piston pumps.

Open Power Fluid System (OPF)

The system in Figure 9-3 is a typical open power fluid system. This arrangement is not recommended for dewatering a gas well as the path for gas is through the pump. For dewatering a gas well, the OPF system shown in Figure 9-4, which is the coiled tubing jet pump system, allows the gas to bypass the pump via the casing–tubing annulus. The CT jet pump is shown in Figure 9-5.

In all OPF systems, only two downhole fluid conduits are required to operate the pump. One conduit (normally the tubing) contains the pressurized power fluid and directs it to the pump. The other conduit (normally the casing-tubing annulus) returns both the spent power fluid and produced fluid to the surface (see Figure 9-3). It is by far the most commonly used.

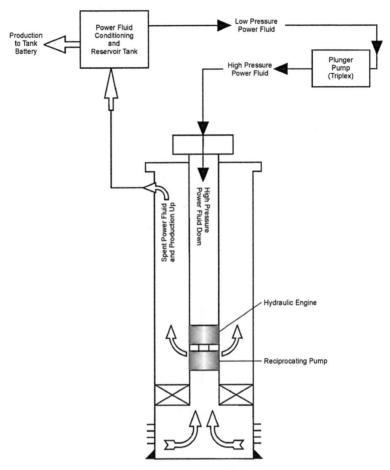

Figure 9-3: Open Power Fluid System

Besides the simplicity and economic advantage of the OPF system, the intermingling of the power fluid and the produced fluid has some other advantages:

- The circulated power fluid can carry chemical additives. Corrosion, scale, paraffin inhibitors, and emulsion breakers can be added to extend the life of the subsurface equipment.
- The commingled power fluid has a diluting effect. Where highly corrosive production fluids are being lifted, the clean power fluid can reduce the concentration of the corrosive elements by approximately 50 to 80 percent. The viscosity of heavy oils can be reduced.

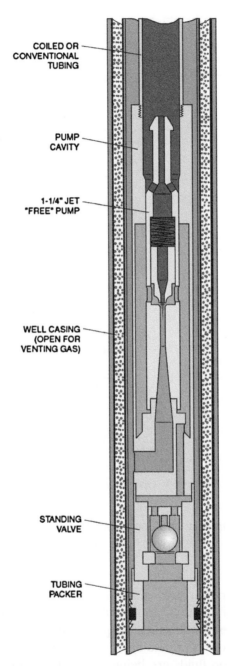

COILED OR
CONVENTIONAL
TUBING

PUMP
CAVITY

1-1/4" JET
"FREE" PUMP

WELL CASING
(OPEN FOR
VENTING GAS)

STANDING
VALVE

TUBING
PACKER

Figure 9-4: 1-1/4 in Coiled Tubing Jet Pump Installation

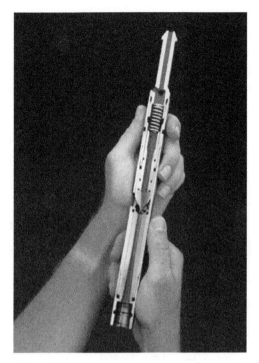

Figure 9-5: 1-1/4 in Coiled Tubing Jet Pump

- In production fluids with a high paraffin content, the OPF system allows the circulation of heated liquids or dissolving agents. This will remove waxy build-ups that might otherwise hinder or halt production.

A drawback to a typical OPF system is that all the gas must go through the pump. Piston pumps have a tendency to gas lock. Throats of jet pumps have a tendency to become choked, inhibiting production.

These problems are overcome with the system in Figure 9-4, which was discussed earlier.

Closed Power Fluid System (piston pumps only)

The discussion of this system is included for completeness, as it is not for dewatering most gas wells.

In a closed power fluid system (CPF), an extra tubing string is required both downhole and on the surface. The extra downhole string is used to bring just the spent power fluid back to the surface. On the surface, the extra string is for carrying just the spent power fluid to the power fluid tank for recirculation and repressurization (see Figure 9-6). The CPF system is used less than the OPF configuration.

A closed system may also find preference on platforms or where available space is at a premium as the size of surface facilities are smaller.

In most downhole pumps used in a CPF system, the pump end is lubricated by the power fluid. The engine piston is designed to have

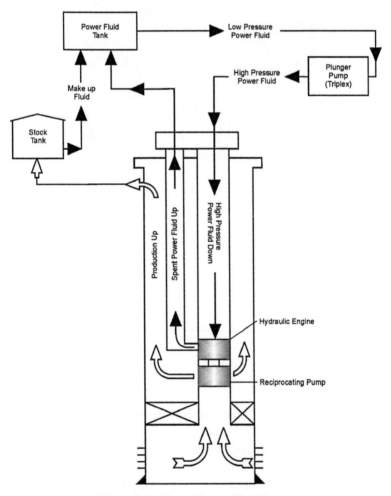

Figure 9-6: Closed Power Fluid System

some leakage so ≈10 percent of the power fluid is lost into the production. This amount of fluid must be fed back into the power fluid system from the production line.

Even in a closed system, the power fluid cannot remain clean since the pipes, fittings, pumps, tanks, and the like are not completely free of contaminates. When a liquid containing solid material passes through a close fit (such as slippage past the engine piston), the material will tend to be held back.

9.1.6 Types of Subsurface Pump Installations

There are three basic subsurface pump systems: free-type, fixed-type, and wireline-type.

Free-Type Installations

The free-type system does not require a pulling unit to run or retrieve the pump. The pump is placed inside the power fluid tubing string and is "free" to move with the power fluid to the bottom of the well and back out again when the power fluid direction is reversed (see Figure 9-7). This may be the primary advantage of hydraulic pumping systems.

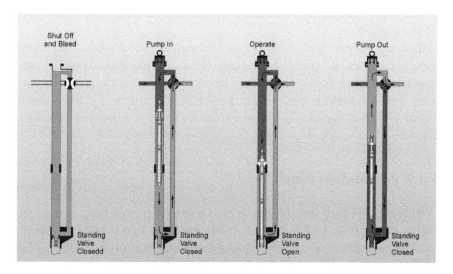

Figure 9-7: Installing/Removing a Free-Style Pump

There are two main types of free-pumping installation designs: the casing-free design and the parallel-free design.

Casing-Free Installations

The OPF is a casing-free installation and is discussed for completeness. It is not recommended for dewatering a gas well because it does not have a separate path for the gas. The exception to this is the configuration shown in Figure 9-4.

The casing-free OPF consists of a single tubing string, a pump housing (bottomhole assembly), and a packer. Power fluid is circulated down the tubing string where it operates the subsurface hydraulic pump and mixes with the produced liquids and gas. This mixture returns to the surface via the casing annulus (casing return).

Figure 9-8 shows a casing-free installation (using casing return) for an open power fluid system. In this design, all the gas must pass through the pump. Any gas that is produced adversely affects the liquid displacement efficiency of a piston pump. This is also true for a jet pump but it is possible for jet pump performance to improve due to the reduction of the discharge pressure from the gas in the tubing. This can reduce the required injection pressure more than the increase from having to use a larger throat to accommodate the gas passing through the pump. Because of the simplicity and cost benefits of the casing-free open power fluid system, there is more of this design than any other type of installation.

The casing-free OPF with a gas vent can be used where displacement efficiency is affected by high GLR. A parallel string is run to a dual packer below the pump to provide a separate path for the gas. The production plus spent power fluid returns by the casing annulus. This is useful when liquid volumes to be produced create high friction.

9.1.7 Parallel-Free Installations

The parallel-free installation is also an open power fluid system in which a free style pump is used. It incorporates two strings of tubing and a bottomhole assembly but no packer (see Figures 9-8 and 9-9). The BHA is attached to the main tubing string and has a landing bowl that receives the spear on the bottom of the parallel string.

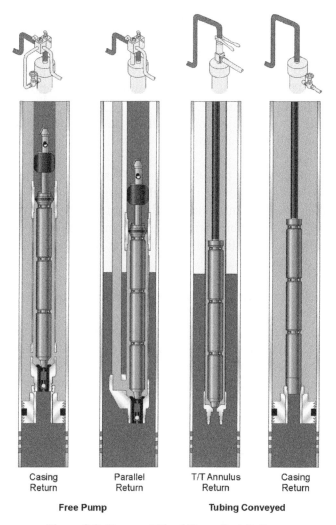

Casing Return	Parallel Return	T/T Annulus Return	Casing Return
Free Pump		**Tubing Conveyed**	

Figure 9-8: Free and Fixed Pump Installations

The power fluid goes down the main tubing string and operates the pump. The spent power fluid then mixes with the produced fluid and the mixture returns up the parallel string to the surface. This subsurface design permits gas to be vented up the casing annulus. With this type of installation, the annulus can be used for gas production. The liquid production can be the water or condensate lifted to unload a gas well.

The drawback is the extra string. Usually the parallel string is smaller than the primary string, which contains the hydraulic pump. This second string must carry both the production and power fluid to the surface. This can create high return friction loss resulting in the need for more horsepower. However, this concern is minimized in a gas well if low volumes of liquid are produced.

Casing size dictates the size of both strings and the pump. This concern is also minimized when dewatering a gas well through the use of the system in Figure 9-9.

It is possible to have a closed power fluid system/parallel installation, but this would require having three tubing strings.

In parallel-free type installations, the main string should always be anchored to minimize tubing stretch as an unanchored string could actually unseat the parallel string(s) and disable it (them) from functioning as a return conduit(s).

9.1.8 Fixed-Type (or tubing conveyed) Installations

The information on fixed-type systems is limited to discussion only (no installation figures with the exception of Figure 9-8) as they are not applicable to dewatering gas wells.

Fixed-pump installations have the bottomhole pump attached to the end of the tubing string. These are considered to be permanent-type installations and may be used for high production rates.

9.1.9 Fixed-Insert Installations

In a fixed-insert installation (or fixed-concentric installation), a large tubing string is run to bottom. The pump is then run on a string of macaroni tubing inside the main tubing string and seated in a seating shoe (see T/T Return in Figure 9-8). The macaroni string carries the power fluid. The returned power fluid/produced fluids travel up the tubing–tubing annulus. Gas can be vented up the annulus.

9.1.10 Fixed-Casing Installations

In fixed-casing installations, the pump is run on the tubing string with a packer below the pump. The tubing string carries the pressurized power fluid down to the pump. The spent power fluid and the production are returned through the casing annulus.

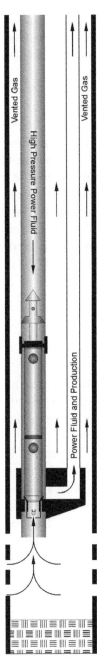

Figure 9-9: Parallel Free, Open Power Fluid Installation

This type of installation generally is used where high production rates require that large pumps are run, but any produced gas must pass through the pump. They are not applicable to dewatering gas wells and included only for completeness.

9.1.11 Wireline-Type Systems

In a wireline-type system (for jet pumps only), the subsurface pump can be installed anywhere there is access to the casing. Typically, it is installed in a sliding sleeve but can also be installed straddling a gas-lift mandrel, a chemical-injection mandrel, or a hole shot in the tubing. The pump is run into the well and pulled from the well on a wireline, and can be operated in standard flow or reverse flow.

This type of installation generally is used when the operator does not wish to pull the tubing string to install the normal bottomhole assembly (see the following section).

9.1.12 Ancillary Equipment

Bottomhole Assemblies

A bottomhole assembly (BHA) is a housing where the hydraulic pump is located when operating. It has an internal seal sleeve(s) that matches with a seal(s) on the outside diameter of the pump when the pump is landed.

The casing type BHA and the parallel BHA both provide the same sealing/operating functions. With the casing type BHA, only a single tubing string is required along with a packer. In a stab-in parallel installation, the main tubing string is run by itself. The smaller, parallel string is run separately and stabbed into the shoe assembly of the BHA. All parallel strings require the use of parallel tubing string clamps. A packer is optional with the parallel type installation.

Figure 9-10 shows some casing free and parallel free BHAs for single displacement free style pumps.

Retrievable Tubing Standing Valve

A retrievable tubing standing valve is required when using a free-type pump (see Figure 9-11). This standing valve lands in the lower end of the BHA and under the pump. The purpose of the SV is to prevent the

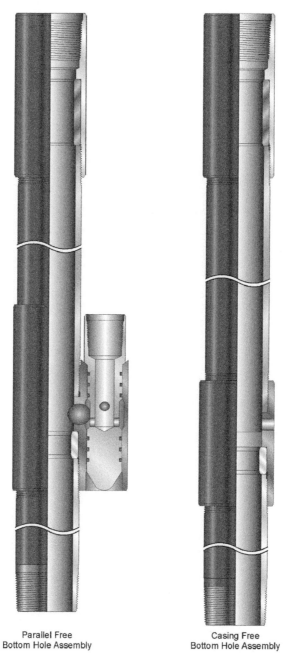

Parallel Free
Bottom Hole Assembly

Casing Free
Bottom Hole Assembly

Figure 9-10: Free Style Bottomhole Assemblies

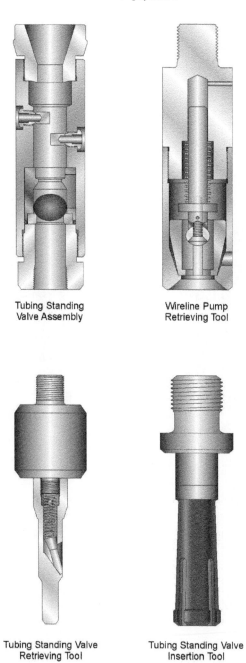

Tubing Standing
Valve Assembly

Wireline Pump
Retrieving Tool

Tubing Standing Valve
Retrieving Tool

Tubing Standing Valve
Insertion Tool

Figure 9-11: Retrievable Tubing Standing Valve and Wireline Tools

power fluid from falling out of the tubing string when installing or retrieving the pump and can be wireline retrieved.

9.1.13 Power Fluid Choices

The predominate liquids used for power fluid are those produced by the well—water or oil.

Power Oil

As oil has lubricity that water does not, the service life of the equipment is usually longer using oil. The compressibility of oil is greater than water thereby reducing fluid hammer effects.

Drawbacks to power oil are the potential fire hazard and pollution damage.

Power Water

During recent years an increased number of hydraulic systems have changed from using power oil to using power water. Many of these changes were due to ecological reasons, code restrictions, town site locations, increased water cuts, or because the produced crude oil had a high viscosity.

A high-viscosity power fluid can mean excessive friction losses in the system. This in turn increases the operating pressure and the horsepower requirements for lifting the well. In some cases, it would be prohibitive to use the produced crude as a power fluid and water, with a lower viscosity, would be preferable. There are hydraulic installations where produced water is heated and then used as a power fluid (and as a diluent or thinner) for heavy oil.

Water has lower lubrication qualities than oil and sometime requires a chemical additive for lubrication when using hydraulic piston pumps. Frequently, the chemicals used will include oxygen inhibitors and agents to combat corrosion that are easily added at the multiplex suction via a chemical pump. Improvements in lubricants as well in the designs of both surface and subsurface equipment have expanded use of power water.

Salt crystals will occasionally be a problem in systems using power water. This problem can usually be solved through the use of fresh water.

A fresh water blanket on the bottom of the power fluid tank or injection of fresh water into the power fluid can be used.

Surface multiplex pump modifications for converting from oil service to water service are limited mainly to the fluid end of the pump and its plungers/liners. A power oil fluid end can be made of ductile iron or forged steel, but a fluid end for power water will be of aluminum/bronze to resist the corrosive effects of water. It has now become common to use the aluminum/bronze fluid end for both power oil and power water. This eliminates the possibility of using a multiplex pump in the wrong application. Metal-to-metal pistons and liners can be used for power oil service, but metal pistons against soft packing are needed for water service. As with the fluid end metallurgy, it has become common to use soft packing for water and oil.

9.2 JET PUMPS

9.2.1 Theory

The key components of a jet pump are the nozzle and throat. The ratio of the nozzle to throat areas is referred to as the area ratio and it determines the performance characteristics of the pump. Pumps with the same area ratio have the same performance and efficiency curves.

The power fluid and production flow rates must be within the design parameters of the physical nozzles and throats being used in order for them to function correctly. It is not uncommon for someone to focus on the ratios used in jet pumps, such as the area ratio, and forget about the actual sizes of the parts. This can lead to misapplications and failure to perform.

Power fluid is pumped at a given rate (Q_S) to the downhole jet pump where it reaches the nozzle with a total pressure, designated as P_N (see Figure 9-13). This high pressure liquid passes through the nozzle where it is converted from a low velocity, high static pressure flow to a high velocity, low static pressure flow (P_S). The low static pressure (P_S) allows well fluids to flow from the reservoir at the desired production rate (Q_S) into the well bore and pump. The volume of power fluid used will be proportional to the size of the nozzle.

Whenever a high velocity jet of liquid is introduced into a stagnant or slowly moving liquid, a dragging action occurs at the boundary between the two liquids due to the interaction of the high velocity particles with the low velocity particles. The mixing of the two liquids is initiated by this dragging action and the transfer of momentum accelerates the slow liquid in the direction of flow. The mixing of the two streams at this point is minimal at most as the slow moving liquid at the boundary is able to move away from the high velocity jet.

The slow liquid then enters a region of decreasing area, the annulus between the mixture stream and the inner walls of the throat. At the throat entrance, that annular area is the difference between nozzle exit area and throat area. As the two flows progress, a thorough mixing of the two streams takes place because the slow moving liquid at the boundary is not able to move away due to the walls of the throat. The area of the mixture stream progressively spreads while the area of the core of the high velocity jet progressively decreases until it disappears (see Figure 9-12).

At or before the throat exit the mixture stream has spread until it touches the walls of the throat. At that point, all the slow liquid has been mixed with the primary jet. The flow then exits the pump through a diffuser section and is converted to a high static pressure, low velocity flow. This high discharge pressure (P_d) must be sufficient to lift the combined flow rate (Q_t) to the surface.

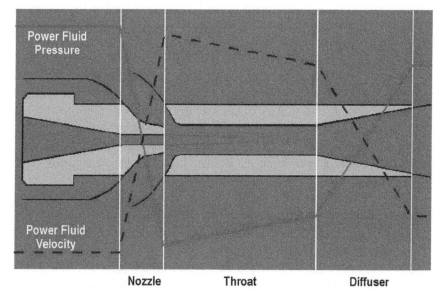

Figure 9-12: Schematic of Flow Velocity and Static Pressure in a Jet Pump

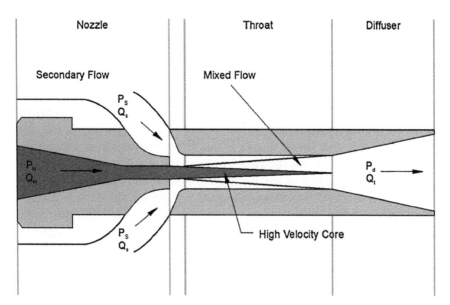

Figure 9-13: Flow Rates/Pressures Entering and Leaving a Jet Pump

The area of the throat must be able to pass the power fluid as well as the liquids and gas being produced. The area in the pump that must accommodate just the produced fluids (liquid and gas) is the annular area between the nozzle and the throat, and it is this area that determines the cavitation characteristics of the pump.

For high flow installations the size of the nozzle is chosen such that the annulus area in the throat is maximized. The resultant area ratio is excellent for high flow/low lift requirements. The reverse is true for low production rate installations. The area of the annulus is minimized. The resultant area ratio for this case is excellent for high lift/low flow installations. However, care must be taken when using the high lift ratios as they are more susceptible to cavitation than the low lift ratios (see Figure 9-14).

9.2.2 Cavitation

Cavitation occurs when the local static pressure is equal to or less than the vapor pressure of the gas dissolved in the liquid. Typically, this is a problem whenever too much fluid is forced through the area that is available for it—that is, throat area minus nozzle area. The higher the

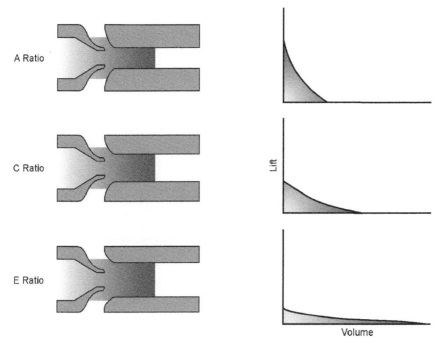

Figure 9-14: Volume/Pressure Relationships for Different Area Ratios

volume for a given flow area, the higher the velocity and the lower the static pressure.

Cavitation is also possible whenever there is too little production. This situation is commonly called "power fluid" cavitation. As is always the case, the power fluid accelerates the produced fluid to a high velocity but the velocity difference is at a maximum when the production rate approaches zero. The shearing action between the two flows will generate vortices and the cores of the vortices may be at sufficiently low enough pressures that cavitation bubbles will form. These bubbles will travel into the throat and cause cavitation damage in either the constant diameter section of the throat or in the diffuser. More commonly, the damage is located in the diffuser just past the constant diameter section.

As with all cavitation bubbles, the damage is mitigated if any gas or oil is present due to their cushioning effect. The damage is greatest whenever there is a high percentage of water with little or no oil or gas present (see Figure 9-15). It is possible to have a similar condition at

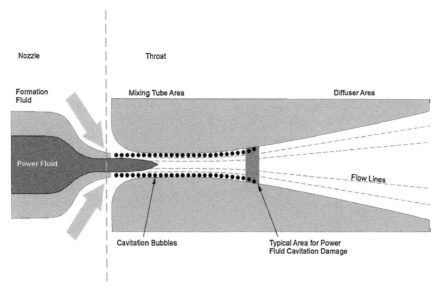

Figure 9-15: Schematic of Cavitation Bubbles and Damage Due to "Power Fluid Cavitation"

start-up. The reason that such damage is not common at that time is due to the cushioning effect of the produced fluid that is present, although it has been known to happen. This effect would not be present when a well is "pumped-off" and cavitation is unavoidable if operations are continued for too long after the well is "pumped-off."

9.2.3 Emulsions

It has long been *assumed* that jet pumps create emulsions, especially if the water percentage is in the range of 60 to 70 percent. However, a verifiable case has yet to be found where that has happened irrespective of the percentage of water. Whenever an emulsion enters the jet pump, it will also exit the jet pump as long as no effort has been made to break it. This typically happens because the emulsion has been created elsewhere (such as acidizing a calcium carbonate reservoir). It is possible to create an emulsion in the jet pump by accident due to using incompatible chemicals in the power fluid where they essentially become emulsifiers. Breaking an emulsion usually involves simply adding an emulsion breaker to the power fluid.

9.2.4 Sizing Considerations

A jet pump must be capable of producing at the desired rate and in accordance with the well's capabilities. The required surface horsepower must be kept at a reasonable level. The first part of the process involves matching the jet pump performance curves with the well productivity (PI/IPR). The balance of the process involves staying within the operating limitations for a particular installation. The most common limitations are power fluid injection pressure and/or rate, and space limitations (such as for offshore installations). The backpressure (discharge pressure) imposed on a jet pump should always be as low as possible.

A computer is required for the calculations.

9.3 PISTON PUMPS

The subsurface production pump is the heart of a hydraulic pumping system. The piston pump is driven by a reciprocating hydraulic engine piston that is connected to the production pump piston (see Figure 9-16). Although the stroke length of a particular pump is fixed, various pump bore sizes are available for different volume and depth requirements. A wide operating speed range adds further flexibility.

The basic components of all hydraulic piston pumps include an engine piston and barrel, an engine reversing valve (which controls the up and down motion by directing the power fluid to the appropriate areas), and the pump piston and barrel. Conventional valves (usually balls and seats) control the production fluid intake and discharge. The arrangement of these components in the pump is based on the specific designs of the individual pumps.

The two most common pump end designs are:

- The "single-acting" production pump end that displaces produced fluid only during the upstroke or only during the downstroke
- The "double-acting" production pump end that displaces produced fluid on both the upstroke and downstroke

The engine end may be designed to displace equal volumes of power fluid on each up and down stroke (double displacement pump) or to displace a greater power fluid volume during one or the other of the two strokes (single displacement pump).

The power fluid used to drive these pumps is clean crude oil or clean water drawn from the top of a settling tank or from a well site unit such as the Unidraulic®. Usually the spent power fluid is mixed with the produced well fluids at the pump and both come to the surface together (an open power fluid system).

The power required to return the spent power fluid to the surface is what is required to overcome mechanical and fluid friction and the difference in piston areas as seen by the power fluid and return fluid. Even though the static pressure head of the incoming power fluid and the static pressure head of the returning power fluid column are not equal, they are similar.

9.3.1 Operation

Hydraulic subsurface piston pumps are composed of two basic sections—a hydraulic engine and a piston pump. They are directly connected with a middle rod. As the engine piston moves upward, the pump piston also moves upward, causing the barrel chamber under the pump piston to fill with production fluid. When the hydraulic engine makes a downstroke, the pump piston also makes a downstroke, displacing the production fluid in the pump barrel.

The arrangement of the pump end is the same as with a sucker rod pump in that there is a barrel, a piston, a piston traveling valve, and a standing valve. Since there is no mechanical linkage to the surface because the sucker rod string is replaced by a column of high pressure power fluid, many limitations of sucker rod pumping are eliminated.

9.3.2 Single Displacement Pump

The pump in Figure 9-16 is a single displacement pump and its operation is as follows. It has a mechanically assisted, hydraulically shifted reversing valve that makes it ideal for dewatering gas wells. This arrangement makes it impossible for the pump to stall no matter how slowly it operates. Additional pump models are available.

The hydraulic engine end of the Power Lift I® series basically is composed of an engine barrel, an engine piston, and a reversing valve mechanism. The reversing valve is shifted by hydraulic pressure but it is initiated mechanically. The assembly is mounted in and moves with the engine piston. The bottom of the engine piston is exposed to the power fluid during the upstroke and to discharge pressure on the down-

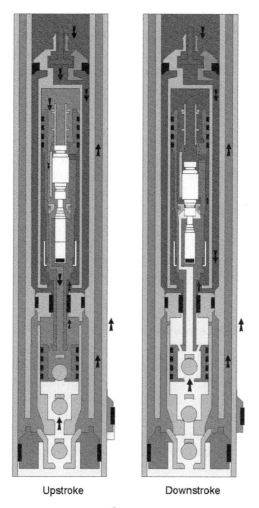

Upstroke   Downstroke

Figure 9-16: Power Lift® I Single Displacement Pump

stroke. The pressures acting on the top of the engine piston are the reverse of those on the bottom of that piston. During the downstroke, the reversing valve is in the down position. At the end of the down-stroke, the mechanical assist mentioned earlier "bumps" the valve off its seat and pressure forces cause it to shift to the up position. With the reversing valve in the up position, a pathway is opened that exposes the top of the engine piston to discharge pressure. The resulting pressure imbalance creates a net force up on the engine piston. At the end of the upstroke, a mechanical assist again "bumps" the valve off its seat and reopens the pathway for the top of the engine piston and reversing valve to be exposed power fluid again. Because there is no middle rod on top of the engine piston, more area on top is exposed to the power fluid than on the bottom, and this results in a net force downward.

The arrangement of the pump end is the same as with a sucker rod pump in that there is a barrel, a piston, a piston traveling valve, and a standing valve. The sucker rod string is replaced with a column of high pressure power fluid, which supplies the energy needed to move the engine piston. As the engine piston moves upward, the pump piston also moves upward, causing the barrel chamber under the pump piston to fill with production fluid. When the engine piston moves downward, the pump piston also makes a downstroke, displacing the production fluid in the pump barrel from below the pump piston to above it.

9.3.3 Double Displacement Pump

The pump in Figure 9-17 is a double displacement pump and its operation is as follows. As is the case with the single displacement pumps, additional pump models are available that utilize different designs but the concept holds true for all of them.

When the pump starts the upstroke, the reversing valve is positioned at the top of the valve body. Power fluid enters the pump through a port in the center of the valve body. The reversing valve directs the power fluid into the barrel above the reversing valve housing and under the upper piston. Since the pressure below the upper piston is greater than the pressure above it, the result is an upward movement of the piston.

The fluid above the upper piston is production that entered the upper barrel during the downstroke. It entered through the valve section of the pump located between the inlet valve and discharge valve. The upward movement of the piston forces the inlet valve to close and the discharge valve to open, allowing the production to exit the pump and BHA and flow to the surface.

The lower piston is also moving up, causing the spent power fluid above that piston to enter the lower end of the valve body. This spent power fluid passes by the reversing valve as it leaves the pump and returns to the surface.

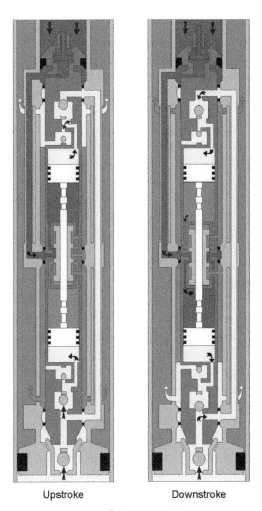

Upstroke   Downstroke

Figure 9-17: Power Lift® II Double Displacement Pump

Below the lower piston, the pressure is reduced inside the lower barrel and between the lower valves by the upward movement. This reduced pressure opens the inlet valve and the higher pressure on the return side closes the discharge valve. This allows production to enter the lower barrel so it can subsequently be discharged on the downstroke.

On the downstroke, production in the lower barrel is discharged. Production also enters the pump below the lower production inlet valve but is directed to the upper valve section through side tubes in the BHA.

This fluid then enters the upper barrel, above the piston, to be subsequently discharged on the next upstroke.

The upper piston is also moving down, causing the spent power fluid below it to enter the upper end of the valve body. This spent power fluid passes by the reversing valve as it leaves the pump and returns to the surface.

Another upstroke begins as the pump reaches the end of the downstroke. Again, a reduced O.D. section of the middle rod enters the I.D. of the reversing valve. This undercut creates a path that allows the bottom of the reversing valve to be exposed to the power fluid. With a higher pressure below, the reversing valve is made to shift to the top of the reversing valve body. This shift causes the power fluid to be directed to the upper half of the pump and the upstroke begins.

9.3.4 Piston Velocity

A common cause of failure among all piston pumps is excessive piston velocity on the displacement stroke. This usually occurs when a pumped-off or fluid pound situation occurs.

The speed of the pump in Figure 9-16 is controlled by the size of the hole, or orifice, in the part that "bumps" the reversing valve off its seat at the end of the upstroke. That part is known as a pushrod. The orifice functions as a choke and reduces the volume of power fluid that can drive the engine piston downward at a given injection pressure. The smaller the orifice, the slower the pump will stroke for a given injection pressure. If the orifice is changed to a smaller size, that would further reduce the volume of power fluid to the engine piston and but will tend to increase the pressure of the power fluid at the surface. However, the bypass in the surface power fluid line should be adjusted so that injection pressure is not increased, but rather there is an increase in the amount of fluid being bypassed back to the power fluid tank.

In this way, the downward velocity of the pump piston is controlled and any damage due to pumped off conditions is minimized.

The pump in Figure 9-17 controls piston velocity through a series of sensing holes in the reversing valve. These holes sense whenever the flow of power fluid through the reversing valve/reversing valve body is excessive and directs the reversing valve to restrict the flow.

9.3.5 Fluid Separation

There must be some means of keeping high-pressure power fluid and low-pressure return fluid separated both inside and outside of the pump.

Internally, separation is accomplished by a close-fitting metal-to-metal seal around the middle rod. This seal prevents the power fluid inside the pump from bypassing the pump engine and mixing with the produced fluids. A leak or loss of this seal results in reduced pump speed and engine efficiency.

Externally, separation is accomplished by elastomeric seals on the outside diameter of the pump and one or more seal collars, which are part of the BHA. These seals prevent the power fluid outside of the pump from bypassing the pump engine and mixing with the produced fluids. A leak or loss of this seal also results in reduced pump speed and engine efficiency.

9.3.6 Piston Size

The engine piston diameter for a given size of hydraulic pump will always be the same. The pump piston, however, can be reduced by several sizes from the size of the engine piston. For example, a 2-1/2-inch hydraulic pump (the maximum size that can be used in 2-1/2-inch tubing) will always have a 2-inch diameter hydraulic engine piston. The pump piston, however, can have a diameter of 2 inches, 1-3/4 inches, 1-5/8 inches, 1-1/2 inches, 1-1/4 inches, or 1-1/16 inches. The displacement rate chart (Table #1) shows the maximum pump end displacement rates for different pump piston sizes of the single displacement Powerlift® I pumps (see Figure 9-16). The displacement rate chart (Table #2) shows the maximum pump end displacement rates for different pump piston sizes of the double displacement Powerlift® II pumps (see Figure 9-17).

Varying the piston size primarily permits two things:

- Sizing the pump end to the actual well requirements
- Sizing the pump to have the lowest operating pressure possible

The smaller the pump piston size in relation to the size of the engine piston, the lower the pressure required to operate the hydraulic unit.

9.3.7 Selecting a Pump

Pump/Engine Ratio

The pump/engine ratio (P/E) is an important factor to consider in selecting a pump because of its relationship to the surface pump that provides the high pressure power fluid. The P/E for pumps with unbalanced middle rods is determined by dividing the engine piston area (A_E) less the middle rod area (A_{MR}) into the pump piston area (A_P):

$$\frac{P}{E} = \frac{A_P}{A_E - A_{MR}}$$

For pumps with balanced middle rods, the equation is

$$\frac{P}{E} = \frac{A_P - A_{MR}}{A_E - A_{MR}}$$

A *higher* P/E requires a *lower* power fluid volume and a *higher* multiplex pump pressure. A *lower* P/E requires a *higher* power fluid volume and a *lower* multiplex pump pressure.

Ensure that pump selection calculations are done carefully. If the P/E is too low, the increase in power fluid volume will cause increased friction losses in the system, resulting in higher multiplex pump pressure.

9.3.8 Surface Power Fluid Conditioning System

The purpose of a surface power fluid conditioning system is to provide a constant and adequate supply of suitable power fluid to operate the subsurface pump. The success and economical operation of any hydraulic pumping fluid installation is dependent on the effectiveness of the surface conditioning system in supplying clean power fluid for the surface power pump and downhole pump.

There are two types of power fluid conditioning systems for hydraulic pump installations: the central power fluid conditioning system and the well-site, self-contained power fluid conditioning system (Unidraulic®).

9.3.9 Central Power Fluid Conditioning System

A central tank battery (see Figure 9-18) is one in which the power fluid for one or more wells is treated to remove gas and solids at a large centralized facility.

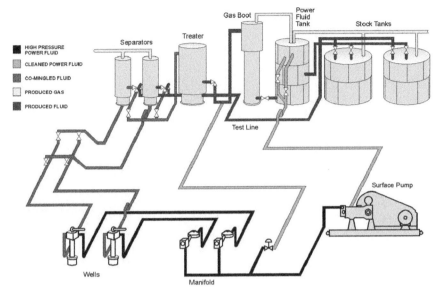

Figure 9-18: Central Tank Battery

Figure 9-19 shows a typical central power fluid treating system that has been proven through years of experience. This power fluid treating system design assumes that the normal lease separators and heater treaters have delivered "stock tank" oil, essentially free of gas, to the treating facility.

The power fluid settling tank in this system is usually a 24-foot-high, three-ring, bolted steel tank that will provide an adequate head for gravity flow of fluid from the tank to the intake of the charge pump.

The purpose of the power fluid settling tank is to allow separation of solids that the lease separator has not removed in the continuous flow system.

In a tank of static fluid, all foreign material that is heavier than the fluid will fall or settle out to the bottom. Some of the particles, such as fine sand, would fall more slowly than the heavier solids. These factors, plus viscosity-related resistance factors, influence the rate of separation. In time, however, all solids and heavier liquids will settle out, leaving a layer of clean fluid.

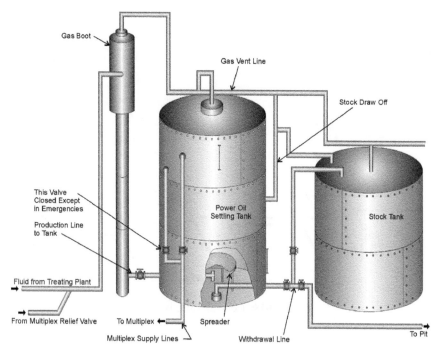

Figure 9-19: Central Power Fluid Treating System

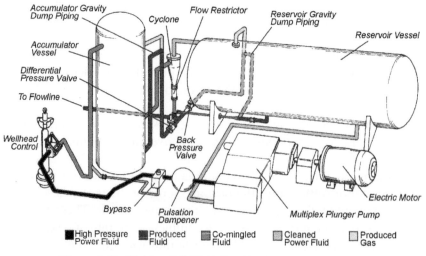

Figure 9-20: Unidraulic® Well Site Fluid Conditioning Unit

In an actual power fluid system it is not practical, nor is it necessary, to furnish tank space to allow settling under perfectly still conditions. Sufficient settling can be accomplished when the upward flow through the settling tank is maintained at a velocity just slower than the velocity at which the contaminating material will fall. It has been found through tests and experience that an upward velocity of 1 foot per hour is low enough to provide gravity separation of entrained particles in most crude oils.

An example of a unitized surface equipment package for hydraulic pumping applications is the Weatherford Unidraulic® Fluid Conditioning Unit. The unit provides complete fluid conditioning as well as a surface pump to supply pressurized power fluid to the downhole pump (see Figure 9-20).

9.3.10 Troubleshooting

Table 9-1
Troubleshooting Jet Pumps

Indication	Cause	Remedy
Sudden increase in operating pressure—power fluid rate constant or reduced.	(a) Paraffin build-up or obstruction in power oil line, flow line, or valve. (b) Partial plug in nozzle.	(a) Run soluble plug or hot oil, or remove obstruction. Unseat and reseat pump. (b) Retrieve pump and clear nozzle.
Slow increase in operating pressure with constant power fluid rate, or slow decrease in power fluid rate with constant operating pressure.	(a) Slow build-up of paraffin. (b) Worn throat or diffuser.	(a) Run soluble plug or hot oil. (b) Retrieve pump and repair.
Sudden increase in operating pressure and power fluid rate essentially stopped.	(a) Fully plugged nozzle.	(a) Retrieve pump and clear nozzle.
Sudden decrease in operating pressure with power fluid rate constant, or sudden increase in power fluid rate with operating pressure constant.	(a) Failure in power fluid tubing string. (b) Blown pump seal or broken nozzle.	(a) Check tubing for leaks and pull and repair if leaking. (b) Retrieve pump and repair.

Continued

Table 9-1 (*Continued*)

Indication	Cause	Remedy
Drop in production while all surface measurements conditions normal.	(a) Worn throat or diffuser. (b) Plugged standing valve or pump. (c) Leak or plug in gas vent. (d) Changing well conditions.	(a) Increase operating pressure. Replace throat and diffuser. (b) Retrieve pump and check. Retrieve standing valve. (c) Check gas vent system. (d) Run pressure recorder and resize pump.
No production increase when operating pressure is increased.	(a) Cavitation damage in pump or high gas production. (b) Plugging of standing valve or pump.	(a) Lower operating pressure or install larger throat. (b) Retrieve pump and check. Retrieve standing valve.
Throat worn as seen by one or more dark, pitted zones.	(a) Cavitation damage.	(a) Check pump and standing valve for plugging. Install larger throat or reduce operating pressure to reduce velocity.
Throat worn—its cylindrical shape changed to barrel shape, smooth finish.	(a) Erosion.	(a) Replace throat, preferably with a premium material throat. 　Install a larger nozzle and throat to reduce velocity.
New installation does not meet prediction of production.	(a) Incorrect well data. (b) Plugging of standing valve or pump. (c) Tubular leak. (d) Side string in parallel installations not landed.	(a) Run pressure recorder and resize pump. (b) Check pump and standing valve. (c) Check tubing and pull and repair if leaking. (d) Check tubing and restab if necessary.

Table 9-2
Troubleshooting for Reciprocating Pumps

Indication	Cause	Remedy
Significant increase in operating pressure while pump speed is significantly reduced.	(a) Pump intake pressure significantly reduced. Pump is stalling. (b) Paraffin build-up or obstruction in power oil line, flow line, or valve. (c) Pumping heavy material, such as salt water or mud. (d) Pump beginning to fail.	(a) Reduce pump speed, retrieve and resize pistons. (b) Run soluble plug or hot oil, or remove obstruction. (c) Keep pump stroking—do not shut down. (d) Retrieve pump and repair.
Gradual increase in operating pressure while pump is stroking.	(a) Gradually reducing pump intake pressure. Standing valve or formation becoming plugged. (b) Slow build-up of paraffin. (c) Increasing water production.	(a) Retrieve pump and check. Retrieve standing valve and check. (b) Run soluble plug or hot oil. (c) Increase pump speed and watch pressure.
Sudden increase in operating pressure but pump is not stroking.	(a) Pump stuck or stalled. (b) Sudden change in well conditions requiring operating pressure in excess of triplex relief valve setting. (c) Sudden presence of emulsion in power oil, etc. (d) Closed valve or obstruction in production line.	(a) Alternately increase and decrease pressure. If necessary, unseat and reseat pump. If this fails to start pump, retrieve and repair. (b) Raise setting on relief valve. (c) Check power oil supply. (d) Locate and correct.
Sudden decrease in operating pressure while pump is stroking. Speed could be increased or reduced.	(a) Rising fluid level—pump efficiency up. (b) Failure of pump so that part of power oil is bypassed. (c) Gas passing through pump. (d) Tubular failure—down hole or in surface power oil line. Speed reduced. (e) Broken plunger rod. Increased speed. (f) Seal sleeve in BHA washed or failed. Speed reduced.	(a) Increase pump speed if desired. (b) Retrieve pump and repair. (c) Consider installing vent string or parallel BHA. (d) Check tubulars. (e) Retrieve pump and repair. (f) Pull tubing and repair BHA.

Continued

Table 9-2 (*Continued*)

Indication	Cause	Remedy
Sudden decrease in operating pressure and pump is not stroking.	(a) Pump not on seat. (b) Failure of production unit or external seal. (c) Bad leak in power oil tubing string. (d) Bad leak in surface power oil line. (e) Not enough power oil supply at manifold.	(a) Circulate pump back on seat. (b) Retrieve pump and repair. (c) Check tubing and pull and repair if leaking. (d) Locate and repair. (e) Check volume of fluid discharged from triplex. Valve failure, plugged supply line, low power oil supply, excess bypassing, etc., all of which could reduce available volume.
Drop in production but pump speed is constant	(a) Failure of pump end of production unit. (b) Leak in gas vent tubing string. (c) Well pumped off—pump speeded up. (d) Leak in production return line. (e) Change in well conditions. (f) Pump or standing valve plugging. (g) Pump handling free gas.	(a) Surface pump and repair. (b) Check gas vent system. (c) Decrease pump speed. (d) Locate and repair. (e) Resize pistons. (f) Surface pump and check. Retrieve standing valve. (g) Test to determine best operating speed.
Gradual or sudden increase in power oil required to maintain pump speed. Low engine efficiency.	(a) Engine wear. (b) Leak in tubulars—power oil tubing, BHA seals, or power oil line.	(a) Surface pump and repair. (b) Locate and repair.
Erratic stroking at widely varying pressures.	(a) Caused by failure or plugging of engine.	(a) Surface pump and repair.

Stroke "downkicking" instead of "upkicking".	(a) Well pumped off—pump speeded up. (b) Pump intake or downhole equipment plugged. (c) Pump failure (balls and seats). (d) Pump handling free gas.	(a) Decrease pump speed. Consider changing to smaller pump end. (b) Surface pump and clean up. If in downhole equipment, pull standing valve and backflush well. (c) Surface pump and repair.
Apparent loss of, or unable to account for, system fluid.	(a) System not full of oil when pump was started due to water in annulus U-tubing after circulating, well flowing, or standing valve leaking. (b) Inaccurate meters or measurement.	(a) Continue pumping to fill up system. Pull standing valve if pump surfacing is slow and cups look good. (b) Recheck meters. Repair if necessary.

9.4 REFERENCES

1. Clark, K. M. "Hydraulic Lift Systems for Low Pressure Wells," Petroleum Engineer International, February 1980.

2. Christ, F. C. and Petrie, H. L. "Obtaining Low Bottomhole Pressures in Deep Wells with Hydraulic Jet Pumps," SPE 15177.

3. Peavy, M. A. and Fahel, R. A. "Artificial Lift with Coiled Tubing for Flow Testing The Monterey Formation Offshore California," SPE 20024.

4. Hrachovy, M. J., McConnell, M. L., Damm, M. W., and Wiebe, C. L. "Case History of Successful Coiled Tubing Conveyed Jet Pump Recompletions Through Existing Completions," SPE 35586.

5. Anderson, J., Freeman, R., and Pugh, T. "Hydraulic Jet Pumps Prove Ideally Suited for Remote Canadian Oilfield," SPE 94263.

USE OF BEAM PUMPS TO DELIQUIFY GAS WELLS

10.1 INTRODUCTION

Beam pumps are likely the most common method used to remove liquids from gas wells. They can be used to pump liquids up the tubing and allow gas production to flow up the casing. Their ready availability and ease of operation have promoted their use in a variety of applications.

Beam pump installations typically carry high costs relative to other deliquifying methods. The initial cost of a beam pump unit can be high if a surplus unit is not available. In addition, electric costs can be high when electric motors are used to power the prime movers, and high maintenance costs often are associated with beam pumping operations. Due to the expense, alternative methods to deliquify gas wells should be considered before installing beam pumps. In the event that these costs are minimal for a particular application, beam pumps can provide a good means of removing liquids from gas wells.

If beam pumps are to be used for gas well liquid production, the beam system often will produce smaller volumes of liquids. Because of the usually low volumes required to deliquify gas wells, and the fact that beam pumps do not have a lower limit for production and efficiency as do other pumping systems such as ESPs, they often are used for gas well liquid production. Figure 10-1 shows an approximate depth-volume range for the application of beam pump systems.

The presence of high gas volumes when deliquifying gas wells means that measures often are required to keep gas from entering the downhole pump or to allow the pump to fill and function with some gas present.

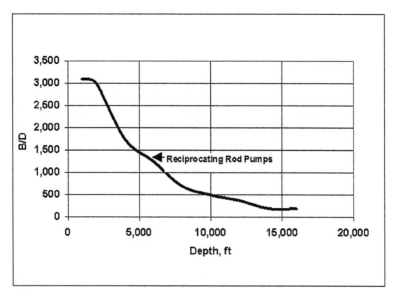

Figure 10-1: An Approximate Depth-Rate Application Chart for Beam Pumping Systems

Figure 10-2 shows a typical beam pumping system.

This chapter discusses the primary concerns associated with the use of beam pumps to deliquify gas wells. Some concerns include:

- Pump-off control of the pumping system to avoid effects of over-pumping
- Gas separation when necessary
- Special pumps to handle gas-induced problems
- The possible use of injection systems to inject water below a packer in a water zone so gas can flow upward more easily

10.2 BASICS OF BEAM PUMP OPERATION

The beam pumping unit changes rotary motion from the prime mover into reciprocating motion. If the prime mover is electric, it usually is a motor with a synchronous speed of 1200 rpm. Under load, it might be rotating at possibly 1140 average rpm. A beam pump is a high efficiency device that makes good use of input electrical energy. A formula for the efficiency of a beam pump unit is:

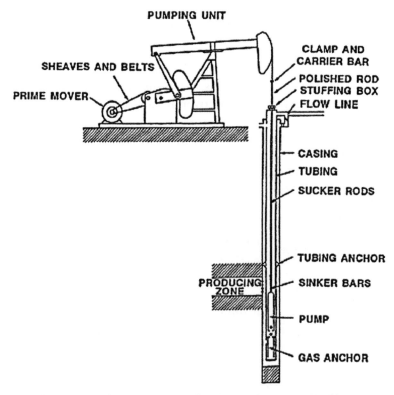

Figure 10-2: Schematic of Beam Pumping System (courtesy, Harbison Fischer)

$$\eta = \frac{.00000736\gamma QH}{kW/.736} \tag{10-1}$$

where:

η = the overall electrical efficiency of the pumping unit.

γ = specific gravity of the fluid

Q = the additional production from pumping the well, bpd

H = the vertical lift of the fluid from approximately the fluid level in the casing to the surface, ft

KW = the electrical power input to the motor at the surface, kilo-Watts

Equation 10-1 can be used for PCPs, ESPs, hydraulic pumping units, and other pumping systems. However, this formula cannot directly be used for gaslift. Typically, PCPs and some hydraulic systems may have better efficiency than beam pump systems, and ESP systems are usually less.

A good beam pump installation can have an efficiency of more than 50 percent. However, for gas wells, often the gas interference into the pump downhole may reduce the overall electrical efficiency to much less.

To reciprocate the sucker rods and pump, the high-rpm motor speed must be reduced to the required SPM for the pump. The speed of the motor is reduced by the motor sheave and the gearbox sheave and a gearbox speed reduction of usually 30 : 1. The pump SPM is calculated from

$$\text{SPM} = \text{Motor RPM} \times \frac{\text{motor sheave diameter}}{\text{gear box sheave diameter}} \times \frac{1}{\text{gearbox ratio}}$$

$$(10\text{-}2)$$

As an example, a beam unit with a motor sheave of 12 in and a gearbox sheave of 37 in, then the speed reduction from the motor to the rods or horses head will give a SPM of

$$\text{SPM} = 1140 \frac{12}{37} \frac{1}{30} = 12.3 \, \text{spm}$$

The belts in the sheaves carry the power from the motor to the gearbox. The gearbox slows the RPM by 30 : 1 and increases the torque to the output shaft of the gearbox by 30 : 1, discounting some inefficiencies. The crank turned by the slow-speed shaft of the gearbox, rotates and, through a pitman arm connected to the crank, moves the back end of the long walking beam up and down at the back. The up and down motion is translated to the front of the walking beam and to the rods to impart reciprocating motion. Counterweights on the crank arm balance one-half of the fluid load plus the weight of the rods in fluid.

The rods are connected to a polish rod at surface to pass through the stuffing box to seal the well. The polish rod is clamped on the top of the carrier bar hanging by two cables from the horses head end of the walking beam. The rods, usually connected with couplings, are connected all the way from surface to the pump near the perforations. The rods are usually 25 ft in length (30 ft in California) and come in different grades of metallurgy. The rod string usually has a section of larger rods at the top and one or more sections of smaller diameter rods to the pump.

The pump strokes up and down with motion imparted by the rods to affect a downhole stroke usually less than the surface stroke length due to rod stretch. The formula for volumetric production at the pump is

$$BPD = .1165\ D^2 L\ SPM \qquad\qquad (10\text{-}3)$$

where:

  D = diameter of downhole pump, in
  L = downhole stroke length at the pump, in
SPM = reciprocating cycles per minute

The pump is rarely full of liquid and has leakage so the preceding formula should be multiplied by some factor, perhaps 0.8 or so to better match a good installation. A poor installation with leaky valves or gas interference may show much less production than the formula would predict.

The downhole pump consists usually of a plunger connected to the rods with a traveling valve on the end of the oscillating plunger. The barrel has a standing valve on the bottom. The pump is connected to the tubing end by a top or bottom hold-down for insertable pumps (pumps that can be removed by the rods), whereas tubing pumps have the barrel screwed into the bottom of the tubing.

The pump works much better if free gas is kept from the intake of the pump. This is accomplished best by setting the pump below the pay zone. Special gas separators can be used if the pump is set above the pay. If neither of these options is successful, special pumps are available to better handle gas.

10.2.1 Comments on API Pumps

By Benny Williams, HF Pumps
Benny Williams, PE, BSME, MBA has worked in the sucker rod pumping industry more than 25 years and has introduced several new products to the industry. He has chaired API committees and coauthored papers in this industry. He is Vice President of Engineering for Harbison-Fischer Manufacturing Company.

Sucker rod pumps are available in two broad categories: standard American Petroleum Institute (API) configurations, and non-API sucker rod pumps. Non-API sucker rod pumps are sometimes categorized as special sucker rod pumps since they are proprietary designs by sucker rod pump companies or minor modifications of API pumps. Both categories of sucker rod pumps are application specific for certain down-hole pumping conditions, fluid types, and fluid quantities.

API sucker rod pumps are standardized by API specification API 11AX and thus, regardless of the manufacturer of the pump, all the pump parts are interchangeable. This API specification also standardizes the materials and quality control procedures for this group of sucker rod pumps.

Non-API, or special sucker rod pumps, have become more commonplace in the last 20 or 30 years due to increased competition in the industry and due to the perceived need for specialty pumps for certain pumping conditions. Materials and quality control procedures, although not controlled by API 11AX, are generally the same, and most special pump parts are API 11AX parts due to their wide availability.

It is interesting to note that simple changes in the parts of API pump assemblies put slightly modified API pumps into the special pumps category. For example, for heavy crude applications the traveling valve may be moved from the bottom to the top of the plunger, giving the pump less fluid restriction and earning it special pump status. However, there is always a tradeoff, and this minor change makes it less effective in gas production conditions.

Sand and scale can stick the bottomhole pump and must be accounted for in solids producing wells using special pumps or filters.

10.3 PUMP-OFF CONTROL

Often if a beam pump is used to dewater a gas well, then relatively small amounts of liquid must also be produced to allow the gas to flow. The usual procedure is to pump liquids up the tubing and allow gas to flow up the casing. Because small rates of liquids may be produced, it is not unusual for the beam system to pump at a rate higher than the well can deliver liquids over time. When a beam pump is operated at a rate beyond the capacity of the reservoir to produce liquids, the liquid level in the well is pumped below the pump intake and the pump is said to "pump-off." This condition allows gas to enter the pump barrel and potentially damage the pump.

There is considerable literature [1, 2] concerning beam pump systems on pump-off control. With gas in the pump barrel, the pump plunger initially compresses the gas on the downstroke of the pump before contacting the liquid. If sufficient gas is allowed into the barrel, the plunger can contact the fluid causing "fluid pound" with sufficient force to ultimately damage the pump and rod string. This is of primary concern in gas wells due to the relatively high volumes of gas produced with typical low volumes of liquid.

The pump-off controller enables the beam pump to operate with sufficient liquid levels to prevent damage while operating the pump at a high efficiency. The controller essentially stops the pump when the well has been pumped off. However, some pumping systems often are allowed to operate in the pumped-off condition with continual gas interference at the pump. This results in poor efficiency and can result in "fluid pound" as the plunger contacts the fluid in gas/liquid filled barrel on the downstroke. Fluid pound can lead to mechanical damage to the system.

10.3.1 Design Rate with Pump-off Control

The beam pump system should be designed to be able to pump the fluid level in the annulus down to the minimum value consistent with efficient pump operation and prevention of fluid pound.

To achieve this design objective, the pump should be designed to pump at a rate given by

$$\text{Design Rate} = \frac{\text{Maximum Inflow Capacity} \times 24\,\text{hrs/day}}{\text{Pump Volumetric Efficiency} \times \text{hrs pumped/day}}$$

$$(10\text{-}4)$$

The pump volumetric efficiency is essentially the percent fillage of liquids in the pump barrel. For effective pump-off control, 20 hours/day pumping time is a good rule of thumb. The maximum reservoir inflow capacity should be used for the desired daily rate. Example 10-1 illustrates this equation.

Example 10-1: Design System Pumping Rate for POC (pump-off control)

A gas well with maximum liquid flow capacity of 300 bfpd is to be put on beam lift to pump-off the liquids. For what rate should the pump be designed, assuming a pump volumetric efficiency of 80 percent?

$$\text{Design Rate} = \frac{300 \times 24}{0.80 \times 20} \text{ bfpd} = 450 \text{ bfpd} \qquad (10\text{-}5)$$

Using this technique, the pump is designed to operate about 20 hours/day with an 80 percent volumetric efficiency. The pump-off controller will turn the well off when it reaches fluid pound conditions. The operator usually sets the downtime based on production considerations.

Using a typical volumetric efficiency of 80 percent and 20 hours/day pumping time, a simpler rule of thumb is simply to design the beam pump system to deliver a rate equal to 1.5 times the reservoir maximum inflow capacity.

Design Rate = 1.5 × Maximum Inflow Capacity

10.3.2 Use of Surface Indications for Pump-off Control

An important tool for diagnosis of beam pump problems is the surface dynamometer card, or load vs. position at the top of the rod string. Usually the shape of the surface dynamometer card is used by a computerized system to determine when the well is beginning to pump-off. Other systems include using a calculated downhole dynamometer card shape, cycle time, vibration, and other techniques to determine pump-off.

The surface dynamometer card is a plot of load and position on the polished rod just above the top rod. The shape of the surface card and especially the calculated card for the load and position in the rod just above the pump can be indicative of problems or good operation at the downhole card.

Figure 10-3 shows that the surface card can be indicative of the condition of the pump. The two left-most cards show the surface dynamometer card. The surface card is shown for a full pump and the pump card

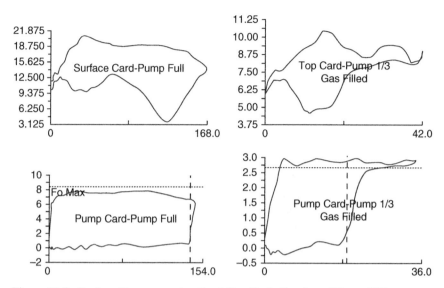

Figure 10-3: Surface Dynamometer Card Can Be Indicative of Pump Fillage (bottom hole card)

below it shows that the pump card is almost rectangular, indicative of a full pump with few or no problems.

However, for the right-most two cards, the bottom pump card shows that the bottom right of the pump card is not outlined by a load-position line. This is because the TV (traveling valve load) has not been released from the rods until about one-third into the downstroke. This is because the gas and/or fluid below the TV has not built up enough pressure below the TV to open it and drop the load on the TV. The load can be dropped gradually (gas interference) or quickly when the TV hits the fluid. The worst situation is for the TV to hit fluid for the first time somewhere near the middle of the downhole stroke when the plunger is traveling much faster than at the beginning and end of the stroke.

Since, as shown earlier, the surface card can be indicative of what is happening at the pump, then pump-off-control can use the surface dynamometer card to control on since it can indicate when the bottomhole pump card is full and when it is beginning to fill with some gas.

Figure 10-4 shows a surface card with a computerized set point, indicated by the "+", that indicates the point at which the POC will shut the pumping action off and allow the well to build up liquids. The latter stages of lesser pump fillage are not allowed to occur.

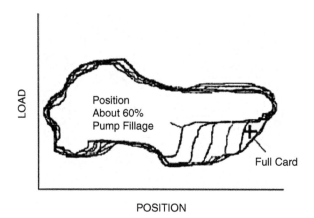

Figure 10-4: Surface Dynamometer Card Showing Various Degrees of Pump Fillage

It is usually desirable to shut the pumping system in when the barrel becomes no less than perhaps 80 to 85 percent full, although conditions vary. Pump-off-control is simply a method of oversizing the pumping action of the pumping system and then shutting off the system when gas interference begins at the pump. Another, but harder to control, method that would achieve about the same results for production would be simply to maintain a small fluid level over the pump at all times.

10.4 GAS SEPARATION TO KEEP GAS OUT OF THE PUMP

When removing liquids from a gassy well using beam pumps, it is quite possible that the pump will be subjected to gas interference. Measures may be needed to separate the gas [4,5] from the liquid stream prior to its entering the pump to prevent gas locking, low efficiency, reduced production, and possible damage from fluid pound.

Before outlining some guidelines, let us first identify what is meant by gas interference and fluid pound. For fluid pound the pump intake is at a low pressure, and the barrel is partially full of liquid and partially full of gas. The plunger comes down on the downstroke and passes through gas and then suddenly impacts the fluid in the barrel, causing fluid pound. This can cause rods to unscrew, rod and tubing damage as the rods bend to the tubing, and other bad effects. This is an indication of the well being pumped off or pumping too fast. A pump-off controller may prevent this. It could be that when pump-off-control is installed or the pump is slowed, that the fluid pound will cease, but gas interference due to gas from the formation coming with the production will still take place.

For gas interference, the pump intake is at a higher pressure, and the pump is volumetrically filled with part gas and part fluid. The gas or a mixture of liquids and gas with higher pressure gradually build up the pressure in the barrel below the plunger and help to cushion the impact of the plunger with the fluid on the downstroke. This still causes low pump fillage or efficiency and low production but may not cause the mechanical damage that the fluid pound causes. There is usually a fluid level above the pump for gas interference and the gas is coming with the fluid from the perforations.

Figure 10-5 shows what the calculated bottomhole dynamometer looks like for fluid pound or for gas interference.

The following rules are based around the height of fluid over the pump in the annulus of the well. This height of fluid measured by acoustic means should be corrected for gas content in the fluid level [4].

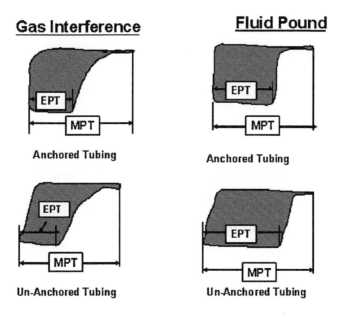

MPT= max pump travel, EPT = effective pump travel

Figure 10-5: Surface Dynamometer Card Showing Various Degrees of Pump Fillage

If the fluid level is low:

- No gas interference in the downhole pump is indicated, then "good job."
- If some gas interference present, but no fluid pound, then still acceptable.
- If gas interference with possibly damaging fluid pound present, then consider gas separation.

If the fluid level is high:

- No gas interference is present, then pump at a higher rate to lower well pressure and produce more gas up the annulus. This would be a high priority.
- Gas interference is present, then consider gas separation so you can pump liquids at a higher rate and allow more gas to be produced. This would be a high priority.

The following are some methods of separating gas from the downhole pump intake.

10.4.1 Set Pump Below Perforations

One of the simplest and best methods of separating the gas from the liquid at the pump is to set the pump below the perforations. The slow downward velocity of the liquid in the casing/tubing annulus down to the pump intake allows the gas to separate from the liquids and migrate freely up the annulus. At the same time, the liquids migrate downward to the pump intake carrying a minimal amount of gas through the pump. If this downward velocity is less than about 0.5 ft/sec, then the amount of gas being carried downward is minimal, especially if only water is being pumped. If the pump cannot be set below the perforations, then other types of gas separation techniques can be considered.

10.4.2 "Poor-boy" or Limited-Entry Gas Separator

One leading type of gas separator is the so-called "poor-boy" gas separator. Various modifications of the "poor-boy" separator have been used widely in the industry over the past 20 years. Figure 10-6 shows a

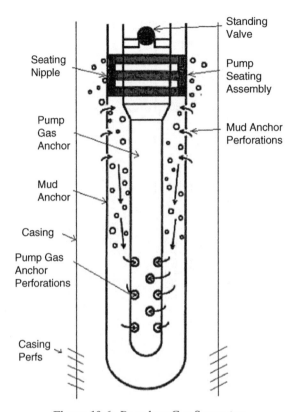

Figure 10-6: Poor-boy Gas Separator

schematic of the "poor-boy" separator, also referred as the limited entry separator.

The device is named "limited entry" because the entry for the fluid is also the entry for stray bubbles, which if entrained has no place for escape. The poor-boy separator is designed so that the down flow in the annular area inside of the separator is less than 0.5 ft/sec so any bubbles in the flow will not be carried into the pump intake through the dip tube. In gassy wells, however, the free gas component makes it difficult to determine when the actual velocity is below 0.5 ft/sec relative to the surface production. In addition, bubbles that migrate inside the separator are unable to escape and can eventually gas-lock the separator.

A rough rule of thumb is that if a gassy well is producing over 200 bpd it will gas-lock a poor-boy separator.

A simple modification that has been made to the natural separator, making it more applicable to gassy wells, is shown in Figure 10-7. The

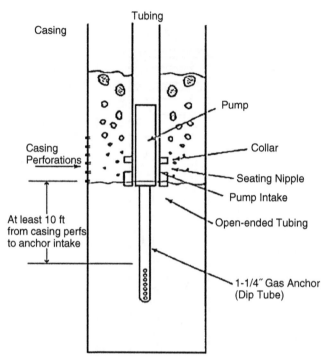

Figure 10-7: Separator Using a Dip-Tube to Allow Intake below the Perforations

modification is to use a stinger to set below the perforations but not the entire pump body. This modification allows a very slow velocity down to the intake, allowing gas to come up the annulus. The inlet of the stinger is positioned below the perforations, allowing the gas to separate from the low velocity fluid in the annular region. The length of the stinger should be kept to a minimum. If the stinger is too long, then the combination of the frictional pressure drop and pressure head due to elevation change can bring gas out of solution in the stinger and defeat this system.

10.4.3 Collar-sized Separator

Another separator is the collar-sized gas separator [4] shown in Figure 10-8. It is fairly inexpensive, has large intake and discharge ports, and can be expected to give good results at fairly low pressures (less than a few 100 psi).

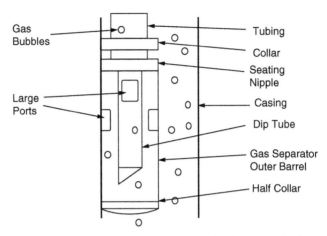

Figure 10-8: Collar-sized Separator (Echometer, Inc.)

A collar-sized gas separator should be selected that is the same size as the tubing unless the pump capacity exceeds the gas separator capacity. In this case, a larger gas separator should be selected that has a liquid capacity equal to or greater than the pump capacity. At high liquid and gas rates, even an optimum size gas separator in limited size casing may not have the capacity to separate all the free gas from the liquids at low pump intake pressures [4].

It is imperative that a beam pumping system operating in a gassy environment have some sort of effective gas separation downhole. Although a wide variety of gas separation systems are given in the literature, those discussed earlier are found to be among the most successful.

10.4.4 Benefits of Downhole Gas Separation in Dewatering Gas Wells and in Low Pressure Oil and Gas Reservoirs

By Dr. Augusto L. Podio, U of Texas, Austin
Dr. Podio is Professor of Petroleum Engineering, University of Texas, specializing in well control and drilling optimization, computer-based production systems, multiphase flow, and beam pumping optimization.

Introduction

Coal bed methane wells have several characteristics that make downhole gas separation one of the tools to achieve more efficient operation while maximizing gas production:

1. Water production initially occurs at relatively large rates but then declines to low rates.
2. Completions involve the use of screens and liners to control production of fine solids.
3. Pumps generally are set in the slotted liner with no rat hole. FBHP is low at 50 psi or less, making the height above the pump very low and typically are pumped off.
4. Pumping systems for dewatering are sized based on initial water rates, resulting in oversized capacity when water production declines. Water rates are typically less than 50 BWPD.
5. Pumping systems are operated with POCs or timers.

The use of a downhole gas separator has the objective of maximizing the efficiency of the system by maintaining a high water fraction at the pump intake even as water production decreases from a high rate to a low rate.

Although the focus is mainly CBM wells, the following discussion is extended to low pressure oil and gas reservoirs where flow rates are limited due to relatively high gaseous liquid columns in spite of the use of pumping systems.

Typical Coal Bed Methane Operations

The following information summarizes the types of bottomhole assemblies in pumping wells in CBM leases. The vast majority of these are wells that produce water with the coal bed methane. The water rates vary from less than 1/2 bwpd to 250 bwpd, with most being less than 10 bwpd. Data from a recent lease review show the following distribution for rod pumped wells is shown in Figure 10-9.

The gas rates in these wells vary from 150 mcfd to 3700 mcfd with an average of less than 1000 mcfd. Some wells have wellhead compressors pulling at 15 psig or produce into gathering system pressures of up to 350 psig. Most produce against less than 150 psig.

Many of the wells have pump-off controllers, so pump intermittently. Generally pumps smaller than 1-1/4 in are not used. Units may have up to 120 in stroke lengths, but most are between 74 and 48 in. The SPM vary from just under 5 to 11 in.

Most wells have 7 in casing set at the top of the coal and are completed open hole with uncemented 5-1/2 in perforated liners across the

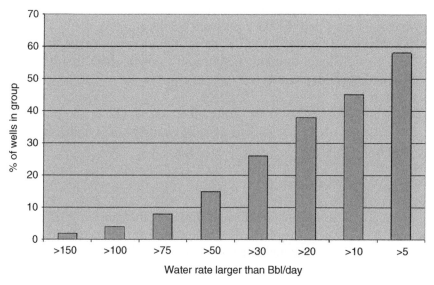

Figure 10-9: Distribution of Water Production Rate in CBM Wells

coal section. The pumps are set down in those liners as deep as possible. However, 50 percent of the wells do not have a rathole.

Flow Mechanics

Simultaneous production of gas and water from the formation results in the creation of a gaseous liquid column in the annular space. This results in two types of multiphase flow:

Above the pump intake. The liquid in the gaseous liquid column is permanently resident in the annulus while the gas percolates through the liquid and is produced at the surface. This is defined as a "zero net liquid flow" condition, although the liquid is in continuous motion but recirculates in place. Depending on the cross-sectional area of the wellbore annulus (casing–tubing) and the gas flow rate, the multiphase flow pattern that develops in the vicinity of the pump intake could range from bubble to annular flow. In the majority of the cases the flow pattern is likely to be the highly turbulent churn flow regime.

The height of the gaseous column, its liquid fraction, the liquid density, and the casing-head pressure determine the magnitudes of the pump intake pressure and the back pressure on the formation.

Below the pump intake. From the perforations to the intake, there is both flow of liquid and gas. Liquid and gas will enter the pump at various ratios depending on the two-phase flow pattern that is developed.

When the pump is set in the slotted liner the intake of the pump or the gas separator (if installed) can also experience impinging flow from the formation.

Steady State Conditions

This is a condition of equilibrium between the mass rate of fluids flowing *into* the well from the formation and the mass rate of fluids *leaving* the wellbore through the pump and through the annulus. Steady state implies that there is no accumulation of material in the wellbore.

At steady state conditions the liquid rate entering the pump intake matches the liquid rate that is entering the wellbore from the formation. The gas rate entering the pump depends on the liquid fraction that is present at the pump intake.

At steady state conditions the height of the gaseous liquid column will remain constant and also the casing-head pressure will be constant as a function of time. With pump capacity exceeding water production the production never reaches true steady state but is intermittent corresponding to the on/off time of the POC or interval timer. However, if the on/off intervals are short (such as when using a 15 minute % timer) the wellbore conditions oscillate about what may be considered a steady state.

Initial Phase of Gas Well Dewatering

During this phase the objective is to produce as much water as possible in order to reach the maximum gas rate as quickly as possible. In this example well, producing from a zone at 2800 feet, a 3-1/4 inch pump was set at 2550 feet and operated at 6.7 SPM with a stroke of 84 inches.

Figure 10-10 shows the result of a fluid level survey that indicates that about 72 MSCF/day of gas flowing up the annulus generate 1560 feet of gaseous column above the pump intake. The well is completed with 2-7/8

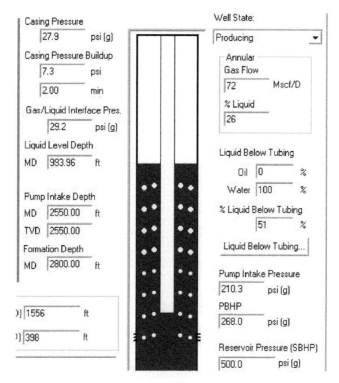

Figure 10-10: Annular Fluid Level Survey in Gas Well with Downhole Gas Separator

inch tubing inside 5-1/2 inch casing. In this annular space the gaseous column contains only 26 percent liquid. Below the pump intake, inside the 5-1/2 inch casing, the percentage of liquid is greater at about 51 percent, due to the larger flow area.

The completion *includes* a downhole gas separator so that although the percent of liquid at the depth of the pump intake is over 50 percent, the percentage of liquid that enters the pump is greater than 87 percent, as can be seen in the dynamometer cards in Figure 10-11.

Note that the pump card shape clearly shows the effect of operating with a pump plunger diameter larger than the tubing ID, which causes a marked increase in load during the first half of the upstroke due to the friction and inertia required to force the fluid into the smaller area of the tubing (bottled-up pump). The effective pump displacement, which corresponds to the 61.8 inch effective plunger stroke, is computed at 507.4 Bbl/day.

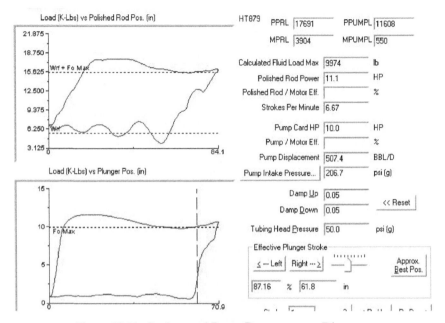

Figure 10-11: Surface and Pump Dynamometer Diagrams

If the same well were to be produced *without* a downhole gas separator, it is estimated that the liquid fillage of the pump would be less than 50 percent, considering that the liquid fraction at the pump intake is about 50 percent as computed from the acoustic fluid level measurement.

Correspondingly, the effective pump displacement would be reduced to (507*50/87 = 291 Bbl/day).

In this instance the benefits of using the downhole gas separator would be:

- Larger water production rate at the same stroke length and SPM
- Time to final dewatering reduced
- More efficient operation
- Less wear of TV and SV due to ball rattling caused by gas interference

Final Phase of Gas Well Dewatering

At this point the objective is to maintain the minimum bottomhole pressure so as to maximize the gas production. Ideally the fluid level should be below the formation perforations, but this is generally not possible due to mechanical problems that may arise by setting the pump intake below the perforations. Therefore, the pump intake is set as low as feasible and the pumping system operates close to a "pumped-off" condition.

The volume of water that needs to be pumped is much lower than during the initial dewatering phase (less than 50 Bbl/day) so that the pump displacement (unless the pumping unit is outfitted with a variable speed drive or a strap jack) is much larger than the formation liquid production rate. For this reason the pumping system generally is controlled by a timer or a pump-off controller so as to minimize rod and tubing damage due to fluid pound.

It would seem that for these conditions a downhole gas separator would not provide many advantages since the liquid flow rate is low and timers and POCs should be able to do a good job of controlling pump operation. To illustrate the effects of using a downhole gas separator the following discussion first describes the mechanics of operation without a separator, and then considers the same conditions when using a separator.

Operation without a Downhole Separator

The pump intake is set above the perforations and includes a perforated nipple or strainer to prevent large solids from entering the pump. Gas is produced through the casing annulus at relatively high flow rates so that a gaseous column is present in the annulus above the pump intake. The height of the column could be several hundred feet or in some cases several thousand feet, but generally the percentage liquid will be less than 30 to 40 percent due to the high gas flow velocity.

These are the conditions existing in the following example well:

Initial completion without downhole gas separator.
Casing 4-1/2 inch
Perforations at: 1227–30, 1613–14, 1622–23, and 1915–18
PBTD: 1946
Tubing 2-3/8
EOT: 1923 ft tailpipe end
Rathole: 23 ft
Pump: 2 in × 1.5 in × 10 ft Brass Cpid 20 ring PA

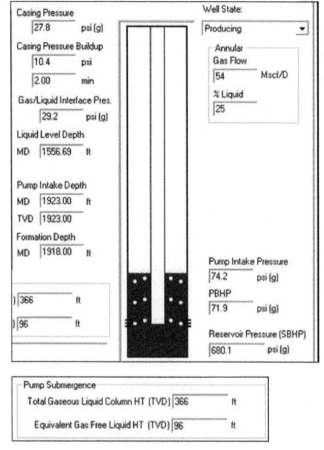

Figure 10-12: Annular Fluid Level Survey in Gas Well without Downhole Gas Separator

Seating nipple: 1891 ft
Average gas sales: 103 MSCF/D

Note that the tubing intake (end of tail pipe below seating nipple) is set just below the bottom of the bottom perforations.

Figure 10-12 shows the result of the fluid level survey indicating that the top of the gaseous column is measured at 1556 feet, which is below the topmost perforations but above the other two sets of perforations. The 366 ft high gaseous column consists of 25 percent liquid for an annular gas flow rate of 54 MSCF/D.

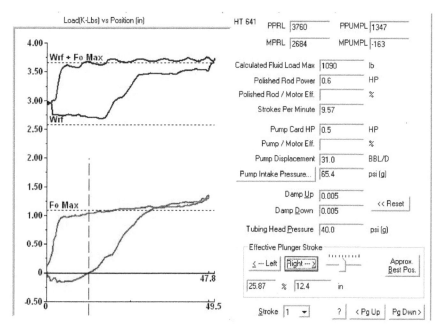

Figure 10-13: Surface and Downhole Dynamometer for First Stroke

The dynamometer survey was recorded using a horseshoe-type load cell and indicates consistently that gas interference is present in the pump. Figure 10-13 shows the surface and pump dynamometer cards for the first stroke of a set of 19 recorded strokes.

The effective plunger stroke is 12.4 inches or 25.9 percent of the plunger travel of 47.8 inches. This corresponds to a pump displacement of 31 Bbl/D. The computed pump intake pressure of 65.4 psi checks favorably with the value of 74 psi computed from the acoustic survey. The downstroke portion of the pump dynamometer shows the characteristics of gas interference.

Figure 10-14 shows the dynamometer cards for the nineteenth stroke. Note that the effective plunger stroke has decreased to only 11.2 inches corresponding to a pump displacement of 13.4 Bbl/D. The pump card shows that the pump is close to being gas locked.

Observations

- Even though the total plunger travel corresponds to a rate of 120 Bbl/D, the effective displacement of the pump varies between 13 and 31 Bbl/D.

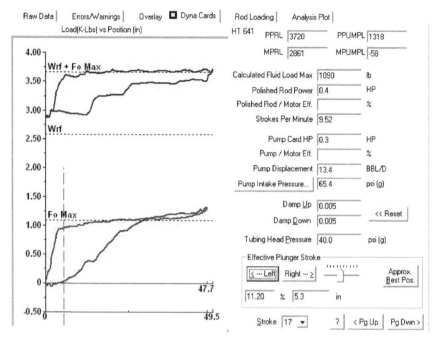

Figure 10-14: Surface and Downhole Dynamometer for Last Recorded Stroke

- Liquid fillage of the pump is probably much less than the effective plunger displacement due to the large gas fraction that is present at the pump inlet.
- Whether using the shape of surface or pump cards as the reference criterion, setting the shut-off point of a pump-off controller is made difficult due to the gas interference and the changing fillage of the pump.
- A large volume of gas is flowing through the pump potentially causing high rate of wear of the valve assembly. Based on the value of annular gas flow (54 MSCF/D), over 50 percent of the gas sold is flowing through the pump.
- Very low polished rod power (0.3 to 0.5 HP) is computed compared to the installed motor of 10 HP, indicating a very low system efficiency.
- In spite of the fact that the tubing intake is set below the perforations and the large capacity of the pump, the system is not able to remove all the liquid that is present in the annulus.

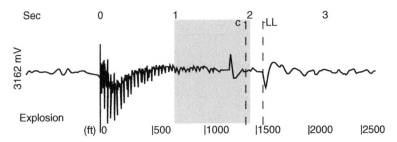

Figure 10-15: No Downhole Gas Separator Record That Corresponds to Figure 10-12

Acoustic Liquid Level Record

Note echo (up-kick) from perforations at 1227 to 1230 feet.

Operation with Downhole Separator

Perforations at: 1227–30, 1613–14, 1622–23, and 1915–18
Removed tail pipe and installed downhole gas separator in its place
PBTD: 1946 ft
EOT: 1937′ to end of bullplug
Rathole: 9 ft
Pump: 2 in × 1.5 in × 10 ft Brass Cpid 20 ring
Seating Nipple: 1928 ft
Average gas sales: 124 MSCF/D

Liquid level has been drawn down to the separator ports. Figure 10-17 shows the dynamometer record that corresponds to pump operation after 15 hours of continuous operation to draw the fluid level down to a minimum.

The liquid fillage corresponds to about 37 percent of the downhole stroke of 47.4 inches resulting in an effective pump displacement of 28.1 Bbl/day. Computed pump intake pressure of 31.1 agrees with acoustic PIP = 35.3 psi.

Figure 10-18 shows the superposition of 23 surface dynamometer cards showing that operation and fillage of the pump is quite consistent and that a steady state condition has been achieved.

The previous figure indicates that setting a POC using the surface dynamometer would be feasible without problems.

The pump dynamometer indicates that in this well the pump should operate only 37 percent of the time, so that a percent timer could

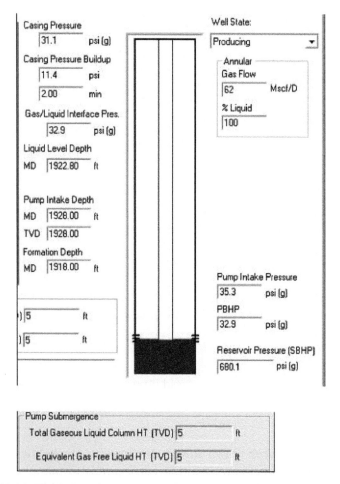

Figure 10-16: Fluid Level Survey after 15 Hours of Operation after Installing the Downhole Gas Separator (corresponding acoustic record shown in Figure 10-19)

be used to control operation and minimize problems due to fluid pound.

Note echoes (up-kicks) from perforations at 1227 and at 1613 feet.

Observations

- Installing the downhole gas separator allowed the removal of most of the liquid from the wellbore and pump down the fluid level to the separator intake.

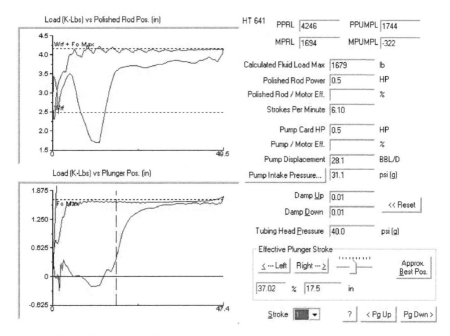

Figure 10-17: Surface and Pump Dynamometer Cards after Separator Installation

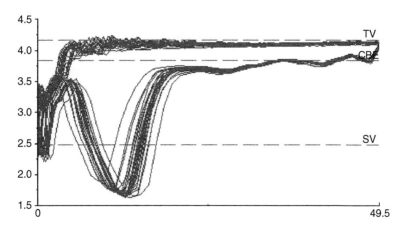

Figure 10-18: Superposition of 23 Surface Dynamometer Cards

- Pump operation and liquid fillage is repeatable from stroke to stroke indicating a steady state condition.
- Either a POC or an interval timer should have no problem in controlling operation of the pumping system to minimize fluid pound.

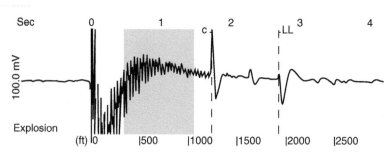

Figure 10-19: Acoustic Liquid Level Record after Installing Downhole Gas Separator

• With a POC or timer the pump would operate with a high percentage liquid fillage from the beginning of start-up until automatic shut down.

Application of Downhole Gas Separator in Well Producing from Depleted Oil Reservoir

The following example shows the effect of installing a downhole gas separator in an oil well producing from a low pressure reservoir.

Operation without Downhole Gas Separator

The well was rod pumped with intake above the perforations as shown in Figure 10-20, which shows 148 feet of gaseous fluid column in the annulus above the pump. Gas flow rate is only 5 MSCF/D, giving a liquid percentage at the pump intake of about 81 percent.

The operator was having difficulty in properly setting a POC to operate this well reliably and requested that a fluid level and dynamometer survey be taken to analyze the situation.

The pump was stopped for 5 minutes and a dynamometer record was taken immediately after restarting the pump. Figure 10-21 shows the superposition of 61 surface dynamometer cards. Note that the pump fillage is less than 50 percent for the first stroke and continually decreases to about 10 to 15 percent for the last stroke.

The fact that even the first stroke exhibits partial liquid fillage is an indication of the presence of gas at the pump intake due to the continued flow of annular gas from the perforations during the time that the pump is stopped.

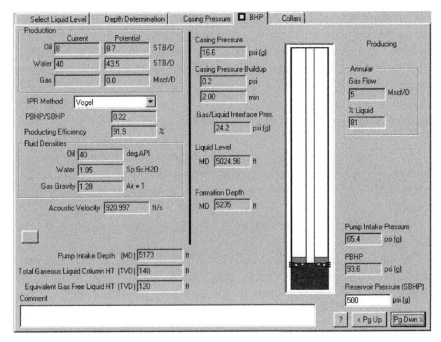

Figure 10-20: Fluid Level Survey without Downhole Gas Separator

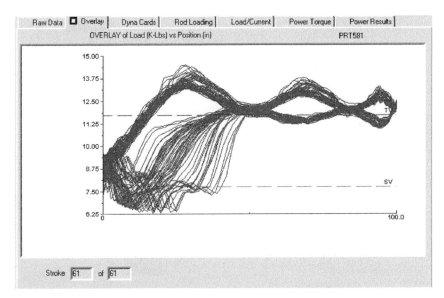

Figure 10-21: Superposition of 61 Surface Dynamometer Cards Recorded after Pump Was Stopped for Five Minutes

The fact that the liquid fillage was much less than the liquid percentage in the wellbore was probably due to the presence of a double standing valve and a perforated strainer at the pump intake. These restricted the flow of liquid and caused additional pressure drops that would allow additional gas to evolve from the liquid.

Operation after Installation of Downhole Gas Separator

The tubing was pulled and a downhole gas separator installed below the landing nipple. The pump was replaced with one with a single standing valve. The pull rod was replaced due to damage caused by fluid pound.

Figure 10-22 shows the result of the fluid level survey taken after the workover.

The survey shows that the liquid level is at the pump intake and very near the perforations. Producing bottomhole pressure has dropped to 36 psi compared to 94 psi when the well was produced without the separator.

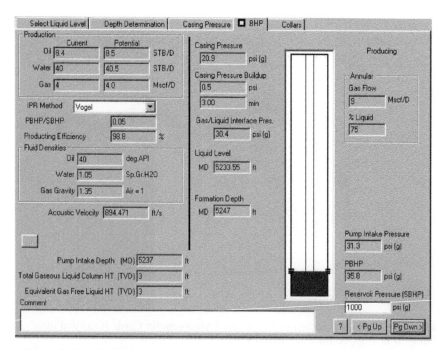

Figure 10-22: Fluid Level Survey after Installation of Downhole Gas Separator

After stabilization of flowing conditions, the pump was stopped for five minutes, then restarted to investigate the liquid fillage that was obtained after the rest period. Figure 10-23 shows the surface and pump dynamometer for the second stroke of the recorded series.

Effective pump stroke is 84 percent of the pump stroke of 82.5 inches due to the tubing stretch. The shape of the pump card indicates that liquid fillage is better than 95 percent. The effective pump displacement corresponds to about 110 Bbl/day, which is over twice the well test rate of 48 Bbl/day.

The dynamometer record was continued for about 10 minutes and over 60 strokes recorded. Figure 10-24 shows the superposition of all the recorded dynamometers and shows that high liquid fillage was maintained during 28 strokes until the liquid level is drawn to the intake ports of the separator since the pump displacement is twice the liquid production from the reservoir.

Eventually the pump liquid fillage settles at about 40 percent of effective plunger displacement.

The well was then operated intermittently using a 15 minute interval timer adjusted to operate 40 percent of the time.

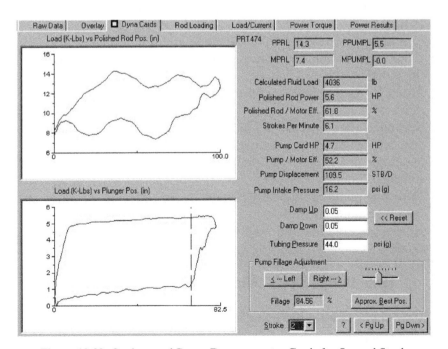

Figure 10-23: Surface and Pump Dynamometer Cards for Second Stroke

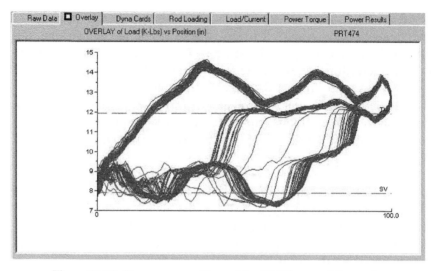

Figure 10-24: Dynamometer Record for 10 Minutes of Operation

Observations

- Operation without a separator precluded reliable operation of POC.
- Continuous operation resulted in pump and rod damage.
- Even for continuous operation a gaseous column was present above the pump.
- Producing bottomhole pressure was high compared to low pressure in reservoir.
- Separator allowed removing all annular liquid above the pump and reducing the producing bottom hole pressure from 94 to 31 psi.
- Separator maintained high pump liquid fillage as long as there was sufficient liquid in wellbore.
- Detecting pump-off is clearly possible using either surface or downhole dynamometer card.
- Efficient operation of the well was achieved with 15 minute interval timer.

10.5 HANDLING GAS THROUGH THE PUMP

If separators are not successful in eliminating gas from the pump, then special pumps or pump construction will assist in handling gas through

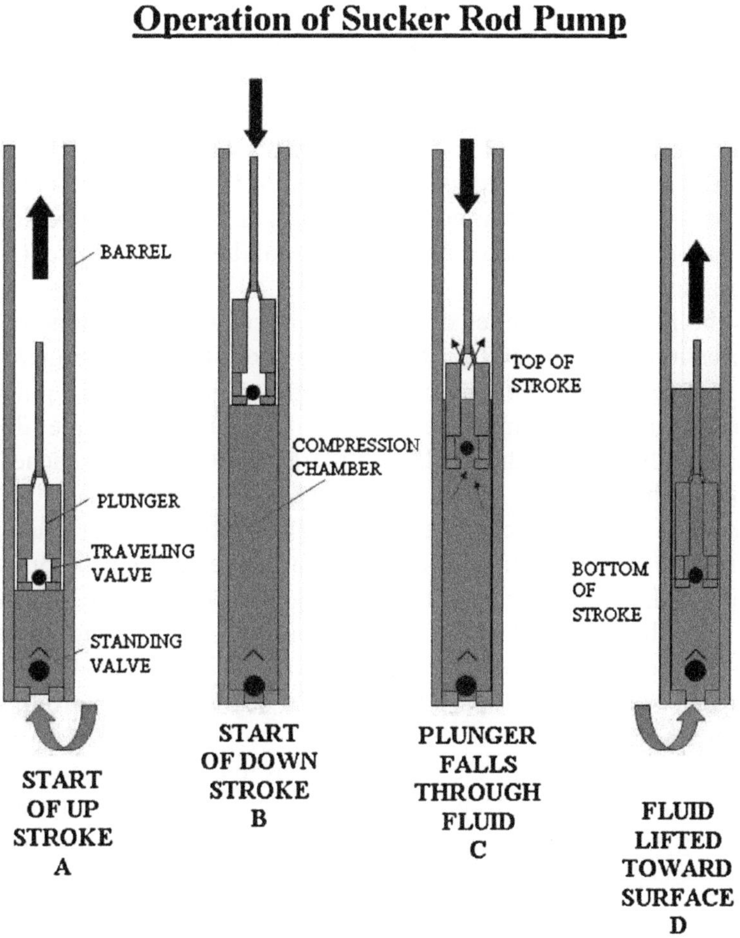

Operation of Sucker Rod Pump

BARREL

COMPRESSION
CHAMBER

PLUNGER

TRAVELING
VALVE

STANDING
VALVE

TOP OF
STROKE

BOTTOM
OF
STROKE

START
OF UP
STROKE
A

START
OF DOWN
STROKE
B

PLUNGER
FALLS
THROUGH
FLUID
C

FLUID
LIFTED
TOWARD
SURFACE
D

Figure 10-25: Normal Pump Cycle

the pump as a second resort. Before discussing special pumps, the following discussion on normal pump operation is contributed by B. Williams, HF Pumps, Ft. Worth, TX.

Figure 10-25 shows a schematic of a stationary barrel sucker rod pump. At the start of the upstroke (see Figure 10-25A), the plunger begins moving upward. This movement increases the volume of the compression chamber between the traveling valve located in the plunger and the standing valve in the barrel. As the compression chamber volume increases, the pressure decreases until it is lower than the hydrostatic pressure in the casing/tubing annulus. At this point the standing valve opens and admits fluid into the compression chamber. For solid fluid with no gas this happens almost immediately when the plunger starts upward.

Fluid continues to fill the compression chamber until the plunger reaches the top of the stroke. As the plunger begins to travel downward at the start of the downstroke (see Figure 10-25B), fluid tries to escape from the compression chamber through the standing valve. The flow of fluid past the standing valve ball pulls the ball back onto the seat, thus closing the standing valve.

As the plunger continues downward it decreases the volume of the compression chamber and raises the pressure. The pressure increases until it is higher than the fluid pressure in the tubing above the plunger. At this point the traveling valve opens and the plunger continues downward and falls through the fluid (see Figure 10-25C), until it reaches the bottom of the stroke.

When the plunger reaches the bottom of the stroke and starts upward, fluid tries to flow back through the traveling valve into the compression chamber. This fluid flow past the traveling valve ball causes it to move back onto the seat, sealing the fluid pressure and closing the traveling valve. As the plunger moves upward in the barrel it lifts the fluid column toward the surface (see Figure 10-25D) and begins the fluid production cycle again.

And before solution to the problem of gas in the pump, let us define gas locking as problems to be avoided in the pump along with just the fact that gas in the pump takes space and reduced the liquid volume per cycle produced.

10.5.1 Gas Locking

By B. Williams, HF Pumps, Ft. Worth, TX

The gas locking pumping characteristics include fluid and gas production along with intermittent acceptable fluid production and unacceptable periods of no fluid production. The API 11AR recommended method of diagnosis for gas locking is to lower the pump plunger until the pump tags, then raise the plunger to its normal position after the gas lock is broken by the mechanical shock. Sometimes this restores the pump to functioning but often the pump must be left tagging in order to maintain production at an acceptable level. This type of mechanical abuse is hard on the pump, the sucker rods, and the tubing. Thus, if a pump can overcome gas locking without tagging, then an operator can save repairs and troubleshooting.

There are several types of gas locking and all cause the pump's performance to be less than expected:

1. Insufficient compression in the pump compression chamber to open the traveling valve on the downstroke, resulting in the pump not "grabbing a bite of fluid" on the downstroke. The result is that the pump does not produce any fluid on that stroke. The cause can be too high of a gas-to-fluid ratio, poor pump compression ratio due to the valve rod being cut too short, or the pump being spaced too high at the well site.
2. Insufficient uncompression in the pump compression chamber on the upstroke to open the standing valve and allow fluid to flow into the pump compression chamber. The result is that when the plunger starts down there is not any new fluid in the compression chamber for the plunger to capture and lift on the next upstroke, and therefore no fluid is produced on the next upstroke. Insufficient uncompression is due to the same reasons as insufficient compression except that in addition there can be a lack of adequate fluid level over the pump to open the standing valve or a restriction in the pump intake.
3. The well is flowing through the pump valves, holding them open and thus not allowing the pump to operate. This is generally not a pumping problem, since the well is producing gas and some fluid, but other problems such as a dry stuffing box and subsequent leakage can occur.

10.5.2 Compression Ratio

In some cases, the produced gas volume is so high that most of the gas cannot be separated. In this case, the pump must be designed to minimize the effects of the free gas that will enter the pump.

As discussed in Section 10.2, the traveling valve must open on the downstroke in order for the pump to work effectively. When pumping gas through the pump, this means that the pump must compress the gas in the pump on the downstroke to a pressure greater than the pressure above the traveling valve in order to force the traveling valve off its seat. If the traveling valve does not open, the pump action will continue but the pump cannot pump liquid. This condition is called *gas locking*.

Beam pump installations can be designed so that they are not susceptible to gas lock regardless of the amount of gas passing through the pump [6]. If the compression ratio of the downhole pump is high enough to always open the traveling valve, it will not gas lock even if it contains 100 percent gas. This certainly will not improve the volumetric efficiency of the pump, but the pump will not gas lock.

The compression of the pump discussed in this section is how much the fluid below the traveling valve is compressed on the downstroke. This pressure should be sufficient to build up enough pressure to open the traveling valve on the downstroke. If the traveling valve always opens on the downstroke, then the pump will not gas lock.

The definition of compression ratio (CR) is given by (see Figure 10-26):

$$CR = \frac{\text{Downhole Stroke} + \text{Spacing Clearance} + \text{Dead Space}}{\text{Spacing Clearance} + \text{Dead Space}} \quad (10\text{-}4)$$

The key to attaining a high compression ratio is to maximize the pump stroke while minimizing the spacing clearance and dead space. Typically little can be done to increase the stroke but careful spacing can drastically increase the compression ratio.

The pull rod must be cut in the shop so that the clearance between the traveling valve and the standing valve is less than approximately 0.5

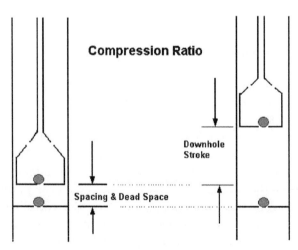

Figure 10-26: Beam Pump Compression Ratio [6]

inch when the pump is at its down-most position. In the well, the pump must be spaced to a bare minimum taking care that the pump does not "tag" or strike bottom on the downstroke, but the traveling valve assembly comes close to the standing value assembly on the downstroke to minimize the dead space in the pump.

The traveling valve seat plug can be specified as a "zero clearance" seat plug. This is an "all thread" seat plug, which does not extend below the traveling valve cage and thus saves about 1" of length, enabling the plunger assembly to be spaced 1" lower during pump assembly. It requires a special hex or square wrench to tighten it on the inside of the seat plug since the outside of the seat plug is no longer available (Source: B. Williams, HF Pumps).

Several standing valve cage designs that reduce the wasted space inside the cage are available from pump manufacturers. Some of these designs significantly lower the unswept volume, especially when combined with a zero-clearance seat plug and properly selected valve rod length (Source: B. Williams, HF Pumps).

If this is done, many pump gas handling problems will be solved. It is easily overlooked because you cannot see how long the pull rod is until you disassemble the pump.

10.5.3 Variable Slippage Pump® to Prevent Gas Lock

The Harbison-Fischer Variable Slippage Pump® shown in Figure 10-27 is primarily for gas locking conditions. This pump has eliminated gas lock in each field test to date.

Leakage is allowed to occur from over the plunger to under the plunger at the end of the upstroke due to a widened or tapered barrel. This reduces pump efficiency, but the liquid allowed to leak below the plunger insures that the traveling valve opens on the downstroke and that gas lock will not occur.

10.5.4 Pump Compression with Dual Chambers

The pump of Figure 10-28 works by holding back the hydrostatic pressure in the tubing on the downstroke while still allowing fluid and gas to enter the upper chamber. The fluid is compressed once on the downstroke into an upper smaller chamber. Then it is compressed on the upstroke into the tubing. If the upper and lower compression ratios are 20 : 1 then the overall compression ratio is 400 : 1.

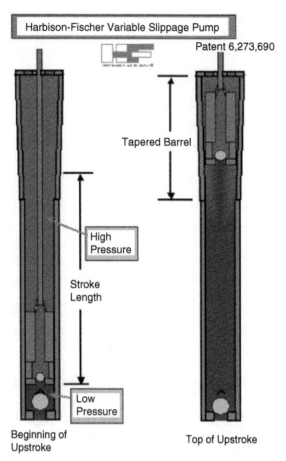

Figure 10-27: Example of a Pump That Uses Designed Leakage to Prevent Gas-Lock (Harbison-Fischer)

10.5.5 Pumps That Open the Traveling Valve Mechanically

There are several pumps that involve a mechanism to open the traveling valve automatically on the downstroke, thereby preventing gas lock. Some have sliding mechanisms and others have devices that directly dislodge the traveling valve from its seat if not already dislodged by pressure. The pump assembly in Figure 10-29 is an example of the latter, using a rod to force the traveling valve ball off the seat on the downstroke.

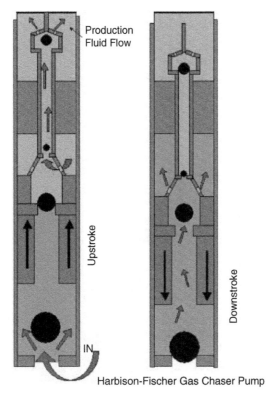

Figure 10-28: Example of a Pump Adding Compression to the Fluid on the Upstroke (Harbison-Fischer)

10.5.6 Pumps to Take The Fluid Load off the Traveling Valve

Figure 10-30 shows a slide above the pump that seals the pressure above the pump from being on the top of the traveling valve on the downstroke. There are other pumps that use this concept.

10.5.7 Gas Vent Pump® to Separate Gas and Prevent Gas Lock (Source: B. Williams, HF Pumps)

A patent-pending H-F Gas Vent Pump® [11] is shown in Figure 10-31. This unusual sucker rod pump was introduced recently and has been able to pump without gas locking by separating the gas from the fluid before it gets into the pump. A strategically placed hole in the

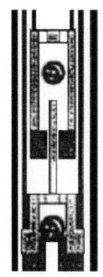

Hart Gas Lock Breaker
Standing Valve Assembly

Figure 10-29: Example of a Pump That Mechanically Opens the Traveling Valve on the Downstroke with a Rod That Lifts the Traveling Ball off Its Seat

barrel, or in a coupling joining two barrels, allows gas to escape from the compression chamber into the casing/tubing annulus during the top part of the upstroke. It also allows fluid to enter the compression chamber with minimum required pressure, allowing the well to be pumped down further than with other sucker rod pumps. This positive fill feature gives the operator the option of slowing down the pumping rate substantially, thus saving energy and wear on the pumping equipment.

There are many other specialty pumps. First try to separate the gas using completion techniques with the pump below the perforations. If this fails, try a gas separator. If this still fails then try the more exotic pumps to handle gas.

10.6 INJECT LIQUIDS BELOW A PACKER

In recent years, methods have been developed to separate the liquid and gas phases downhole and then reinject the liquids back into the formation below a packer. This eliminates both the need for disposal of the liquids (water) at the surface and the means required to lift the

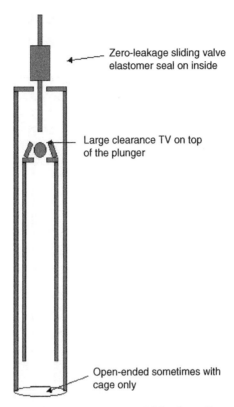

Figure 10-30: Quinn Multiphase Flow Pump: Slide above Pump Closes on Downstroke to Take Fluid Load off of the Valve; for Usual Pump with TV and SV, the Load off the TV Will Allow the TV to Open with Gas and Liquids Below the Pump, and Reduce Fluid Pound (Quinn Pumps, Canada)

liquids to the surface. Once the liquids are reinjected, the gas can flow freely up the casing/tubing annulus.

There are several commercial devices available [7–9] to do this. The concept shown in Figure 10-32 uses gravity as both the separation mechanism and the injection mechanism. The water is pumped up the tubing. The bypass seating nipple allows water pressure and flow to bypass the pump. The pressure exerted on the formation below the pump injects the water. The higher the fluid column in the tubing generated by the pump, the greater the pressure on the formation. If a larger pressure is needed than a full column of liquid in the tubing, then a back pressure regulator can be placed on the surface of the tubing. Cases of 300 psi and greater are reported to inject at the desired rate.

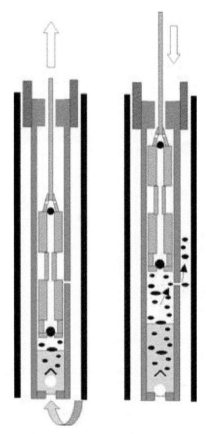

Figure 10-31: Gas Vent Pump® (Harbison-Fischer)

10.7 OTHER PROBLEMS INDICATED BY THE SHAPE OF THE PUMP CARD

In the previous materials, figures were shown of the surface dyna-mometer cards and the bottomhole pump cards and how the shape of the cards can be used to help diagnose gas problems such as fluid pound and gas interference.

There are other problems, however, that can also be diagnosed by the shape of the cards, as shown in Figure 10-33.

Figure 10-33 shows pump cards related to several problems.

Unanchored Tubing. The top two cards show a pump that is full of liquid, but the one the right has the tubing un-anchored. This allows

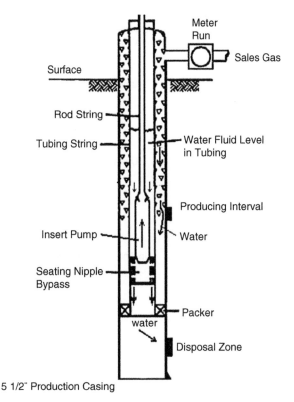

5 1/2" Production Casing

Figure 10-32: Example of Beam Pump System to Inject Liquids below Packer so Gas Can Flow Unobstructed (Harbison Fischer bypass seating nipple)

the tubing to travel upward some on the upstroke and the pick-up of the load takes place over a distance of upstroke, resulting in a slanting of the sides of the card with the un-anchored tubing.

Leaky Traveling Valve. The second row of two cards illustrate a leaking traveling valve (TV). The cards with the leaky TV show a rounding of the top of the card. The load is not immediately picked up and as the top of the upstroke is neared, the load begins to fall off again. This results in low or no production.

Leaky Standing Valve. The third row of two cards illustrate a leaking standing valve (SV). The cards for the leaky SV show a rounding of the bottom of the card. The leaky SV lets fluid out below the TV delaying the loss of load and opening of the TV on the rods above the pump. At the end of the downstroke the loss of fluid allows the

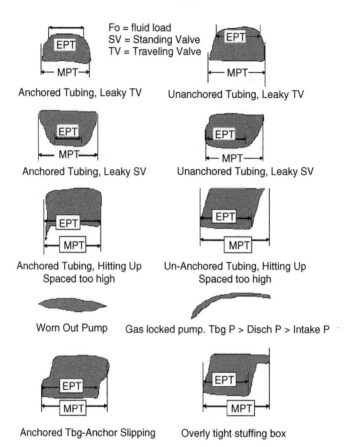

Figure 10-33: Shapes of Pump Downhole Dynamometer Cards in Response to Various Pumping Situations and Problems

TV to close prematurely, bringing up the load again on the rods prematurely.

Improper Spacing. The fourth row of cards shows improper pump spacing. The pump can be spaced too high or too low resulting in the pump "tagging" on the upstroke (spaced too high) or on the downstroke (spaced too low). This can cause rods and pump damage, although it is good practice to space the pump as low as possible without tagging the pump.

Worn Pump. The worn out pump is a combination of the leaky TV and leaky SV. It may be producing no fluid at the surface.

Gas Locked Pump. The gas locked pump is in a situation such that on the downstroke, the TV never opens, and no fluid is being pumped.

As fluid in the annulus builds up the gas lock will clear. Spacing the pump as low as possible without tagging will usually prevent this from occurring for all but the gassiest wells. A long stroke will also help to prevent gas lock.

Tubing Anchor Slipping. The card shown above for a jagged pick-up and release of the load is for the case of the tubing anchor slipping. This is because although the anchor is placed and set, it is slipping on the upstroke and the downstroke, leading to an erratic load pick-up and release. This is a bad situation because the slipping anchor can lead to casing wear as well as rod and tubing wear.

Tight Stuffing Box. The card for the overly tight stuffing box (last card in this figure) should show and extra amount of fluid load ((Pabove-Pbelow) × Area of Pump) on the card that is exhibited as an extra vertical thickness for card. The extra friction usually is released at the top of the stroke leading to the shape of the card.

There are many other problems working with beam pumps including leaky tubing, slipping belts, worn sheaves, worn gearboxes, improperly sized motors or prime movers, incorrect design such as pumping too fast with a small diameter pump, and other factors too numerous to mention. Many of the factors dealing with gas have been mentioned in this chapter since gas problems at the pump are often a concern when trying to pump liquids off of gas wells.

10.8 SUMMARY

Beam pumps often are used for dewatering a gas well but special methods may be required to prevent gas interference. Gas interference is handled most often and easily by setting the pump below the perforations and flowing the gas up the annulus.

If the pump does not fit below the perforations for gas separation, then separators or, as a last resort, specialty pumps may be required to combat pump gas interference.

To handle produced water, the beam pump system may be incorporated with a system to inject water below a packer to a water zone. This method eliminates water hauling charges and leaves a free path for gas to flow to the surface.

Since gas flows up the casing, even if the pump is set below the perforations and all fluid is off the formation, there is still more than the surface CHP pressure on the formation. Therefore for beam pumping

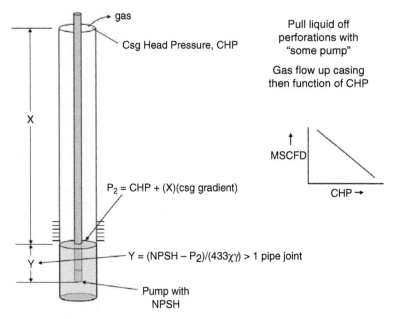

Figure 10-34: Formation Producing Pressure with Pumping System Is a Function of CHP if the Fluid Level is below the Perforations

and any pumping system that pumps liquid up the tubing and flows gas up the casing, the casing pressure must be low if low formation pressures are to be achieved. See Figure 10-34, where NPSH is the pressure at the pump intake required to allow the pump to produce effectively.

10.9 REFERENCES

1. Lea, J. F. "New Pump-Off Controls Improve Performance," *Petroleum Engineer International*, December, 1986, 41–44.

2. Neely, A. B. "Experience with Pump-off Control in the Permian Basin," SPE 14345, presented at the annual technical conference and exhibition of the SPE, Las Vegas, NV, September 22–25, 1985.

3. Dunham, C. L. "Supervisory Control of Beam Pumping Wells," SPE 16216, presented at the Production Operations Symposium, Oklahoma City, OK, March 8–10, 1987.

4. McCoy, J. N. and Podio, A. L. "Improved Downhole Gas Separators," Southwestern Petroleum Short Course, Lubbock, TX, April, 7–8, 1998.

5. Clegg, J. D. "Another Look at Gas Anchors," Proceedings 36th Annual Meeting of the Southwestern Petroleum Short Course, Lubbock, TX, April 1989.

6. Parker, R. M. "How to Prevent Gas-Locked Sucker Rod Pumps," *World Oil*, June 1992, 47–50.

7. Enviro-Tech Tools Inc. brochure on the DHI (Down Hole Injection) tool.

8. Grubb, A. and Duvall, D. K. "Disposal Tool Technology Extends Gas Well Life and Enhances Profits," SPE 24796, presented at the 67th annual SPE conference in Washington, DC, Oct. 4–7, 1992.

9. Williams, R. et al. "Gas Well Liquids Injection Using Beam Lift Systems," Southwestern Petroleum Short Course, Lubbock, TX, April 2–3, 1997.

10. Elmer, W. and Gray, A. "Design Considerations when Rod Pumping Gas Wells," First Conference of Gas Well De-Watering, SWPSC/ALRDC, March 3–5, 2003, Denver, Co.

11. Williams, B. J. and Brown, T. L. "New Gas Vent Pump® Sucker Rod Pump for Gas Locking and Gas Interference Conditions," Southwestern Petroleum Short Course, Lubbock, TX, April 20–21, 2005, 189–191.

GAS LIFT

11.1 INTRODUCTION

Gas lift is an artificial lift method whereby external gas is injected into the produced flow stream at some depth in the wellbore. The additional gas augments the formation gas and reduces the flowing bottomhole pressure, thereby increasing the inflow of produced fluids. For dewatering gas wells, the volume of injected gas is designed so that the combined formation and injected gas will be above the critical rate for the wellbore [1], especially for lower liquid producing gas wells. For higher liquid rates, much of the design procedure may more closely mirror producing oil well gas lift techniques.

Although gas lift may not lower the flowing pressure as much as an optimized pumping system, there are several advantages of a gas lift system that often make gas lift the artificial lift method of choice. For gas wells in particular, when producing a low amount of liquids, the producing bottomhole pressure with gas lift may compare well with other methods of dewatering. For higher liquid rates, the achievable producing BHP may be higher than pumping techniques.

Of all artificial lift methods, gas lift most closely resembles natural flow and has long been recognized as one of the most versatile artificial lift methods. Because of its versatility, gas lift is a good candidate for removing liquids from gas wells under certain conditions. Figure 11-1 shows the approximate depth-pressure ranges for application of gas lift, developed primarily for oil wells.

The most important advantages of gas lift over pumping lift methods are:

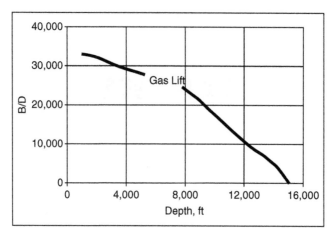

Figure 11-1: An Approximate Depth-Rate Feasibility Chart for Conventional Continuous Gas Lift

- Most pumping systems become inefficient when the GLR exceeds some critical value, typically about 500 scf/bbl (90 m³/m³), due to severe gas interference. Although remedial measures are possible for conventional lift systems, gas lift systems can be applied directly to high GLR wells because the high formation GLR reduces the need for additional gas to lower the formation flowing pressure.
- Production of solids will reduce the life of any device that is placed within the produced fluid flow stream, such as a rod pump or ESP. Gas lift systems generally are not susceptible to erosion due to sand production and can handle a higher solids production than conventional pumping systems.
- For some applications, a higher pressure gas zone may be used to auto-gas-lift another zone.
- In highly deviated wells it is difficult to deploy some pumping systems due to the potential for mechanical damage to deploying electric cables or rod and tubing wear for beam pumps. Gas lift systems can be employed in deviated wells without mechanical problems.
- New techniques (discussed in this chapter) allowing gas lift gas to help lift long pay intervals below the usual packer in a gas lift installation.

Gas lift has features to address these production situations.

Another advantage that gas lift has over other types of artificial lift is its adaptability to changes in reservoir conditions. It is a relatively simple matter to alter a gas lift design to account for reservoir decline or an increase in fluid (water) production that generally occurs in the latter stages of the life of the field. Changes to the gas lift installation can be made from the surface without pulling tubing by replacing the gas lift valves via wireline and reusing the original downhole components. However, many onshore lower volume gas well gas lift installations may choose to use conventional mandrels where the tubing must be pulled to access gas lift valves and to replace valves.

The two fundamental types of gas lift used in the industry today are *continuous flow* and *intermittent flow*. This is the conventional breakdown. However, one could say there are gas lifted gas wells and there are gas lifted oil wells. Gas wells can also be lifted by continuous or intermittent gas lift so the conventional discussion will be presented, although many gas wells are being lifted by continuous flow.

11.2 CONTINUOUS GAS LIFT

In continuous flow gas lift, a stream of relatively high pressure gas is injected continuously into the produced fluid column through a downhole valve or orifice. The injected gas mixes with the formation gas to lift the fluid to the surface by one or more of the following processes:

- Reduction of the fluid density and the column weight so that the pressure differential between the reservoir and the wellbore will be increased.
- Expansion of the injected gas so that it pushes liquid ahead of it, which further reduces the column weight, thereby increasing the differential between the reservoir and the wellbore.
- Displacement of liquid slugs by large bubbles of gas acting as pistons.

11.3 INTERMITTENT GAS LIFT

Often in gas wells as the bottomhole pressure declines, a point is reached where the well can no longer support continuous gas lift and the well is converted to intermittent gas lift. This conversion can also employ the identical downhole equipment (mainly the gas lift

Table 11-1
Maximum Flow Conditions for Intermittent Lift

Tubing Size (inch)	Maximum Flow Rate for Intermittent Lift
2-3/8	150 bpd
2-7/8	250 bpd
3-1/2	300 bpd
4-1/2	Not Recommended

valve mandrels) yet fully adapt the well to intermittent flow. In this case, the unloading valves are replaced with dummy valves to block the holes in the mandrels and prevent injection gas from passing into the production stream. The operating valve then is replaced with a production pressure valve with a newly set pressure capacity reflecting the desired fluid level to be reached in the tubing before the well is lifted.

Fitting the operating valve with the largest possible orifice will greatly improve the efficiency in an intermittent gas lift system. The large orifice diameter exerts a minimum restriction to the flow of the injection gas. The injection gas will then quickly fill the tubing below the fluid, ultimately lifting the slug of liquid to the surface with the minimum amount of lift gas.

The optimum time to convert a gas lift well from continuous lift to intermittent lift is a function of the reservoir pressure, the tubing size, the GLR, and the flow rate of the well. The individual well conditions will dictate the optimum time for conversion; Table 11-1 lists some good rules of thumb to use to estimate the best time to convert to intermittent lift.

It is becoming common practice to use a plunger (see Chapter 7) to increase the production from wells on intermittent lift. The lift gas is injected below the plunger and the plunger acts as a physical barrier between the lift gas and the fluid to reduce the fluid fallback around the gas slug that is characteristic of intermittent lift operations. The plunger extends the life of the well by more effectively removing water from the formation. A plunger with extensions can be used so that it can pass by gas lift mandrels if needed, as discussed later. When a plunger is used over a standing valve and the gas lifts the plunger and liquid slug above a standing valve, this is more similar to a gas-powered long stroke pump or plunger assisted chamber lift than conventional intermittent lift.

11.4 GAS LIFT SYSTEM COMPONENTS

Figure 11-2 shows a typical continuous gas lift system that includes:

- A gas source
- A surface injection system, including all related piping, compressors, control valves, etc.
- A producing well completed with downhole gas lift equipment (valves and mandrels)
- A surface processing system, including all related piping, separators, control valves, etc.

The gas source is often reservoir gas produced from adjoining wells that has been separated, compressed, and reinjected. A secondary source of gas may be required to supply any shortfall in the gas from the separator. The gas is compressed to the design pressure and is injected into the well through the gas lift operating valve, where it enters the tubing string at a predetermined depth.

For conventional gas lift, valves or orifices should be used to port gas to the tubing, rather than holes or simply the end of the tubing string, so that the gas stream is well dispersed within the liquid column and the flow continues smoothly.

However in this chapter, *gas cycling* [4] is discussed, which is a method to flow additional gas down the annulus and into the bottom of the

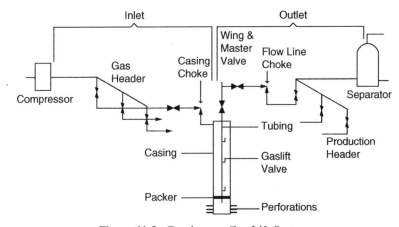

Figure 11-2: Continuous Gas Lift System

tubing. This is possible because the amount of gas is high in the tubing relative to the fluids so severe slugging does not occur as it would with a lower operational GLR, which is typical for gas lifting oil wells.

It is recognized that there are single well compressors for gas wells where small compressors (reciprocating or screw for example) are used to lower the surface pressure of a flowing or plunger lifted gas well to solve liquid loading issues.

11.5 CONTINUOUS GAS LIFT DESIGN OBJECTIVES

Gas lift increases well production by reducing the density of the produced fluid and thereby decreasing the flowing bottomhole pressure. This is accomplished by introducing the injection gas, at an optimum (usually maximum) depth, pressure, and rate, into the produced fluid stream. The valve through which the gas is injected into the wellbore fluid stream under normal operating conditions is called the *operating valve*.

Nodal Analysis (see Chapter 2) can be used to evaluate several tubing sizes and GLRs to determine possible production increases for the different tubing sizes and GLRs. The well becomes a candidate for gas lift when the artificially increased GLR significantly increases the well production. Another way of thinking of gas lift for gas wells is to inject a sufficient additional volume of gas to keep the total gas velocity (from produced + injected gas) above the critical velocity for the well.

The efficiency of a gas lift system is highly influenced by the depth of the operating valve. As the depth of the operating valve is increased, more and more of the hydrostatic pressure of the heavier fluid (and gas) column is taken off of the formation, reducing the bottomhole pressure and increasing production. Typically, before a gas lift well is brought online, it is filled or partially filled with kill fluid from the workover operation. To bring the well into production the well must first be "unloaded" by injecting high pressure gas into the annulus to displace the kill fluid in the annulus down to the operating valve.

To push the liquid level to the depth of the operating valve requires an extremely high surface injection pressure. In most installations, this high injection pressure is not available. Several gas lift valves are required to allow the available surface pressure to feed gas to the well at increasing depths until the operating valve at maximum depth is reached. This process is called "unloading the well" and the additional valves are called "unloading valves."

The series of unloading valves are placed at various depths and have different opening/closing pressures to step the injection gas down to the design injection depth. These unloading valves are designed to have a particular port size and set to specific opening pressures to allow the annular fluid level to pass from one valve to the next. The design of the gas lift system includes the size, pressure rating, depth and spacing of the unloading valves, the optimum depth of the operating valve to maximize recovery, the size of the operating valve orifice, and the injection rate and pressure of the lift gas.

The correct spacing of the unloading valves is critical. Valves spaced too far apart for the injection parameters will not allow the well to unload completely. In this case, injection gas will enter the production stream too high in the well, significantly lowering the system efficiency and, more importantly, the well's production.

Determining the best gas lift design requires considerable knowledge of the well conditions, both present and future. These calculations usually are performed by sophisticated commercial software packages or design charts supplied by gas valve manufacturers. The complete fundamentals of gas lift design and optimization are beyond the scope of this text although field applications [2] of gas lift technology for gas wells are presented in this chapter.

11.6 GAS LIFT VALVES

The key to a properly designed gas lift system is the proper choice of gas lift valves. Gas lift valves fall into one of three major categories:

- Orifice valves
- Injection pressure operated valves
- Production pressure operated valves

Schematic examples of injection and production pressure operated valves are given in Figure 11-3. By far, the Type 1 and possibly Type 2 are used to dewater lower pressure gas wells.

11.6.1 Orifice Valves

Strictly speaking, orifice valves are not valves because they do not open and close. Orifice valves are simply orifices, or holes, providing a communicating port from the casing to the tubing. Because they do not

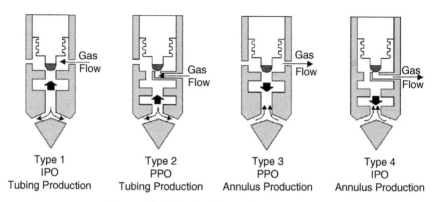

Type 1	Type 2	Type 3	Type 4
IPO	PPO	PPO	IPO
Tubing Production	Tubing Production	Annulus Production	Annulus Production

Figure 11-3: Typical Gas Lift Valve Types

actually function as valves, orifice valves are used only as operating valves to provide the correct injection flow area as required by the valve design and to properly disperse the injected gas to minimize the formation of slugs. Orifice valves typically are used only for continuous flow applications. The valve includes a check to prevent tubing to casing flow.

11.6.2 Injection Pressure-Operated (IPO) Valves

Injection pressure operated (sometimes called casing pressure operated) valves are the most common valve used in the industry to unload gas lift wells. Although somewhat influenced by the pressure of the flowing production fluid, injection pressure operated valves are controlled primarily by the pressure of the injection gas.

Figure 11-4 shows a schematic of an IPO gas lift valve where the injection pressure is applied to the base of the bellows and the produced fluid pressure is applied to the ball (stem tip) through the valve orifice area. Since the bellows area is much larger than the orifice area, the injection pressure dominates control of the valve operation.

Injection pressure valves act like backpressure regulators and close when the backpressure (casing pressure) reaches a predesignated "minimum" value. Typically this minimum value is designed to be when the kill fluid in the casing/tubing annulus, being pushed downward by the injection gas during the unloading process, just reaches the next lower valve. This allows the upper valve to close to the flow of injection gas, forcing the pressure to continue to push the fluid level further down the annulus to eventually reach the operating valve.

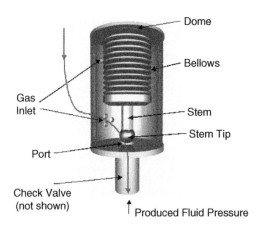

Figure 11-4: Schematic of Gas Lift Valve

11.6.3 Production Pressure Operated (PPO) Valves

Production pressure operated valves (sometimes called tubing pressure valves) are operated primarily by changes in pressure of the production fluid. Unloading is then controlled primarily by the reduction in hydrostatic pressure in the production stream by injecting lift gas.

PPO valves are used typically:

- Where the production fluid is produced through the annulus
- In dual completions where two gas lift systems are installed in the same well to produce two differently pressured zones
- For intermittent lift

PPO valves are ideal for intermittent lift applications since the valve is designed to remain closed until a sufficient fluid load is present in the tubing, at which time the valve opens, producing the liquid.

Once an unloading valve closes during the unload process, it should remain closed. Both injection and production pressure valves use a charged bellows (typically pressurized with nitrogen), a spring, or sometimes both to obtain the valve closing force. The nitrogen charged bellows is the most common. The bellows type valves are set to the design pressure in a controlled laboratory environment by the valve shop.

All gas lift valves are equipped with reverse flow check valves to prevent backflow of fluid through the valve. For subsea completions, where minimal intervention is a design objective, the spring-loaded valve may provide the most reliability since in the event of a bellows rupture the spring will keep the stem on seat and the valve will remain closed. The spring-loaded valve is also not sensitive to temperature variations as is the nitrogen charged bellows.

11.7 GAS LIFT COMPLETIONS

The heart of a gas lift installation is the gas lift valves. Their placement in the tubing string is fixed during the installation of the tubing by the gas lift mandrels. Gas lift mandrels are placed in the tubing string to position each gas lift valve to the desired depth.

There are two basic "conventional" gas lift systems in use today. These are systems using conventional mandrels with threaded nonretrievable gas lift valves and systems using side pocket mandrels (SPM) with retrievable gas lift valves.

Conventional mandrels accept threaded gas lift valves mounted on the outside of the mandrel. These valves can be retrieved and changed only by pulling the tubing, and usually are not run where workover costs are high.

Side pocket mandrels (SPM) allow the gas lift valves to be retrieved using slickline from the surface without the need to pull the tubing. These mandrels are most commonly used today. Both systems are discussed next.

11.7.1 Conventional Gas Lift Design

A schematic of a gas lift system using conventional mandrels is shown in Figure 11-5. With this system gas lift mandrels and valves are installed at the surface when the tubing is run in the well. The valves are threaded into the mandrels and therefore cannot be removed without removing the entire tubing string. Gas lift designs using conventional mandrels are among the lowest cost gas lift designs available.

An added benefit of using conventional mandrels, particularly when removing liquids from gas wells, is that they can readily integrate with plunger lift systems. This is not the case for installations using side pocket mandrels. The ID of a conventional mandrel is relatively uniform but the internal pocket of a side pocket mandrel is eccentric to permit the insertion of gas lift valves via slickline. This presents a problem for plunger operations because as the plunger assembly enters the SPM's

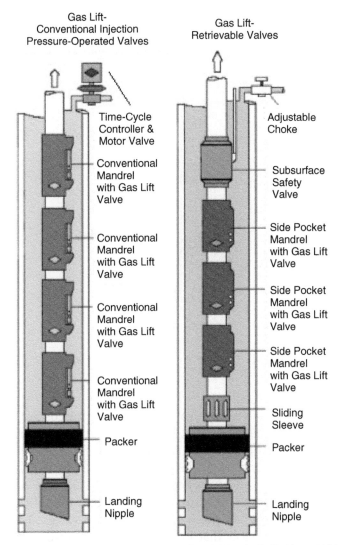

Gas Lift-
Conventional Injection
Pressure-Operated Valves

Gas Lift-
Retrievable Valves

Time-Cycle
Controller &
Motor Valve

Conventional
Mandrel
with Gas Lift
Valve

Conventional
Mandrel
with Gas Lift
Valve

Conventional
Mandrel
with Gas Lift
Valve

Conventional
Mandrel
with Gas Lift
Valve

Packer

Landing
Nipple

Adjustable
Choke

Subsurface
Safety
Valve

Side Pocket
Mandrel
with Gas Lift
Valve

Side Pocket
Mandrel
with Gas Lift
Valve

Side Pocket
Mandrel
with Gas Lift
Valve

Sliding
Sleeve

Packer

Landing
Nipple

Figure 11-5: Gas Lift Design Using Conventional Mandrels (left) and Side Pocket Mandrels with Wireline Retrievable Valves (right) (courtesy Schlumberger-Camco)

eccentric pocket it allows gas to bypass liquid. This typically results in a loss of plunger velocity and in some cases makes it difficult for the plunger to reach the surface. Some operators have successfully adapted extensions to the plunger to effectively straddle the pocket, but these have succeeded only for shallower wells.

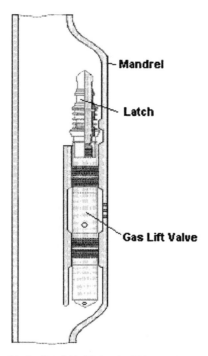

Figure 11-6: Gas Lift Valve in Side Pocket Mandrel

Side pocket mandrels were developed to reduce the costs of changing a gas lift system to maintain a gas lift valve design that optimizes production as well conditions change. A schematic of a side pocket mandrel (SPM) is shown in Figure 11-6. The primary feature of side pocket mandrels is the internally offset pocket that accepts a slickline retrievable gas lift valve. The pocket is accessible from within the tubing using a positioning or kickover to place and retrieve the valves. The gas lift valves use locking devices that lock into mating recesses in the SPM. Both conventional and SPM mandrels are installed in the well in the same manner, but only the SPM system is serviceable with slickline operations for post-completion repair or well maintenance.

The high-pressure gas in a gas lift system usually is supplied by a central compressor that compresses the gas produced by the field for reinjection into those wells on gas lift. If the field gas supply is insufficient to meet the needs of the artificial lift system, more gas generally is obtained from the sales line.

Gas lift compression can also be supplied for individual wells when one or two wells in a field are being lifted with gas lift. These small well

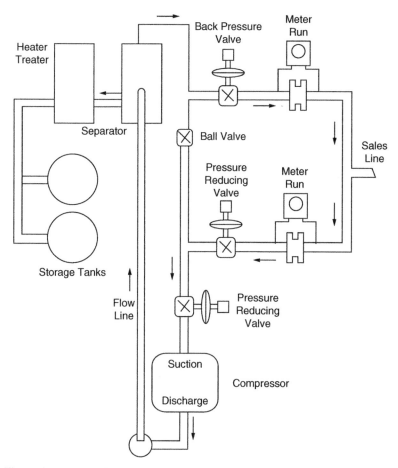

Figure 11-7: Typical Compression System for Low Pressure Gas Lift System

site compressors are typically skid-mounted for easy mobilization when it becomes necessary to move the system from one well to another. Figure 11-7 shows a typical system for an individually compressed low pressure well on gas lift. This might be a system on a gas well to help lift liquids.

11.7.2 Chamber Lift Installations

When the completion configuration prevents the point of injection from achieving the desired depth, or when the volume of gas in an intermittent lift installation is less than acceptable, a chamber lift design is used.

The concept of chamber lift is to create a large diameter volume (chamber) to collect liquids. The larger diameter of the chamber, as opposed to the tubing, allows higher volumes of liquid to accumulate while keeping the liquid column height to a minimum. Lower liquid column heights put less hydrostatic pressure on the formation. Increasing the diameter of the chamber can drastically reduce the hydrostatic head since the bottomhole pressure is reduced by the square of the chamber diameter. For example, increasing the chamber diameter from 2-3/8-inch to 3-inch will drop the hydrostatic pressure at the bottom of the hole by almost half for the same volume of liquid.

Typically, the chamber consists of a portion of the casing as shown in Figure 11-8. Chamber packers isolate the chamber, and a dip tube frequently is used in the top packer to allow the gas collected in the chamber to bleed off into the casing above the packer. Chambers also can be manufactured at the surface and installed in the tubing string.

Chamber lift is one method of producing a relatively high volume of liquids in a low pressure formation without loss of gas production due to excessive liquid head in the tubing.

In Figure 11-8, a chamber is formed between two packers. Well liquids are allowed to enter the space between the packers at low pressure. After the chamber is filled, gas is injected into the top of the chamber, displacing the liquids into and up the tubing. An additional gas lift effect is added to the liquids as rise with gas injected from gas lift valves spaced higher in the tubing. A time-cycle controller is provided to control the cycles.

11.7.3 Horizontal Well Installations

Over the past decade, the number of horizontal wells has ballooned worldwide. Many of these wells are on gas lift to either increase oil production or, in gas wells, to more effectively produce the liquids.

Some operators have attempted to install gas lift in the horizontal section of the hole but found this not to be practical for a variety of reasons.

- Gas lift operates by reducing the hydrostatic head on the formation. In a horizontal, or near horizontal section of the hole, there is very little vertical head. Placing gas lift valves in the horizontal

Gas Lift-
Retrievable Valves
Chamber Installation

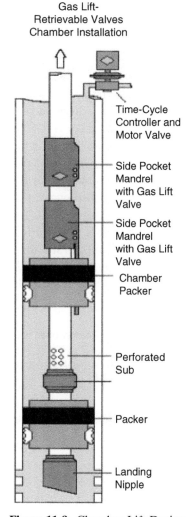

Figure 11-8: Chamber Lift Design

lateral gains little benefit over valves located up-hole in the vertical section.

- In the horizontal section of the well, the two-phase (gas/liquid) flow tends to become stratified, allowing the gas to pass over the top of the fluid without pushing the fluid to the surface. This greatly reduces the efficiency of the gas lift.
- Servicing gas lift valves with slickline becomes increasingly difficult with increased wellbore inclination.

The preferred completion configuration for a horizontal well on gas lift is to use the gas lift mandrels only in the portion of the wellbore where the deviation from vertical is less than 70 degrees. Figure 11-9 shows a horizontal well with inherent slugging in the vertical portion of the well. If gas lift is used, then gas lift valves should be placed in the vertical or near vertical portion of the well and not in the horizontal section. This eliminates the problems discussed earlier while producing the well at an optimum rate and efficiency.

Horizontal wells are also notorious for slugging, which dramatically reduces the overall production. Slugging can also create many problems with pumping systems such as ESPs and beam pumps, where the slugging typically causes intermittent shutdowns in the equipment as well as cooling problems with the ESP motors. Slugging in the near vertical portion of a horizontal flowing well is shown Figure 11-9.

Installing gas lift in a horizontal well can stabilize slugging and thereby increase production. Gas lift removes the liquid head and controls the influx of gas to prevent or drastically reduce slugging, returning the

Loading Problems with Horizontal Wells

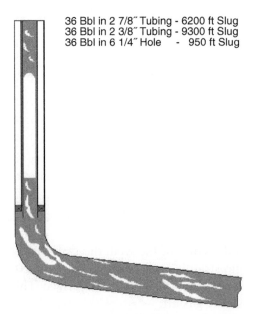

36 Bbl in 2 7/8″ Tubing - 6200 ft Slug
36 Bbl in 2 3/8″ Tubing - 9300 ft Slug
36 Bbl in 6 1/4″ Hole   -   950 ft Slug

Figure 11-9: Flowing Horizontal Well (if gas lift is used to enhance production, install mandrels in the near-vertical portion only)

production stream to more continuous flow. Gas lift also is not affected by slugging as are most other mechanical pumping methods.

11.7.4 Coiled Tubing Gas Lift Completions

To reduce costs while improving the versatility of gas lift systems, coiled tubing suppliers have developed spoolable systems that can be run on coiled tubing. These systems provide complete downhole assemblies that can be installed in small diameter casing or even tubing strings. The system can save initial costs with rigless completions and lower installations times. The cost of the coiled tubing gas lift string is comparable to a jointed tubing installation. The smaller diameter coiled tubing can also improve the efficiency of the lifting process by reducing the overall area of the pipe, but at the expense of the added frictional drag imposed by the smaller cross-sectional flow area.

Coiled tubing gas lift completions have been in use for nearly a decade. Several successful installations have been presented in the literature. Development of the systems continues to improve string reliability and better algorithms to predict the depth of the gas lift valves that include the sometimes significant stretch of the coiled tubing.

Figure 11-10 shows a typical spoolable gas lift system with a close up view of the spoolable gas lift valves. This system has the valves made up inside the CT during run-in and the CT must be retrieved if the valves are to be serviced. Figure 11-11 shows a new system by Nowcam whereby the valves are inside the CT but the valves can be serviced by wireline.

11.7.5 A Gas Pump Concept

A gas pump (see Figure 11-12) is a form of intermittent gas lift where the injection gas does not mix with the produced liquids [3]. Although developed for viscous oil, this method can also be used for dewatering gas wells.

The gas pump is applicable only for shallow gas wells where there is sufficient injection pressure to overcome a hydrostatic gradient to the bottom of the well. The gas pump is a form of chamber lift in that a large downhole chamber is used to collect the liquids prior to being pushed to the surface by the gas. This method requires high-pressure gas at surface, but the volume of lift gas required is small compared to conventional intermittent gas lift systems.

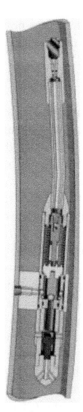

Figure 11-10: Spoolable Coil Tubing Gas Lift System (Schlumberger)

Operation of the system begins with the chamber filling with produced liquids. After a predetermined time, high-pressure gas is injected
rapidly into the chamber, forcing the liquid into the production tubing.
During the injection process, the liquid is pushed into the production
conduit with very little of the injection gas. An intake check valve closes
during gas injection to prevent backflow into the formation. Once the
injection gas begins to break around the bottom of the chamber, the
well is shut in and again allowed to accumulate liquids and the cycle is
repeated.

11.7.6 Gas Circulation

Another method of controlling liquid loading is to inject gas continuously down the casing and up the tubing to keep the gas velocity above

Flow
By-Pass
Area

Valve in Place

Figure 11-11: Valve in CT That Is Wire-Line Retrievable (Schlumberger)

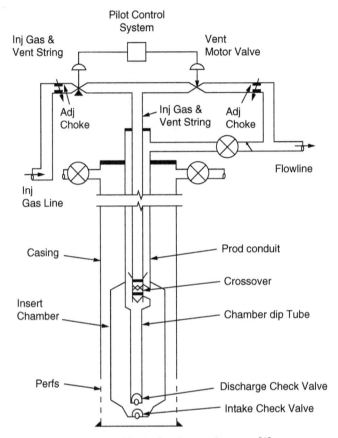

Figure 11-12: A Gas Pump Concept [3]

the critical velocity at all times [4]. Figure 11-13 shows a schematic where this is done by compressing some of the gas back down the annulus.

Figure 11-14 shows how this is done with injection and also using a compressor to reduce the wellhead pressure.

Figure 11-15 shows how a compressor can be used to lower the wellhead pressure and also inject gas downhole to stay above the critical velocity.

11.8 GAS LIFT WITHOUT GAS LIFT VALVES

It is possible to install holes in tubing with a check valve and hard orifice [5] using wireline as shown in Figure 11-16.

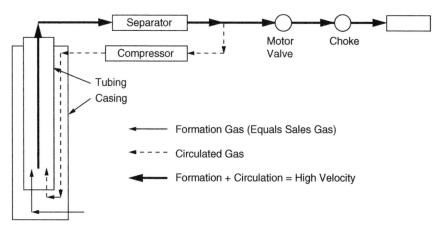

Figure 11-13: Gas Injection to Stay Above Critical Velocity in Tubing

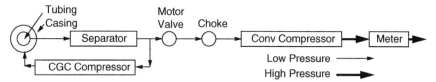

Figure 11-14: Gas Injection to Stay above Critical Velocity in Tubing with Gas Compressor and Wellhead Compression

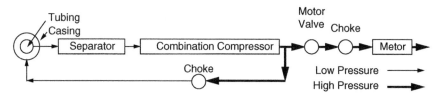

Figure 11-15: Gas Injection to Stay above Critical Velocity in Tubing with Single Compressor

This method allows gas to be injected into the tubing without having to pull the tubing and install mandrels and valves. It is "shot" though the tubing using a charge somewhat like shooting a perforation.

The gas cannot be injected as deep as would be possible using gas lift valves because there is no closing mechanism for the wireline set "seats." However, if one seat is installed, this will allow lightening the tubing gradient above this point when gas is injected and then a second and

Figure 11-16: A Check Valve and Orifice for Gas Lift Installed in Tubing Using Wireline Techniques

possibly a third seat can be installed for deeper injection. However, this will entail multiple point injection as the upper seats cannot be closed. This is an inexpensive method of trying gas lift to see if you can unload a loaded gas well. Later conventional gas lift could be installed if deeper injection is needed for improved production.

11.9 SPECIFICS OF GAS LIFTING GAS WELLS

There are differences in oil well and gas well gas lift. A listing of the differences from B. Rouen, SLB, Denver Gas Well Symposium, 2005, is listed in Table 11-2.

What kind of pressures on the formation can be expected from gas lifting gas wells? As in Table 11-2, the liquid rate is many times lower for gas wells but higher liquid rate wells are subject to dewatering considerations as well. The following graphs show in general how gas wells perform with gas lift for a relatively deep well.

The data for the following graphs are as follows:

Depth: ~10,000 ft
4.5″ casing
2 3/8″ tubing
WHP: Various

Table 11-2
Comparison of Gas Lift for Oil Wells and Gas Wells

Oil Wells	Category	Gas Wells
Water Drive	Reservoir	Solution Gas Drive
Top Down	Design	Bottom Up—Packer Perfs
Usually Higher	Rates	Lower
High	PI	Extremely Low
Shallow, Small Intervals	Perfs	Deeper, Large Intervals
Rare	One Well System	Common
Most are same	Design Methods	Numerous
1000:1	GLR	Extremely High
Not Important	Critical Rate	Important
No Decline	SBHP	Fast Decline
Large	B/L Port Size	Small
Lighter	Fluid Gradient	Small
Low (initially)	Water %	High
Lighter	Fluid Gradient	Small

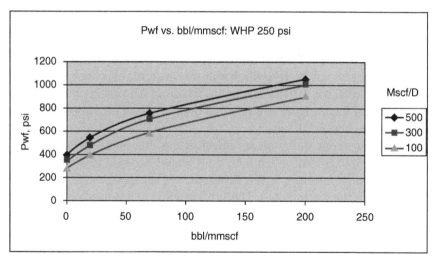

Figure 11-17: Example Gas Lift Tubing Performance in Deep Well with High WHP

LGR: Various
.67 Gas Gravity

As shown in Figures 11-17 and 11-18, gas lift can lower the gradient in the tubing as gas is increased to a point but the pressures achieved

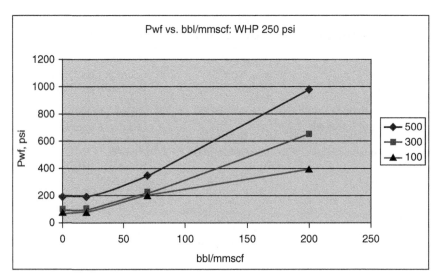

Figure 11-18: Example Gas Lift Tubing Performance in Deep Well with Lower WHP

are diminished if the LGR (liquid-gas ratio) increases. Also the importance of the surface WHP is shown in the two graphs where the WHP is lowered to 50 psi for the results in Figure 11-18 (using compression perhaps) and the pressure predicted at the bottom of the tubing is much less with the low WHP case.

If gaslifting a gas well, you are injecting gas into a gas well, so it may be a problem to optimize production. It would be good to set up a production rate to monitor is the difference of all gas produced up the tubing minus the gas injected. Then this reading of gas produced is what must be optimized and would eliminate confusion of total gas and produced gas.

In summary, gas lift can achieve very low pressures on the formation if the LGR is not too high and the WHP that the stream flows against is low. For higher liquid rates the achievable pressure on the formation increases substantially. As for the effects of the WHP on a pumping system, if the liquids are pumped up the tubing, the gas would flow up the casing and even if the pumping system pulled the fluids below the pay section, if the gas flows up the casing to a high surface pressure, the formation still would not see low flowing pressures.

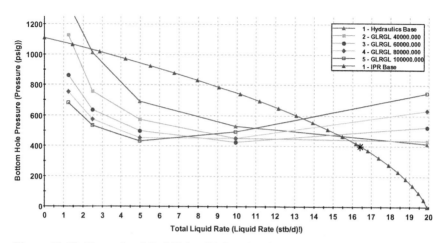

Figure 11-19: Example of Stabilizing Well or Solving Liquid Loading with Injected Gas for a Gas Well: Liquid Production with High GLRs Shown

Regardless of the values of the pressures and production, Figure 11-19 shows how a well that is liquid loaded as predicted by Nodal Analysis (top curve on the left in Figure 11-19) can be unloaded or stabilized by gas injection (see top to curves on Figure 11-19). This simulation shows liquid production with a high GLR to simulate a gas well. Unfortunately multiphase correlations seem to vary widely for gas lifting gas wells, and you must find a multiphase correlation that fits your well conditions to make an analysis such as the following. For low rates of gas and liquids, it could be better and easier just to add enough injection gas to bring the well above critical flow, using a critical flow correlation. For example, if a well is producing 200 Mscf/D at 100 psi WHP, the required critical rate using Turner at the surface could be around 330 Mscf/D depending on how the Turner correlation is programmed. Then you could just add perhaps 160 Mscf/D to the bottom point of injection and the well would once again be producing, with natural flow and gas lift gas, above the critical rate and no liquid loading would be occurring.

For higher liquid rates, it may require a more conventional gas lift design analysis other than simply adding gas to be above the critical flow rate.

Figures 11-20 and 11-21 show a low pressure, low liquid rate producer and a higher pressure, higher liquid rate producer equipped with gas lift

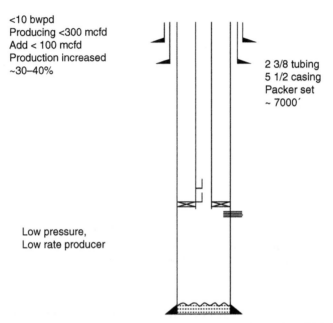

<10 bwpd
Producing <300 mcfd
Add < 100 mcfd
Production increased
~30–40%

2 3/8 tubing
5 1/2 casing
Packer set
~ 7000´

Low pressure,
Low rate producer

Figure 11-20: Schematic of Low Pressure, Low Rate Producer: Few Valves Need to Unload and Inject Gas at Bottom of Well (after Rouen, SLB)

valves. Note only a few deep-set valves are needed to unload and produce the low pressure, low rate well.

Note also that this conventional gas lift has the packer set above the perforated zones so there is no gas lift effect below the packer.

The previously discussed conventional valve designs show gas injected only to the packer that is set above the perforations. New innovations to allow the gas lift effect to extend into the perforations are now available, notably PerfLift™ from Schlumberger and systems from Weatherford. Other methods may exist.

Figure 11-22 illustrates injecting below the packer with a new valve from Weatherford. The well has a bypass packer set at 11,000 ft. Tubing below the packer is 2 7/8 in, and above the packer the tubing is 2 3/8 in. Objectives were to increase production by dewatering. The production was 102–500 mcfd and 50–60 bpd. A target was set for 1 mmscfd and 150 bpd of fluid. Weatherford custom-designed a gas lift valve and manufactured a bypass packer that could run through 26-lb casing and be set in 23-lb casing. Mandrels and a 1-in orifice tubing through the packer

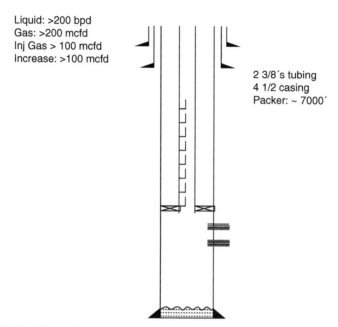

High Liquid Rate Producer
Higher Pressure Well

Liquid: >200 bpd
Gas: >200 mcfd
Inj Gas > 100 mcfd
Increase: >100 mcfd

2 3/8´s tubing
4 1/2 casing
Packer: ~ 7000´

Figure 11-21: Schematic of Higher Pressure, Higher Rate Producer: More Valves Need to Unload to Inject at Bottom of Well (after Rouen, SLB)

took the gas-injection flow from casing to tubing above the packer, and then from tubing to casing below the packer.

Results indicated 800 mcfd and about 100 bpd of fluids.

11.10 SUMMARY

Gas lift for gas wells can be thought of as a method to keep the gas velocity above the critical velocity at all times. If this is done, then no liquid loading can occur and manpower is usually not high for gas lift wells.

Low pressures can be achieved on the formation if WHP is low and if liquid rates are not too high. There is little measured data, however, on actual performance of gas lift on gas wells, such as producing BHPs when gas lift is performing.

New methods of continuous flow gas lift for gas wells with long perforated intervals are available to obtain some gas lift effect below the

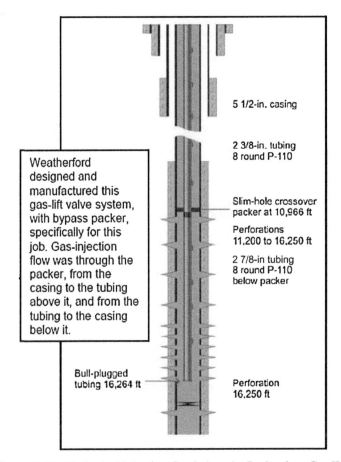

Figure 11-22: Method of Injecting Gas below the Packer in a Gas Well

packer and into the long pay zones. Intermittent methods use a burst of gas to lift liquid slugs from the well. Chamber lift allows liquid accumulations to occur for lifting with a minimum pressure on the formation.

11.11 REFERENCES

1. Trammel, P. and Praisnar, A. "Continuous Removal of Liquids from Gas Wells by Use of Gas Lift," SWPSC, Lubbock, TX, 1976, 139.

2. Stephenson, G. B. and Rouen, B. "Gas-Well Dewatering: A Coordinated Approach," SPE 58984, presented at the SPE International Petroleum Conference and Exhibition in Villahemosa, Mexico, February 1–3, 2000.

3. Winkler, H. W. "Gas Lift Solves Special Producing Problems," *World Oil*, November 1998, 209, No. 11, 35–39.

4. Boswell, J. T. and Hacksma, J. D. "Controlling Liquid Load-Up with Continuous Gas Circulation," SPE 37426, presented at the Production Operations Symposium, Oklahoma City, OK, March 9–11, 1997.

5. Kinley, J. C. Co., 5815 Royalton St., Houston, TX.

CHAPTER 12

ELECTRIC SUBMERSIBLE PUMPS

Significant contributions by Peter Oyewole
Peter O. Oyewole, MSPE U of Houston, BSPE U of Ibadan, Nigeria, is
the Artificial Lift Production Engineer for BP America, in Durango, CO.
He oversees artificial lift performance and optimization for tight gas and
coalbed methane dewatering operations in the North San Juan Asset. He
has 10+ years of experience that span upstream/downstream sectors includ-
ing work with Schlumberger, Valero Energy, and Texaco. He has authored
SPE and IPTC papers. He is an IADC (WellCAP) certified well control
instructor and teaches well control at San Juan College.

12.1 INTRODUCTION

Electric Submersible Pumps (ESPs) typically are reserved for applications where the produced flow is primarily liquid. High volumes of
gas inside an electrical pump can cause gas interference or severe
damage if the ESP installation is not designed properly. Free gas dramatically reduces the head produced by an ESP and may prevent the
pumped liquid from reaching the surface. In gas reservoirs that produce
high volumes of liquids, ESP installations can be designed to effectively
remove the liquids from the wells while allowing the gas to flow freely
to the surface up the casing. For any pumping system for gas wells, the
flow of gas up the casing depends on the casing head pressure (CHP)
once the liquids are minimized or eliminated over the perforations.

The adaptation of existing mature oil-patch technologies, such as ESP
by the petroleum industry, to low liquid gas production has created the
evolution of ESP hybrid systems and low liquid volume ESP systems.
Advanced monitoring and surveillance telemetry used in conjunction

with the systems provides not just an optimization tool, but also equipment protection, in the low liquid gas well environment.

This chapter is concerned with the four main methods used to employ ESPs to dewater gas wells.

1. The first method develops techniques to separate the gas from the intake of the ESP so that primarily liquid enters the pump. The gas separation is accomplished by using completions or special separation devices. In this way the liquid is produced to surface through the tubing and the gas is allowed to flow freely up the annulus between the tubing and casing.
2. The second method is to use special stages at the pump intake to handle the gas. Special stages build pressure from the intake to compress the gas sufficiently so that conventional stages take over and can continue building pressure. They could also change the flow pattern of the fluid or recondition it, to keep from gas locking at the entrance of the conventional stages. This allows the ESP to pump under a packer and still handle a fairly reasonable volume of free gas through the pump with the early special stages.
3. The third method is a technique where the liquid is reinjected into a formation below the packer. In this method, the liquid never reaches the surface. If the pump is well below the gas perforations, the water falls by gravity to the pump intake while the gas flows up the annulus. This system is commercially available and has been used in a number of successful installations.
4. The fourth method is the ESP hybrid systems and low liquid volume ESP. These are techniques developed to produce low liquid rates from gas wells. This includes special adaptations to conventional ESP systems with a centrifugal pump, which enables the pump to handle low liquid production. In some other cases, a completely different type of pump other than centrifugal pump is coupled with electric submersible motors. Progressing cavity pump, twin screw pump, and hydraulic diaphragm pump are all running on downhole electric motors to dewater gas wells.

12.2 THE ESP SYSTEM

This chapter is not intended as a tutorial on the installation, operation, and troubleshooting of ESP systems. Some basic knowledge of ESPs is assumed. Some introductory comments are included describing

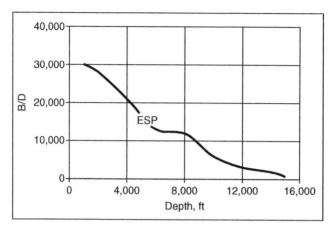

Figure 12-1: An Approximate Lift-Rate Application Chart for ESPs (Weatherford)

only the basics of an ESP installation. See [3], for example, for more details.

Figure 12-1 shows an approximate lift-rate chart for ESP applications. This chart indicates ESP applications can be made inside the envelope and not outside of the envelope. However, there are many exceptions to the chart, possibilities that would extend or reduce the area indicated as possible for applications.

Figure 12-2 shows a basic ESP system. The system consists of a downhole motor connected to a seal section which in turn is attached to a centrifugal pump. A high voltage electric cable connects the motor to the surface where power is transformed from the utility lines or is taken from an electric generator. VSD/VFD (Variable Speed Drive, Variable Frequency Dive) or an inline switchboard are surface controllers used to control ESP operations.

The motor is a two-pole, squirrel cage motor with synchronous speed of 3600 rpm and an operational speed of about 3500 rpm at 60 Hz. It is imperative that the motor be cooled by the produced fluid passing its outer casing. In the event that large quantities of gas pass the motor, the heat transfer from the motor to the produced fluid will be drastically reduced, potentially causing severe motor damage. This can be a frequent occurrence in gas wells.

The seal section houses a pump thrust bearing and restricts the wellbore fluids from entering the motor. The pump has an intake where the fluid enters the pump at the bottom of the pump. The intake can be

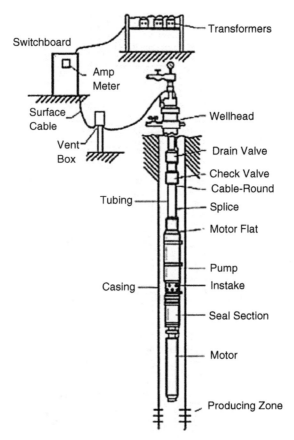

Figure 12-2: Typical ESP System (Schlumberger)

replaced by a rotary gas separator, which separates gas to the annulus while nearly all liquid enters the pump. The pump itself consists of a stack of impeller/diffuser combinations called stages that generate head and pressure. The amount of head required to bring the liquids to surface dictates the numbers of stages to be stacked in the pump housing, and the flow rate required determines what type of stages to use.

The motor controller typically has protective shut-offs, the on/off controls, usually some method of recording motor amps, and other parameters often used to supply diagnostics for the pump operation. The transformer steps up or steps down the voltage to the value needed by the motor after accounting for the cable voltage loss.

A typical pump head curve, for a single pump stage, is shown in Figure 12-3. This curve shows the pump to reach maximum efficiency at a rate

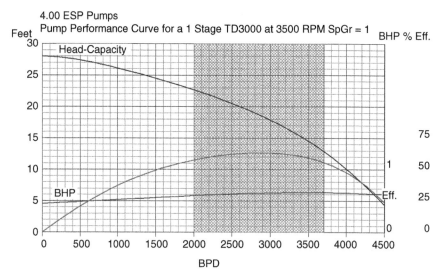

Figure 12-3: Pump Performance Curve Showing the Head Curve, the Brake HP or BHP Curve, and the Stage Efficiency Curve

of approximately 3000 bpd and producing about 18 ft of head per stage at Best Efficiency Point (BEP). Higher head is possible at lower flow rates and more flow is possible at significantly lower values of head.

The single stage head curve shown in Figure 12-3 is for the special case where the pump is pumping 100 percent liquid. If gas is present, the head curve tends to become erratic and will drop to zero head prematurely as the flow rate of gas is increased. It is therefore necessary to keep large quantities of gas from entering the pump intake of an ESP system.

12.3 WHAT IS A "GASSY" WELL?

It is well known that ESP performance can be severely degraded by the flow of excessive gas through the pump. But what constitutes excessive gas? How much free gas can a given ESP handle before performance is affected?

A high GOR or GLR could be an indicator of excessive gas, but a high intake pressure would compress the free gas, making the free gas volume smaller, whereas a low intake pressure would expand the free gas, resulting in higher gas volume through the pump. In addition, if the intake pressure is greater than the bubble point, all the gas will be in solution so that no free gas will be present in the pump.

This section will summarize a method for evaluating the effects of free gas on ESP performance (see [3] for more details).

A useful correlation for evaluating ESP performance with free gas is [2]:

$$\Phi = \frac{667 \times VLR}{P_{ip}} \qquad (12\text{-}1)$$

where:

VLR = vapor/liquid volume ratio at the pump intake
P_{ip} = pump intake pressure, psia

It is found empirically that the effect of free gas on the pump head curve is negligible when $\Phi \leq 1$.

Example 12-1: Calculate Free Gas Percentage and ESP Limitations

Desired Production	1000 STB/d
Pump intake pressure	850 psia
Pump intake temperature	165°F
Produced GOR	430 scf/bbl
Gas gravity	0.65
Water gravity	1.08
Oil gravity	35 API
Water cut	65%

In this example, we will use Standings [6] Black Oil correlation for solution gas R_S and oil formation volume factor B_O.

$$R_S = \gamma_G \left(\frac{P_{ip} 10^{0.0125API}}{18 \times 10^{0.00091T}} \right)^{1.2048} \text{scf/STB-oil} \qquad F = R_S \left(\frac{\gamma_g}{\gamma_o} \right)^{0.5} + 1.25T$$

$$Bo = 0.972 + .000147 F^{1.175} \text{ bbl/}STB$$

1. First calculate the gas in solution at the pump intake from Standings [6] solution GOR, scf/bbl-oil:

$$R_S = .65 \left(\frac{850 \times 10^{.0125 \times 35}}{18 \times 10^{.00091 \times 165}} \right)^{1.2048} = 150 \, \text{scf/STB-oil}$$

2. Calculate the formation volume factor B_O following Standings [6]:

$$\gamma_o = 141.5/(131.5 + 35) = 0.85$$

$$F = 150 \left(\frac{.65}{.85} \right)^{0.5} + 1.25 \times 165 = 337.43$$

$$B_o = 0.972 + .000147 F^{1.175} = 1.11 \, \text{bbl}/STB$$

3. Calculate the gas volume factor at the pump intake using $Z = 0.85$ for the compressibility factor:

$$B_G = \frac{5.04 ZT(^\circ R)}{P_{ip}} = \frac{5.04 \times .85 \times (165 + 460)}{850} = 3.15 \, \text{bbls gas/Mscf}$$

4. Calculate the volume of free gas in in-situ barrels and the oil and water volumes at the pump intake. The total free gas at the pump is then

$$Q_{GAS} = Q_{OIL} \left(\frac{GOR - Rs}{1000} \right) Bg = 350 \times \left(\frac{430 - 150}{1000} \right) \times 3.15$$
$$= 308.7 \, \text{bbls gas/day}$$

Assume 30 percent of the free gas at the pump travels up the annulus, bypassing the pump intake. Then 70 percent of the free gas actually enters the pump:

$$Q_{GAS,Pump} = 0.7 \times 308.7 = 216 \, \text{bbls gas/day}$$

5. Calculate the volume of liquids entering the pump.

$$Q_{OIL,Pump} = Q_{OIL,STB} \times B_O = 350 \times 1.11 = 388.5 \, \text{bbls oil/day}$$
$$Q_{WATER,Pump} = 650 \, \text{bbls water/day}$$

6. Calculate the parameter Φ:

$$\Phi = \frac{666(\text{Qgas/Qliquid})}{P_{ip}} = \frac{666}{850} \times \frac{216}{388.5 + 650} = 0.164$$

If $\Phi < 1.0$, then the ESP is predicted to perform on the nominal head curve, even though some free gas is present.

If $\Phi > 1.0$, then the pump head curve is predicted to be degraded and a gas separator will be needed to reduce free gas through the pump. The gas separator would augment the natural separation of the free gas, bypassing the pump into the annulus.

In this example, $F < 1.0$, so no additional gas separation is needed.

7. Calculate the percent free gas at the pump.

$$\% \text{ Free Gas} = \frac{Q_{GAS}}{Q_{GAS} + Q_{LIQUID}} \times 100 = \frac{216}{216 + 388.5 + 650} 100 = 17\%$$

The percent free gas is fairly high, but the high pump intake pressure of 850 psia reduces the detrimental effects on the pump head curve.

12.4 COMPLETIONS AND SEPARATORS

As we have seen, excessive gas at the pump may require additional gas separation to reduce the free gas into the pump. This section will discuss methods to achieve higher gas separation.

Perhaps the best method of keeping gas from entering the pump is to set the pump intake below the perforations. This configuration would allow the liquids to gravity drain to the pump intake while the lighter gas is diverted into the annulus above the pump. However, this completion locates the ESP motor outside the flow path of the production liquids where it would not normally receive sufficient cooling. To alleviate this problem, the motor can be fitted with a shroud that forces the produced liquids down past the motor before entering the pump intake as illustrated in Figure 12-4.

If the pump must be set above the perforations, then the pump can be fitted with an upward opening shroud on the pump intake, also shown in Figure 12-4. This forces the produced fluid to reverse direction and travel downward before entering the pump, breaking out the larger gas bubbles. In high rate wells, however, where the velocity of the liquid in the annulus between the casing and the shroud is above approximately 0.5 ft/sec, the production fluids can carry significant amounts of free gas to the pump intake even with shrouds.

Although shrouds can increase gas separation, there are several potential problems to consider before running shrouds on ESPs.

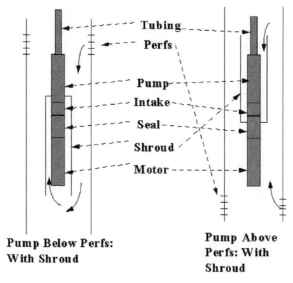

Figure 12-4: Shrouded ESP Installations

- A shroud can substantially decrease the clearance between the pump assembly and the casing. In wells having clearance problems, it is recommended that a full-length gauge section be run prior to running the pump.
- The shroud can accumulate sand and scale. This is particularly true for upward facing shrouds, which act like collectors for heavy particles.

If the annular region between the shroud and the motor (or pump) is small, the increased pressure drop inside the shroud can reduce the pump intake pressure and break gas out of solution just ahead of the pump intake. Other methods exist such as pumps designed to recirculate fluids to the motor when the pump system is set below the perforations.

Early gas separator designs were based on increasing gas separation by changing fluid flow direction to reverse flow in the wellbore. Hence, this type of gas separator is known as a *reverse flow* gas separator. It is also known as *static* gas separator. Well fluid that enters the gas separator at an angle is forced to change direction. Some of the gas bubbles will continue to rise at the annulus instead of turning, while some gas bubbles will rise inside of the gas separator, exit the housing, and continue to rise as in Figure 12-5 and 12-6.

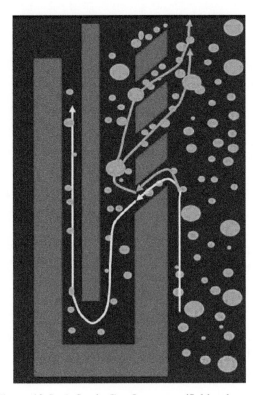

Figure 12-5: A Static Gas Separator (Schlumberger)

Figures 12-7 and 12-8 show variations of the static gas separator, which are bottom feeder intake, also known as gas avoiders, that are used in highly deviated and horizontal wells. A bottom intake feeder works better in low gas rate horizontal wells with a distinctive two-phase flow regime. The eccentricity of the bottom intake feeder enables liquid and some gaseous liquid to be produced into the pump from the low side of the wellbore. Lighter gas bubbles break out and migrate to the high side.

Another common device used to remove gas from the production stream before entering the pump is the *rotary separator*, shown in Figure 12-9. This device is fitted to the pump intake and is attached to the rotating pump shaft. The centrifuge action of the separator causes gas to be diverted to the annulus, leaving mostly liquid to enter the pump. Tests have shown that the rotary separator can be over 90 percent effective. The rotary separator, however, can also be gas choked if gas volumes become too large and rates too high. It may wear with sand since there are large velocity gradients where the fluids are swirled. A

Figure 12-6: REDA 65 GS Static Separator (Schlumberger)

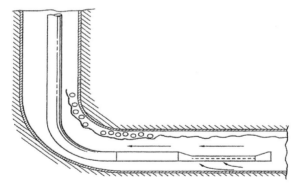

Figure 12-7: A Bottom Intake Feeder or Gas Avoider

diffuser may be added to the rotary separator to help separate gas and liquid phases. It adds velocity to the mixture and creates separation with a vortex action. See [1], [4], and [6] for more information on various gas separation and handling devices applied in the industry.

Figure 12-8: REDA Bottom Intake Feeder (Schlumberger)

12.5 SPECIAL PUMP (STAGES)

Some pump suppliers have stages that are designed to better handle free gas. These stages are installed at the pump intake and build pressure in the presence of free gas. After the gas is sufficiently compressed, additional standard pump stages are used to generate the desired pump discharge pressure.

Examples of these special "gas handling" stages are the Schlumberger Advanced Gas Handler [5] (AGH)™ and Poseidon™, the Centrilift Multi Vane Pump[8] (MVP)™, and Wood Group XGS™ system [6]. A few of these stages in a pump, in a gassy environment, even pumping under a packer, may be able to build pressure in the pump such that gas is not a problem for the remainder of conventional stages in the pump assembly.

The AGH has a recirculation path to keep bubbles from accumulating and to keep some pressure building in the pump. It reconditions the gas and enables the pump to produce gas without gas locking. This is achieved by homogenizing the gas liquid mixture, which reduces gas bubble sizes by putting gas bubble back in the liquid solution. The MVP, which performs similar gas handling as the AGH, has split vane design that prevents gas accumulation in the impeller. The steep vane angle impacts high momentum energy to the fluid leaving the impeller. It also has oversized balance holes that assist in managing thrust and gas handling by creating turbulence to break up the gas bubbles.

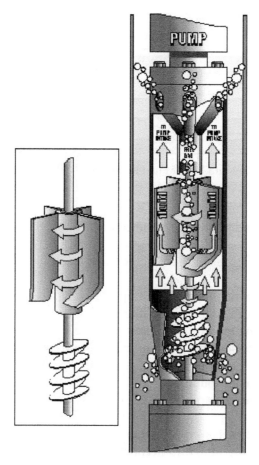

Figure 12-9: A Rotary Separator (Baker Hughes Centrilift)

The XGC system has a compression chamber downstream from the tandem gas separators. Free gas is compressed back into the solution by the compression chamber. It also breaks the bigger gas bubbles into smaller bubbles at the same time. This provides an increasingly homogenized solution, which a submersible pump stage can handle without gas locking.

The Poseidon™ is a multiphase helicoaxial pump that is installed between intake (or gas separator). The Poseidon is capable of handling up to 75 percent free gas. It uses a similar stage design used by Framo and Sulzer to handle up to 95 percent free gas on surface pumps. The special stage design of the Poseidon™ allows for axial flow (see Figure

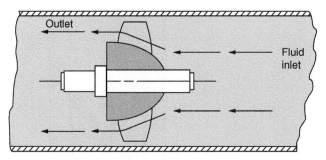

Figure 12-10: Axial Flow of Fluid in Poseidon™ Pump (Schlumberger)

12-10) of fluid and gas, which significantly reduces the possibility of gas locking.

Some main features of the Poseidon are:

- It primes the main production pump and pushes the gas/liquid mixture through the production pump stages.
- It increases the mixture pressure to reduce the gas volume.
- It homogenizes the fluid to ensure good mixture between gas and liquid.
- It breaks large gas bubbles.
- It does not induce separation of gas and liquid.

A simpler option is to use the tapered pump concept of design with larger stages at the intake to accommodate free gas, and switching to smaller stages as compression reduces the total volume can help to handle gas.

12.6 INJECTION OF PRODUCED WATER

ESPs can also be used to reinject liquids back into the formation in a manner similar to that performed by beam pumps (see Chapter 10). In this system, illustrated in Figure 12-11, the ESP is inverted to push fluids downward. The pump generates the necessary head to push the liquids into a water injection zone below the packer while the gas is produced up the casing annulus. No liquids are produced to the surface. The system usually is installed with a downhole pressure sensor that detects when a predetermined level of liquids have accumulated in the hole. Once this level is reached the sensor starts the ESP motor to inject

Centrilift GasPro System

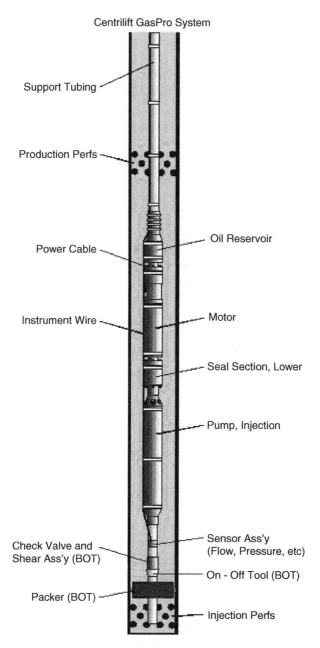

Support Tubing

Production Perfs

Power Cable

Oil Reservoir

Instrument Wire

Motor

Seal Section, Lower

Pump, Injection

Check Valve and
Shear Ass'y (BOT)

Sensor Ass'y
(Flow, Pressure, etc)

On - Off Tool (BOT)

Packer (BOT)

Injection Perfs

Figure 12-11: Inverted ESP Installation for Dewatering Gas Wells by Injecting Water below a Packer into a Water Zone (Baker Hughes, Centrilift)

the liquids. Once the liquids have been injected, the liquid level drops and the sensor automatically shuts off the ESP.

12.7 ESP HYBRID SYSTEMS AND LOW LIQUID VOLUME ESP

12.7.1 Special Adaptation to Conventional Centrifugal ESP

Traditionally, a centrifugal ESP is designed to perform in a high liquid volume environment. Major concerns with operating centrifugal ESP with low liquid includes pump cavitation and gas locking, pump wear due to downthrust stages, and heat damage of electric motor due to inadequate conductive cooling due to low or no flow of liquid past the electric motor. These problems have been mitigated with special adaptation to enable conventional ESP to operate in a gas well with liquid production as low as 40 BLPD. Even lower liquid rates have been achieved in some cases. The previous sections covered in detail methods of separation and gas handling that prevent pump cavitation and gas locking. This is one of the adaptations for low liquid volume ESP.

Pump wear due to downthrust stages is another concern for a low liquid volume gas well if one is forced to operate to the left of the recommended range to achieve low rates. Downthrust cannot be ignored; at low liquid rate, the pump will operate on the left side of the recommended operating range (ROR). As it may be expected, the stages will go into higher downthrust on the left side of the range. In some other cases a balance circulation ring between the impeller and diffuser, created by drilling a balance holes in the upper impeller skirt that recirculate lower pressure fluid over the majority of the upper surface has proven to reduce downthrust. Most often the low liquid rate pump does not benefit from "shimming." Pumps built in compression construction (mostly high liquid rate delivery pump) are shimmed (spaced) to keep every impeller fixed to the shaft rigidly so that it cannot move without the shaft moving. All the impellers are compressed together. The pump shaft mates the protector shaft. This enables all the thrust developed by the pump to be transferred to the protector thrust bearing instead of creating wear between the impeller and diffuser.

To combat inadequate conductive cooling due to low or no flow of liquid past the electric motor, shrouding of motor landed at perforation or in the rathole below perforations has been discussed earlier in Section 12.4. Motor derating and the use of variable rated motor leads to lower internal temperature rise in the motor. Additional features can increase overall motor operating maximum temperature significantly when compared with conventional fixed rated motor. Effective motor temperature monitoring and control is perhaps the most important single factor in motor protection due to heat. A temperature monitor, for example a thermocouple element, is connected to the motor winding to observe the motor winding temperature. Observed motor temperature data is transmitted with other downhole data to the surface. The motor temperature set point can be set in the surface motor controller to automatically shut down the unit, in the event of a high motor temperature violation. This saves the motor from failure due to excessive heat generated as a result of motor internal temperature rise. A recirculation pump designed to pump some liquid below the motor is also an adaptation used to reduce the motor heat rise if the motor is set below the perforations.

12.7.2 Electric Submersible Progressing Cavity Pump (ESPCP) and Electric Submersible Progressing Cavity Pump Through Tubing Conveyed (ESPCP TTC)

ESPCP, also known as Bottom Drive PCP, has been in use in oil producing environments for some decades. This application has been extended to deliquify gas wells. Here, a PCP is coupled with a submersible electric motor. Depending on the system manufacturer and assembly company, the configuration from bottom up includes (see Figure 12-12) a standard ESP motor used to drive the system, and a gear reducer to reduce the downhole motor speed to PCP operation speed range is stacked on the motor. The gear reducer also generates increase torque required to drive the PCP. The seal or protector is then added on top of the gear reducer. The seal separates the motor and gear reducer oil from the wellbore fluid and houses the thrust bearing. A flex shaft and intake component, which is coupled on top of the seal, let fluid into the pump. It also converts the eccentric motion of the PCP to concentric motion of the seal, gear reducer, and motor sections, and transfers the pump thrust to the thrust bearing in the seal. A PCP is then attached to the top of the assembly [9].

The ESPCP system takes advantage of the PCP, which handles a lower liquid rate than conventional centrifugal pump, and additional solid handling capability, without dealing with jointed sucker rod or continuous rod (corod) string problems and challenges. With the ESPCP

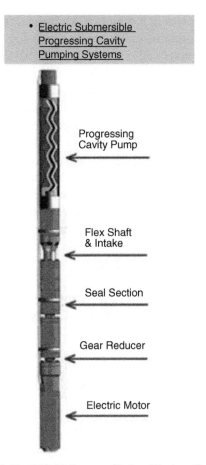

Figure 12-12: ESPCP System (Baker Hughes Centrilift)

configuration, the pump is landed in or through deviated areas of the wellbore where rod and tubing wear with rod-driven systems is a major concern.

The elastomeric stator in PCP is an unwanted limitation that often leads to pump failure. PCPs are very prone to damage if run dry (i.e., without sufficient liquid passing through the pump)—the elastomer will rapidly overheat. This will lead to full rig intervention to replace the damaged pump. Adequate monitoring of key performance data and the use of a pump-off controller can help mitigate this problem. Good design and best operational practices can also help reduce PCP failure. Nevertheless, the pump is still the weak link in the assembled system.

A "through tubing conveyed" (TTC) package is developed to reduce workover cost by eliminating the need for full workover rig intervention to replace the damaged pump. With a TTC package, installation and removal of the PCP pump can be achieved only by slickline, electric wireline, coil tubing, jointed sucker rod string, or corod.

In an ESPCP TTC [9] package, the motor, gear reducer, and the seal sections are run downhole on tubing as in conventional tubing conveyed ESPCP. A tubing intake receptacle crossover is latched on top of the seal. It connects the tubing to the seal and provides a fluid passage to the pump. The rest of the assembly is conveyed through the tubing as in Figure 12-12. From the bottom up, the flex shaft base/coupling, the flex shaft, PCP, tubing pack-off, and tubing anchor are all through tubing conveyed. The flex shaft base/coupling connects the flex shaft coupling to the intake coupling on the tubing. The locking mechanism in the flex shaft base secures the pump stator to the tubing and prevents stator rotation. The flex shaft converts the eccentric motion of the PCP to concentric motion of the seal, gear reducer, and motor sections. PCP has the rotor and stator that pump the fluid. Tubing pack-off separates high pressure discharge from low pressure intake. It also provides flexibility for PCP length change and spacing. Mechanical, electrical, hydraulic, or X-lock system locking set options are available. The tubing anchor is installed to prevent axial movement of the PCP and flex shaft assembly. A slight modification has been made to the assembly to accommodate running all the "through tubing conveyed jewelry" within one installa-

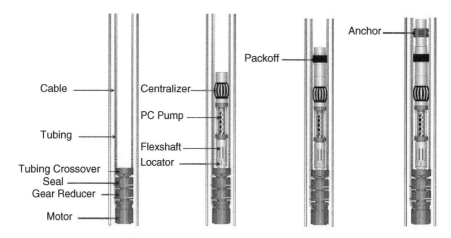

Figure 12-13: ESPCP TTC Installation (Baker Hughes Centrilift)

tion run. This has eliminated or replaced some of the components just described. An ESPCP TTC can be run in 3½ in tubing, as well as 2-7/8 in with low liquid rates.

12.7.3 Hydraulic Diaphragm Electric Submersible Pump (HDESP)

The hydraulic diaphragm pump is coupled together with an electric submersible motor (see Figure 12-14) (www.smithlift.com).

The HDESP™ (Hydraulic Diaphragm Electric Submersible Pump [10]) is designed for use in low volume gas wells. The HDESP™ was designed to efficiently produce low liquid volume gas wells in coalbed methane dewatering operations. The system is similar to a conventional ESP system, but utilizes a positive displacement, double acting diaphragm pump driven by a downhole electric motor. The pump is typically efficient at low flow rates, when compared with traditional centrifugal ESP, generating operational power cost saving. In most cases, the initial equipment cost is less than other forms of lift and requires

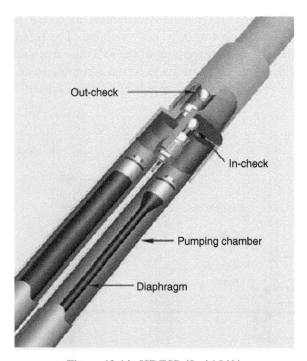

Figure 12-14: HDESP (SmithLift)

no routine maintenance. It may operate at near pumped off condition (i.e., with very low NPSH—Net Positive Suction Head). However, just like any submersible electric motor, it is important to have enough fluid flow the past motor. It can be operated in deviated wells and is capable of handling some solids. The current product offering targets applications less than 2500 feet and less than 200 BFPD.

12.8 SUMMARY

ESPs can be a viable method to dewater gas wells, usually when it is necessary to handle large liquid volumes. But if high rates are needed, they become much more advantageous. Small water well ESPs are used to lift relatively small rates off of coal gas fields.

ESP installations are expensive and usually consume more power per barrel of liquid lifted than a beam pump system. Of course they should be compared only when the rates are well within the good operational ranges for both the beam and ESP systems. The efficiency of an ESP system is significantly reduced (similarly for a beam system and other systems excluding gas lift) when gas is allowed to enter the pump. These shortcomings limit the use of ESPs for gas well dewatering applications.

The use of ESPs to inject water below a packer at fairly high rates is a specialty area for ESPs for gas well operations.

ESP hybrid systems and low liquid volume ESPs are expanding ESP application in gassy wells. This includes special adaptation to a conventional ESP system with a centrifugal pump, which enables the pump to handle low liquid production. In some other cases, a completely different type of pump other than centrifugal pump are coupled with electric submersible motors. Progressing cavity pumps, twin screw pumps, and hydraulic diaphragm pumps are all running on downhole electric motors to unload gas wells. With completion of more deviated, S-shaped, Horizontals, and Multilaterals wells, ESPs are gaining more popularity in gas well dewatering.

12.9 REFERENCES

1. Centrilift Gas Handling Manual.
2. Turpin, J. L., Lea, J. F., and Bearden, J.L. "Gas Liquid Flow through Cen-

trifugal Pumps-Correlation of Data," Proc. 33rd Annual Meeting of SWPSC, Lubbock, TX, 1986.

3. *Centrilift Submersible Pump Handbook*, 6th edition.

4. Dunbar, C. E. "Determination of Proper Type of Gas Separator," REDA Technical Bulletin.

5. Schlumerger-REDA bulletin on the Advanced Gas Handler (AGH™) stages and Poseidon™ pump.

6. Wood Group ESP Inc. bulletin on XGC gas separator and compression chamber.

7. Standing, M. B. "Volumetric and Phase Behaviors of Oil Field Hydrocarbon Systems," Reinhold Pub. Corp., New York, 1952.

8. Baker Hughes–Centrilift Datasheet on the Multi Vane Pump (MVP™).

9. Baker Hughes–Centrilift bulletin on the PCP System.

10. SmithLift bulletin on the Hydraulic Diaphragm Electrical Submersible Pump.

PROGRESSING CAVITY PUMPS

By Tim W. Soltys, Weatherford
Global Sales Support (GSS)—PCP Technical Manager
Weatherford PC Pump Products and Services
Tim Soltys is a graduate of the Northern Alberta Institute of Technology
with a degree in Hydrocarbon Engineering Technology and over 20 years
experience in the oil and gas industry primarily with the design, installation,
and optimization of artificial lift systems. For the last 10 years Tim has
worked exclusively on the global deployment of Progressing Cavity
Pumping Systems, of which a large percentage of applications are the del-
iquification of natural gas and coalbed methane wells. Today Tim spends
much of his traveling globally, leading workshops on the design, installa-
tion, optimization, and operation of Progressing Cavity Pumping
Systems.

13.1 INTRODUCTION

Dewatering CBM wells with Progressing Cavity Pumping Systems
(PCPs) is a relatively routine operation that has been deployed successfully since the mid 1980s. Today there are more than 7000 PCPs extract-
ing water from CBM wells throughout Canada, the United States, the
United Kingdom, Kazakhstan, Russia, China, Australia, and India.

PCPs have the advantage of being able to easily pump solids, liquids,
and gasses. When combined with comparatively the lowest capital cost
and highest operating efficiencies of any ALS, PCPs are the preferred
artificial lift system for many CBM operations. In the United States
alone there are about 6000 PCP systems operating in the Raton, Greater
Green River, Powder River, Black Warrior, Appalachian, and San Juan
basins.

Since natural gas wells other than CBM experience liquid loading problems requiring lift, and also many produce frac sand or perhaps other solids, the information here applies broadly to the dewatering of natural gas wells other than CBM as well.

13.2 PROGRESSING CAVITY PUMPING SYSTEM

The PC pump is comprised of two components, the rotor and the stator (see Figure 13-1). The rotor is manufactured from high strength steel and covered with a chrome layer 0.010 to 0.020 inches thick. The rotor is the only moving component of the pump.

The stator has an internal helix shape molded into an elastomer compound that is chemically bonded to the inside of a steel tube. When the rotor is inserted into the stator, it creates a continuous seal line (compression/interference fit between the rotor and stator elastomer) that extends from the pump suction to discharge. This creates a series of identical but separate cavities that progress from the pump suction to the discharge as the rotor turns. One cavity opens as the other closes, creating a nonpulsing pumping action.

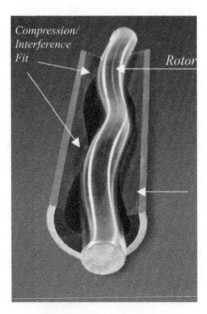

Figure 13-1: PCP Pump

In a conventional installation, the stator is installed on the bottom of the tubing string. The tubing string must be removed from the well to retrieve the pump. Alternatively, there are select models of PCPs (Insert PCP) that can be inserted into 2-7/8 in and larger tubing. An Insert PCP is installed and retrieved with the rod string.

A conventional PCP system (Figure 13-2a, b) will typically have the following components: prime mover, drivehead, stuffing box, rod string, PC pump, and torque anchor (no-turn tool). The torque anchor is to prevent the tubing string from unscrewing during normal pumping operations.

Figure 13-3 highlights the general application envelope for PC pumping systems.

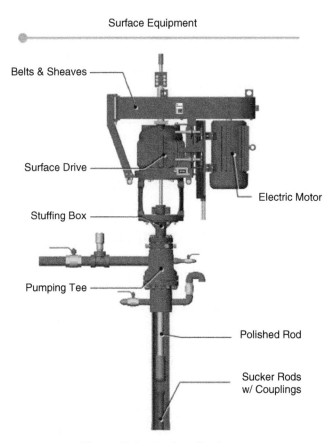

Figure 13-2a: Surface Equipment

Downhole Equipment

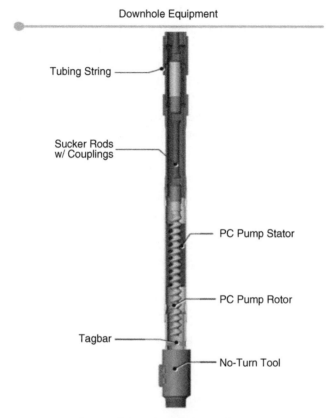

Tubing String

Sucker Rods
w/ Couplings

PC Pump Stator

PC Pump Rotor

Tagbar

No-Turn Tool

Figure 13-2b: Downhole Equipment

13.3 WATER PRODUCTION HANDLING

In Australia, PC pumping systems are producing 5000 BWPD from 2200 ft and 2500 BWPD from 3000 ft with bottomhole temperatures of 175°F. In the U.S. Raton basin, PCPs are operating at 10 rpm to produce 1 BWPD. With pump landing depths of more than 9000 ft TVD, Indonesia's light oil applications represent some of the deepest and hottest (220°F) applications in which PC pumping systems are deployed.

With the largest turn-down ratio of any ALS system, a PCP production rate could range from 500 to 5000 BWPD, but the total system efficiency would be significantly less at the lower production end. High system turn-down ratios are particularly important in CBM applications

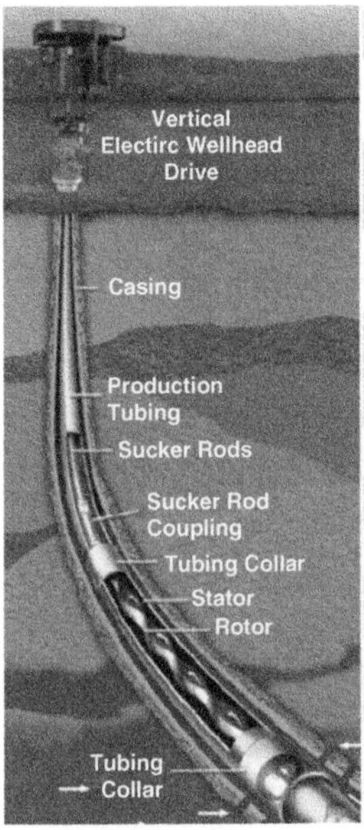

		Typical Range	Maximum*
	Operating Depth	1,000' - 5.000' TVD 330 - 1,550 m TVD	9,800' TVD 3,000 m TVD
	Operating Volume	5 - 2,500 BPD 1 - 400 m³ / day	5,000 BPD 800 m³ / day
	Operating Temperature	75 - 170 °F 24 - 77 °C	300 °F 150 °C
	Wellbore Deviation	N/A	Dogleg Severity less than 15°/100 feet 15°/30m
	Corrosion Handling	Excellent (regarding pump)	
	Gas Handling	Good	
	Solids Handling	Excellent	
	Fluid Gravity	Below 45 °API (highly dependable on aromatics content)	
	Servicing and Repair	Typically Requires Workover or Pulling Rig	
	Prime Mover Type	Electric Motor or Internal Combustion Engine	
	Offshore Application	Good	
	System Efficiency	50% to 75% (up to 90%)	

Figure 13-3: General Application Envelope for PC Pumping Systems

where the initial production rates can decline rapidly over days to months.

13.4 GAS PRODUCTION HANDLING

Because large amounts of gas can be stored at low pressures in coal reservoirs, the reservoir pressure must be drawn down to a very low level to achieve high gas recovery. To accomplish this, the pump is typically landed below the lowest set of perforations to draw the liquid level down into the coal interval.

Production of CBM gas is controlled by a three-step process: gas desorption from the coal matrix, gas diffusion to the cleat system, and gas flow through fractures. If the water production is stopped for an extended period of time, water floods the cleat and fracture system and blocks the gas flow. This increases the hydrostatic head and can stop or significantly reduce gas production. It may take several days or weeks to attain normal gas rates again.

In reciprocating rod installations, there is a tendency for the pump to gas lock. This results in reduced water production rates and gas production declines due to the increase in hydrostatic pressure. Because a PC pump has no internal valves to gas lock, the system will continuously remove water from the wellbore. This is particularly important in mature wells with low producing bottomhole pressures.

PC pumps can produce a significant amount of free gas but there is a trade off with pump performance and life expectancy. Any volume in the pump that is occupied with free gas cannot contain liquid; thus liquid rate production decreases as free gas increases.

The biggest potential concern with high gas production is pump overheating. Rotor rotation generates significant friction heat from contact with the stator. The pump is cooled by the liquid flowing through the pump. If cooling is not sufficient, then the pump will heat up, resulting in expansion of the rotor and stator and producing a tighter rotor/stator fit. This results in more friction and higher temperatures. If not corrected, the pump internal temperature may exceed the elastomer temperature limit. Under conventional design and pump sizing practices, a PC pump will typically have a catastrophic failure in under 30 minutes if operated with no liquid.

Under the right conditions PC pumps can be designed to operate with no liquid production for up to 5 hours before the internal operating temperature exceeds the elastomer rated limit. This is particularly important in operations where flowline pressure increases during pigging operations and forces wellbore liquids back into the formation. If this continues for several hours, the liquid level could fall below the pump intake. Once flowline pressures return to normal and water encroaches into the wellbore the PCP immediately starts to dewater the well.

The impact of free gas also can create a highly nonlinear pressure distribution within the pump, which can significantly stress the elastomer. To maintain a more linear pressure distribution, special pump sizing and lift configurations are used.

13.5 SAND/COAL FINES PRODUCTION HANDLING

The combination of elastomer resilience and viscoelastic properties allow for high continuous production of solids with very little damage or wear to the elastomer. This is of particular importance when producing wells that have been fractured with a sand proppant and/or that produce coal fines. Canadian heavy oil wells routinely produce 2 to 60 percent sand cut with average stabilized cuts of 5 percent.

Sand production can:

- Accelerate equipment wear
- Increase rod torque and power requirements
- Create flow restrictions at the pump intake and in the tubing string

In extreme cases, a rapid influx of sand into the wellbore can plug off the pump intake or discharge and may cause immediate and severe damage. PC pumping systems can be designed to handle moderate, steady rates of sand production. Most problems occur when there is a rapid influx (slugging) of sand over a short period of time. Pumping sand slugs can exceed the rated torque limit of the rod string or available surface horsepower.

Slugging occurs naturally; however any operation that changes the bottomhole or sandface pressure can initiate solids production. Rapid changes in bottomhole pressure can collapse stable sand bridges around the perforations and allow sand to flow into the wellbore. Large adjustments in pump speed to lower bottomhole pressure and increase production rate should be made gradually over a couple of days to allow the well to stabilize. Workovers or remedial operations such as swabbing or washes often are followed by periods of high sand production.

Sand bridging above and below the pump intake are most common in directional and horizontal wells. The ability of the produced fluid to transport sand improves with increased tubing flow velocities and viscosity. The initial system design should consider whether the lowest anticipated production rate will be sufficient to keep the sand moving. One method to increase flow velocities is to use smaller diameter tubing. However, the entire system should be evaluated to determine the magnitude of additional flow losses or if there are restrictions on available sucker rod sizes.

If tubing velocities are insufficient to transport sand to surface, the sand may settle out in the tubing above the pump. As the tubing fills with sand the overall fluid density and hydrostatic head increases. This results in a higher pump discharge pressure and a greater load on the pump. If the hydrostatic head continues to increase, due to the accumulation of sand in the tubing, it may exceed the rated lift capacity of the pump.

In some cases, the sand may bridge inside the tubing and create a total or partial blockage that will increase the pressure differential across the PCP. This increase in pump differential may exceed the rated lift capacity of the pump and could result in immediate and severe damage to the pump elastomer.

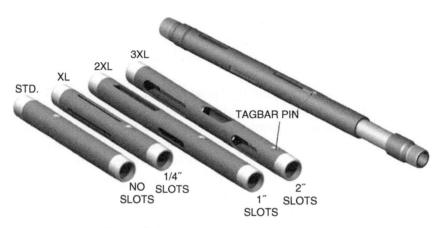

Figure 13-4: Common Tag-Bar Configurations

The build-up of sand into the wellbore can cause production rates to decrease and, in some cases, completely block the pump intake. To minimize sand accumulation around the pump intake, it is important to have a sump below the pump for excess sand to settle out. For sandy applications, pump intakes should be configured so that fluids can flow directly from the wellbore into the bottom of the pump. Most manufacturers offer a number of different tag-bar and rotor options to match specific applications. The more popular design is to go with a slotted tag-bar and extended rotor, which maximize the intake flow area and help to keep the sand fluidized (Figure 13-4 illustrates several common tag-bar configurations).

If the pump is shut down for an extended period of time, sand will settle out on top of the pump and/or create a sand bridge in the tubing. When this occurs, it is strongly recommended that, prior to the pump being started, the rotor be pulled out of the stator and one tubing volume of clean fluid be flushed through the pump to remove all solids. This is a normal operating practice that usually is referred to as a *flush-by operation*.

The key to this operation is to ensure that the polished rod is longer than the rotor, allowing the operator to pull the polished rod up through the drive head without removing any of the wellhead components. Failure to do so may result in high start-up rod string torque and pump discharge pressures that may exceed the lift capacity of the pump and torque limit of the rod string, resulting in a premature pump or rod string failure.

High start-up torque may exceed the available horsepower of the surface drive unit. Once the rotor is free of the stator, clean fluid can be pumped down the tubing. When at least one tubing volume has been displaced, the polished rod can be lowered back down to its original position and the well restarted. Note that depending on the volume of fluid flushed and the amount of sand in the wellbore, this operation may have to be performed multiple times before the system can work the sand slug through without shutting down on high torque. One of the most successful operations to prevent sand from accumulating in the wellbore is to flush the well periodically.

In designing the PC pumping system to handle sand, elastomer wear, rod string torque, and surface horsepower requirements must be carefully considered. Generally the maximum pump rotational speed should be limited to 300 rpm to reduce accelerated elastomer and rod/tubing wear that can occur at higher pump rotational speeds.

High sand production easily can require rod string torque and surface horsepower four or five times the normal operating values. A general rule of thumb is to design the system with four to five times the normal torque and power requirements to allow the pump to work through the sand slugs.

For example, the heavy oil applications in Canada will typically have stabilized sand cuts of 5 percent or less once the well has cleaned up. The time for the well to clean up varies widely from days to months. Once the well has cleaned up, the normal operating torque for these wells might be around 200 ft-lbs, yet when the pump is working through a sand slug the rod string torques may climb as high as 1000 ft-lbs.

The systems are designed to handle the additional torque demands so that the pump does not shut down when working a sand slug through it. If the sand slug is severe enough, the rod string and surface horsepower limits may still be exceeded, resulting in the need for a remedial operation such as a flush-by.

The pump length directly correlates to the amount of torque required when producing a slug of sand through the pump. As pump lengths increase for a specific pump displacement, the more torque is required to move the sand through the pump. To increase the amount of torque available to produce a sand slug, the pump length should be kept as short as possible.

To help keep the sand fluidized and reduce the risk of sand plugging the intake of the pump, it is strongly recommended that, as a minimum, an (XXXL) extended rotor and slotted tag-bar configuration be utilized at the pump intake.

When producing sand, it must be determined whether to produce the sand to surface or allow it to accumulate in the wellbore. Typically, the goal is to prevent the accumulation of sand in the wellbore that may cover the perforations and restrict reservoir flow into the wellbore. Once this occurs, the well is normally shut in until the expensive operation of sand bailing can be performed.

Generally, when trying to produce the sand to surface the pump must be landed below the perforations. However, if the annular velocities are sufficient to lift the sand, the pump intake could be landed above the perforations. The concern with landing any equipment below the perforations is the potential to have sand accumulate around the equipment, preventing it from being easily retrieved to surface.

One of the worst scenarios would be for the well to be shut down for an extended period of time and, depending on the fluid conditions, sand allowed to settle out and accumulate above the pump. Normally when this occurs the pump cannot be restarted nor can the rotor be removed from the stator. This means that the tubing would have to be pulled and the wellbore sand bailed. It is not uncommon to have the PC pumping system shut down on high torque when attempting to work a particular large sand slug through the pump. When this occurs the normal operating procedure is to immediately perform a flush-by operation on the well.

Almost all systems designed for high sand production wells will incorporate a torque-limiting control system that will allow 100 percent of the rod string torque capacity to be exploited. In electrically driven systems, this is typically a VFD with Flux Vector Controller or equivalent. The VFD allows the operator to input the maximum allowable torque to the rod string (based on rod string load limit). When the pump starts to work a sand slug through it, the rod string torque will be allowed to climb until it reaches this maximum. At this time the VFD will start to slow the pump down so that the maximum torque is held. This generally allows the pump to slowly work the sand through without shutting down. Once the sand is produced, the VFD increases the pump speed back to the original set point.

In hydraulically driven systems this is accomplished by setting the hydraulic system pressure via the *pressure compensator screw*. For each hydraulic pump/motor combination, there is a corresponding torque vs. hydraulic system pressure/100 psi constant. For example, the F110/P98 combination with a 5.14 : 1 sheave ratio has a constant of 47.2-ftlbs/

100 psi. The maximum system pressure is 3625 psi; therefore, the output torque at 100 percent efficiency is 1710 ft-lbs. The torque capacity of 1.125 in HS sucker is 1700 ft*lbs; therefore the hydraulic system pressure could be set at the maximum (3625 psi) and still not exceed the rated torque capacity of the rod string.

When working through a sand slug the rod string torque (system pressure) may increase until the set system pressure is reached. If the pressure continues to increase, the *pressure compensator* will open and allow the power fluid to bypass through it, thus maintaining a constant pressure at the set point. As fluid bypasses through the *pressure compensator*, less power fluid is delivered to the hydraulic motor and results in a decrease in the polished rod speed. Once the sand is produced through the pump, the system pressure will decrease, reducing the amount of power fluid bypassed through the *pressure compensator*. As the amount of fluid bypassed decreases, more power fluid is delivered to the motor, increasing the polished rod speed until the preset speed is attained.

13.6 CRITICAL TUBING FLOW VELOCITY

If the tubing flow velocity is less than the critical transport velocity of the sand/coal particles, they will settle out in the tubing above the pump. Particle accumulation can create a partial or complete blockage in addition to increasing the overall fluid density. Any of these situations will cause higher pump discharge pressures that will result in higher rod string torque and horsepower requirements. If the discharge pressure continues to build, it may exceed the rod string or surface horsepower capacities and, in extreme cases, cause immediate and severe damage to the pump elastomer.

In the following example the graph line in Figure 13-5 represents the critical tubing rate for various U.S. standard sheave sizes. If tubing flow rates fall below this line, there is a strong possibility that the velocity will not be sufficient to transport that particular particle size to surface. For example, to lift a U.S. Standard Sieve #30 particle would require a tubing rate of approximately 84 BWPD.

13.7 DESIGN AND OPERATIONAL CONSIDERATIONS

In CBM reservoirs, the coal seam usually is initially fully water saturated with no free gas, and initial production is water only. As the water

CBM—Critical Sand Velocity Example

Chart Parameters

			U.S. Standard	Sieve
Flow Rate from PCP (Q)	100	bbls/day	Sieve Number	Opening
Tubing Velocity (v)	0.23	ft/s		
Critical Tubing Velocity (Vc)	0.29	ft/s		
Reynolds Number (Nre)	75		16	0.0469
Tubing I.D (Dt)	2.441	in.	20	0.0331
Rod O.D (Dt)	0.875	in.	30	0.0232
Partical Diameter (Dp)	0.0331	in.	40	0.0165
Sand Density (ds)	125	lbs/ft3	60	0.0098
Fluid Density (df)	63	lbs/ft3		
Fluid Viscosity (u)	1	cp		

For Reynolds Number 2 < Nre < 500

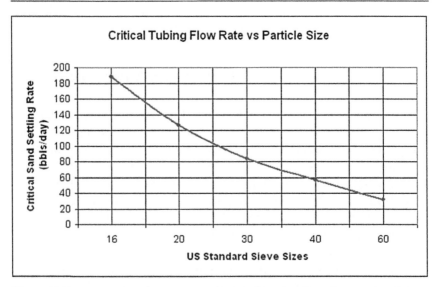

Figure 13-5: Critical Sand Settling Rate for 2 7/8″ Tubing (2.441″ ID) and 7/8″ Rods

is pumped out, the liquid level in the annulus falls, the hydrostatic pressure on the coal seam is reduced, and gas desorbs from the coal, becoming free gas. The free gas then flows into the wellbore via the created hydro-fracture and up the annulus.

If the producing bottomhole pressure is reduced too quickly (by lowering the water level too fast), gas desorption and flow into the wellbore becomes very violent, such that sand (proppant) and any contained coal fines rapidly flow into the well, up the annulus, into the pump, and up the tubing. Slugs of sand/coal fines may plug off the pump intake or bridge across or settle in the tubing/casing annulus and could prevent the removal of the tubing. Solids produced through the pump and up the tubing may settle out and bridge across the inside of the tubing if the flow velocities are insufficient to carry the solids to surface. Any of these scenarios may result in a premature failure of the production equipment.

In some CBM applications, depending on the completion method and characteristics of the coal, the rapid drawdown of the well is desirable as it creates a large cavity (cavity completion) in the coal seam that sometimes helps inflow performance.

To help alleviate the problem of sand and coal-fine production during initial production, a *low delta-P* across the wellbore coalface should be maintained. Every well is different and, depending on such parameters as the static reservoir pressure, ultimate reservoir drawdown and the well inflow performance, this operation can take several days, weeks, or months to achieve. Where feasible the following operating practices should be followed.

1. The producing bottomhole pressure should be lowered a few percent each day relative to the static reservoir pressure.
2. Continue reducing PBHP until a pressure just above the desired PBHP is achieved.
3. After the desired PBHP is achieved, allow the water and gas rate to stabilize.
4. Once the well has stabilized, the bottomhole pressure can then be lowered to the ultimate desired PBHP. In this manner a high delta-P is not created at the coalface in the wellbore.
5. The casing initially is shut in, allowing the build-up of a gas head in the annulus. This can generate high gas pressures that, if not slowly released over the drawdown period, may displace the dynamic fluid level below the pump intake. To prevent this, fluid levels should be monitored during this initial production until the gas and water production rates have stabilized.

Even with these precautions, it is common for the well to produce sand (proppant) and coal fines during the initial production period (weeks or months). As the water level is decreased, some gas will be desorbed from the coal and produced into the wellbore. To help prevent gas from entering the pump and to assist with the removal of solids that accumulate in the wellbore, it is recommended to land the pump below the lowest set of perforations.

13.8 PUMP LANDING DEPTH

CBM wells are typically cased-hole or bare-foot completions. The type of completion can have a significant impact on the pump landing depth, ultimate reservoir drawdown, pump performance, and removal of solids. For maximum reservoir drawdown, the pump intake must be landed below the target coal seam(s). It is recommended to drill 250 to 300 ft (75–90 m) below the target coal seam to provide adequate sump for logging, fracturing, and production operations.

Cased-hole Completion. With sufficient cellar, a PC pumping system can be landed below the coal seam, allowing for maximum drawdown and removal of solids from the well. A cased-hole allows the operator to incorporate a torque-anchor into the PC pumping system. This is extremely useful in high torque applications as it will prevent the tubing from backing off.

Bare-foot Completion. With sufficient cellar, a PC pumping system can be landed below the coal seam, allowing for maximum drawdown and removal of solids from the well. However, in high torque applications there is a potential for the tubing to back off unless a torque-anchor is used. As the torque-anchor has to be set in the cased-hole section of the well, this will typically limit the setting depth of the pump to some height above the coal seam and prevent the dynamic fluid level from being drawn down below this depth. The additional backpressure on the well created by the water column can have a significant impact on the gas desorption rate and inflow (water block) into the wellbore. If the pump is landed above the coal seam a gas separator or tail-joint assembly will have to be installed below the pump to assist with diverting free gas away from the pump intake and up the annulus.

13.9 RESTRICTED OR NO-FLOW SCENARIOS

One of the main concerns with CBM applications are scenarios in which the pump runs for extended time periods either dry or with high free gas production through the pump. This is generally the result of one of three scenarios that can result in a premature pump failure.

1. When the compressor shuts down or the flow line is restricted, causing the line pressure to build. This results in the fluid level being depressed below the level of the pump intake allowing gas to enter the pump.
   In order to prevent excessive build up of line pressure, a pressure control valve should be installed on the flow line. When the compressor shuts down, the control valve will maintain the predetermined line pressure set point by venting the gas to atmosphere or a flare stack. In some applications, if the pump is shut down and water is allowed to encroach into the wellbore, it may take several days for the well to produce gas even after the pumping system is put back on line. This contingency allows the pumping system to operate continually and remove water from the wellbore.
2. When the fluid level is drawn down to the pump intake and free gas enters the pump. This generally occurs when the pumping rate exceeds the in-flow rate.
3. Accumulation of solids into the wellbore restricts flow of fluid into the pump.

Today numerous options are available to prevent pumps from running under restricted and/or no-flow conditions. This typically is accomplished through software or instrumentation process control.

13.10 PRESENCE OF CO_2

The presence of carbon dioxide may cause several operating problems with PC pumping systems. CO_2 has been attributed to a specific failure mechanism in PC pumps called Explosive Decompression (ED). When an elastomeric compound is exposed to high pressure for a period sufficient for gas molecules to diffuse into the compound, subsequent rapid reduction in pressure can cause internal fracturing of the rubber. Internal fracturing may be in the form of cracks or blisters. Crack and blister growth will result in poor volumetric efficiencies and the eventual removal of rubber from the pump. Gas absorption will cause the elastomer to swell and this must be taken into consideration when sizing the pump.

The effects of ED can be minimized by

- Limiting the amount of gas produced through the pump.
- Using elastomers with a high tensile and tear strength value, and that allow gas molecules to diffuse rapidly into and out of the elastomer.
- Limiting the pressure drop across the elastomer face during normal operating and shutdown scenarios.

During normal shutdowns, installing a check-valve on the pump intake can reduce the pressure drop by restricting the movement of fluid through the pump. However, the check-valve ball and seat should be scored to allow some leakage over time to slowly dissipate the tubing hydrostatic head. This will also assist when the rotor needs to be spaced out. Failure to do this may cause the rotor hydraulic inside the stator and prevent proper space-out.

Check-valves should be used in conjunction with a soft-start controller and tubing drain valve. The tubing drain valve is located above the pump and provides the ability to drain the tubing prior to retrieving it to surface. The soft-starter provides a smooth current ramp to the motor. If a soft-start controller is not used, fluid trapped in the tubing will start the motor under 100 percent load and could result in immediate and severe equipment damage to the rod string and drive head.

Not all wells that contain CO_2 will show signs of ED. As fluid properties and wellbore conditions widely vary, it is difficult to predict if ED will occur. In addition, ED is not limited only to wells with CO_2. There have been a number of recorded cases of ED where methane gas was the cause.

CO_2 will combine with water to create a weak acid (carbonic acid), which may corrode the sucker rod and tubing strings. Corrosion inhibitors can be used to inhibit the downhole equipment, but caution must be used to ensure that any chemical that might come in contact with the elastomer is compatible. Coupon samples can be supplied so that the chemical company can conduct an elastomer/chemical compatibility test.

13.11 CORROSION INHIBITORS

The concern with applying corrosion inhibitors in PC pumping systems is that most of them are amine based. Many stators are made from nitrile rubbers (NBR) and amine is one of the worst chemicals for NBR.

Amine inhibitors can be either water- or oil-based, the main difference being the carrier. Water-based inhibitors typically use a methanol carrier, which has little impact on NBR elastomers. Oil-based inhibitors typically use a kerosene or diesel carrier that, in itself, can cause significant elastomer damage.

For these reasons, amine-based corrosion inhibitors are not recommended with PC pumping systems. It is next to impossible to predict precisely how long a PCP will perform in an environment where the elastomer is exposed to amine. As there are so many variables to consider, one cannot know the effect without actually applying the pump in the application and seeing how long it lasts. Depending on how the chemical is applied, the elastomer may see relatively low concentrations. However, since amine has a cumulative effect, prolonged exposure may ultimately result in elastomer deterioration that may lead to a premature failure.

When considering a chemical treatment program with PC pumping systems, the rod/tubing string vs. PC pump life and cost are compared to determine the overall economics on whether it is more viable to live with a shortened rod/tubing or PC pump life.

13.12 CYCLIC HARMONICS

There is an inherent harmonic cycle that can occur in PC pumping systems under certain circumstances. This harmonic cycle is dependent on many variables in the system such as size and grade of the sucker rod drive string, lubricity of the produced fluid, BHT, size of the pump, speed of the pump, rotor/stator fit, and pump landing depth.

This harmonic phenomenon is caused by the relation of the start-up, or breakaway torque, to the normal running torque. The breakaway torque is the torque required to initiate rotor rotation and is almost always greater than the running torque.

In applications where the produced fluid has little or no lubrication, such as CBM or water source wells, the difference between the breakaway torque and running torque can be more significant than in oil wells. As a result these types of wells are more prone to experience the harmonic cycles.

When a PC pumping system is started, the surface drive equipment will rotate and start to twist or wrap-up the sucker rod string, much like if you were to pull a rubber band taut, anchor it at one end, and then continually put twists in one direction into it. For example, 4000 ft of 7/8 in D rod with a 6.3 Kip axial load and 206 ft*lbs or torque will have 28 wraps. As wraps are put into the sucker rod string a torque is developed.

The number of wraps and torque will continue to increase until one of the following conditions is achieved.

1. The breakaway torque of the rotor is reached and the rotor starts to spin inside the stator.
2. The torque-limit of the surface equipment is reached.
3. The drive string breaks.

Once there is sufficient torque in the drive string to overcome the breakaway torque of the pump, the rotor will start turning. If the breakaway torque is extremely high, the sudden release of torsional energy in the drive string can result in tremendous forward acceleration of the rotor. In some cases, this forward acceleration will cause the rotor speed to catch-up to and exceed the normal operating speed.

When the rotor speed exceeds the normal forward speed, torque is no longer transmitted to the rotor and it stops inside the stator. Once the rotor has stopped the drive string must again wrap-up to generate the required breakaway torque, which starts the cycle all over. This harmonic cycle can usually be observed on surface and will likely continue until either:

1. The surface drive speed is increased so that sufficient torque is applied to the drive string to prevent the rotor from stopping during the initial acceleration and subsequent deceleration associated with the pump start up. Obviously there needs to be sufficient fluid production available to support the increased speed.
2. The drive string is replaced with a "stiffer" rod so that the torque is more directly transferred to the rotor.

When designing a PC pumping system for CBM or water source wells, it is strongly recommended that the sucker rod size be upsized over what typically would be used. Because the harmonic cycle also seems to be more prevalent in smaller diameter rod strings, it is recommend that the minimum rod string size used in any PC pump application be equal to or greater than a 0.875 in grade D rod.

13.13 PC PUMP SELECTION

In CBM applications we always assume the dynamic fluid level will be drawn down to or below the perforations (pumped-off).

Under this scenario, our design objective is to select a pump that will operate at a maximum rotational speed of approximately 400 rpm and around 80 to 90 percent of its maximum rated pressure differential. As there are no adverse effects to the elastomer, other than some water swell, the mechanical properties do not significantly deteriorate. This allows the pump to run 90 percent loaded without much risk of elastomer degradation due to hysteresis. Designing the pump to operate initially at higher speeds allows a much greater turn-down ratio of the PC pumping system as fluid rates decline.

Once the well has stabilized under pumped-off conditions, the operator can optimize the PC pumping system to increase the run time and reduce the work-over frequency. This usually results in selecting a PCP that will be operating at speeds and loading of approximately 300 rpm and 80 percent of the maximum rated pressure differential, or less.

Figure 13-6 illustrates two different pump geometries with the same volumetric displacement.

The pump on the left is a long pitched geometry used in CBM, high water cut, low viscosity fluid applications. The pump on the right is designed specifically for high viscous fluids greater than 500 cp and

Figure 13-6: Two Different Pump Geometries with Same Volumetric Displacement

should not be used in CBM applications due to the large rotor/stator interference fits required to generate the seal line and consequently the pump pressure capability.

13.14 ELASTOMER SELECTION

For several reasons the recommended elastomer of choice is a medium nitrile rubber (Buna):

- The highest mechanical properties
- Typically lower water swell than higher nitrile rubbers
- The most economical

Most manufacturers offer a number of medium nitrile elastomers; however, preference should always be given to elastomers that have the lowest water swell. At temperatures below 50°C the difference in water swell between the medium nitrile elastomers is negligible. Below 50°C pumps are typically sized 30 to 40 percent efficient at 300 rpm, tested with 30°C water.

In applications above 50°C, some elastomers have much higher swell rates and need to be considered carefully when selecting and sizing. When using conventional elastomers many different rotor sizes commonly are used to compensate for the thermal and associated fluid swell variance throughout some basins. To reduce the need for multiple rotor sizes, some manufacturers have developed elastomers specifically for CBM application that exhibit negligible water swell (<1% up to 100°C). These elastomers simplify rotor sizing requirements and therefore allow operators to use one rotor size over complete application range. Additionally some of these elastomers have improved resistance to ED.

13.15 SUMMARY

PCPs have advantages such as handling solids and viscous fluids, high power efficiency, and a relatively low surface profile. They are not tolerant of high temperatures.

PCPs are well suited for dewatering coal seam wells and can be used to deliquify gas wells, but care must be taken not to pump the fluid level to the pump and have the pump produce with only gas, even for a short time.

Longer life can be achieved using careful installation procedures, insuring sand transport up the tubing , appropriate fit between the stator/ rotor, correct pump size, and rotation speed for the application and materials selection considering the wellbore fluids and temperatures.

13.16 REFERENCE

1. Weatherford ALS Progressive Cavity Pump Manual.

COAL BED METHANE

By David Simpson
MuleShoe Engineering
David Simpson has 27 years experience in oil and gas and is currently the Proprietor and Principal Engineer of MuleShoe Engineering (www. muleshoe-eng.com). Based in the San Juan Basin of Northern NM, Mule-Shoe Engineering addresses issues in Coalbed Methane, Low Pressure Operations, Gas Compression, Gas Measurement, Oil Field Construction, Gas Well Deliquification, and Project Management. A PE with a BSIM and MSME, David has numerous articles in professional journals and has spoken at various conferences around the world.

14.1 INTRODUCTION

Methane adsorbed to the surface of coal is a very old issue with some new commercial ramifications. This explosive gas has made underground coal mines dangerous both from the risk of explosion and the possibility of an oxygen-poor atmosphere that wouldn't support life. The miner's main concern with coal bed methane (CBM) has been how to get rid of it [1]. Techniques to deal with CBM in mines have ranged from the classic canary in a cage to detect an oxygen-poor atmosphere to huge ventilation fans to force the replacement of a methane-rich environment with outside air, to drilling CBM wells in front of the coal face to try to degas the coal prior to exposing the mine to the CBM. All these techniques have met with some amount of success. None of the techniques to prevent CBM from fouling the air in an underground mine has been totally successful.

Today as expertise in developing and producing CBM increases beyond the Black Warrior and San Juan Basins, it is becoming clear that the CBM is a significant economic resource on its own, and capturing CBM for sale is often profitable even on coal seams that cannot be economically developed.

With the CBM's unique method of gas storage, the preponderance of the gas is available only to very low coal-face pressures. The coal-face pressure is set by a combination of flowing wellhead pressure and the hydrostatic head exerted by standing liquid within the wellbore. Effective compression strategies can lower the wellhead pressure to very low values. Effective deliquification techniques can reduce or remove the backpressure caused by accumulated liquid. Lowering the hydrostatic head creates suction-pressure challenges for most of the deliquification techniques presented in this book, and the successful operator must be very aware of both the minimum suction pressure needs of their deliquification technique and the backpressure requirements of the well. Getting deliquification and compression "right" can result in recovery factors in excess of 90 percent of the original gas in place (OGIP), but getting them wrong can limit recovery to less than half of the OGIP.

14.2 CBM ECONOMIC IMPACT

In 2003 CBM production was 8 percent of total gas production in the United States, 10 percent of gas reserves, and 15 percent of estimated undiscovered gas resources [5].

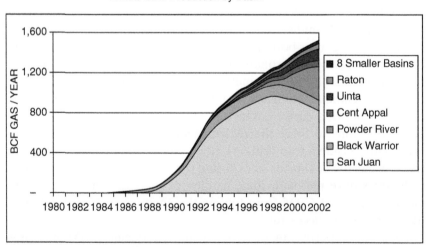

Figure 14-1: The Cumulative Production through Year-End 2002 Has Been Predominantly from the San Juan Basin

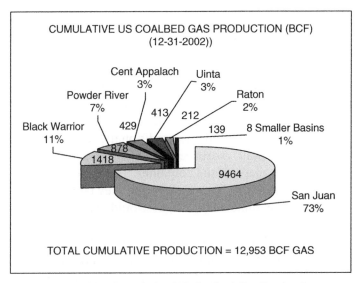

Figure 14-2: Cumulative US Coalbed Gas Production

With this clear dominance from the San Juan Basin, it is obvious that a significant portion of the data available for analysis of the lifecycle of a CBM well must come from this basin. Many things were learned in the San Juan Basin that have proven to be unique to San Juan and not applicable to the other coal plays around the world. Other things were learned through difficult and expensive lessons that can be applied to developing basins.

14.3 CBM RESERVOIRS

14.3.1 Reservoir Characteristics

Rock formations that are typically important to oil and gas production fall into three categories [2]. *Source rock* is a formation containing organic matter whose decomposition has resulted in the formation of complex hydrocarbon products. *Reservoir rock* is a formation with pore volume capable of containing commercial quantities of hydrocarbons. *Cap rock* is a formation that is largely impermeable to the flow of liquids and gases and is located such that fluids that approach it from lower or adjacent formations cannot migrate further.

In conventional reservoirs, oil or gas is made in the source rock, migrates to the reservoir rock, and its migration is stopped by cap rock. CBM reservoirs don't follow this pattern. Coal meets the criteria for source rock since the very matrix of the coal is rich in organic matter. Coal fails the definition of reservoir rock since the pore volume of a coal bed is an order of magnitude less than conventional reservoirs. Coal is often the cap rock for conventional reservoirs.

CBM is adsorbed to the surface of the coal. The adsorption sites can store commercial quantities of gas as part of the coal matrix. This must not be confused with conventional pore-volume storage. Gas within a pore-volume acts as a gas and the traditional pressure/temperature/ volume relationships hold. Adsorbed gas molecules are not gas. They don't conform to the shape of the container. They don't conform to the modified ideal gas laws (i.e., PV ≠ ZnRT) and they take up substantially less volume than the same mass of gas would require within a pore volume.

One effect of CBM not being stored in pore volume is that most of the conventional equations that describe hydrocarbon flow within a reservoir are either outright invalidated or require extensive modification. A simple example is the *Bureau of Mines Method of Gauging Gas Well Capacity* [3]:

$$q = c_p(\bar{P}^2 - P_{BH}^2)^n$$

This equation also is known as the Absolute Open Flow (AOF) equation and can be used to predict how a well's production rate (q) will change with a change in flowing bottomhole pressure (P_{BH}) assuming that reservoir pressure ($\bar{P}$) is relatively constant in the short term and both the nonlinearity term (n), and the flow constant (c_p) are reservoir properties and are constant for all pressure combinations. It has been shown [4] that either c_p or n must be changing dramatically over time to allow changing reservoir pressure to match exhibited CBM performance. Some of the changes recorded have been on the order of 30 percent increase per month.

Although the ideal gas law doesn't apply to CBM while the gas is adsorbed to the coal, there are mathematical relationships that do apply. The most important relationship to describe the way that gas is adsorbed to the coal is the *Langmuir Isotherm*, which assumes that the reservoir is at a constant temperature and defines the quantity of gas adsorbed as:

$$OGIP = 0.031214 AhV_m y\rho \frac{bP_i}{1+bP_i}$$

where:

OGIP = Original gas in place (SCF)
    A = Drainage area in ft^2
    h = Thickness of the coal in ft
  Vm = Gas content of coal (SCF/ton)
    y = Mineral-matter free mass fraction of total coal (fraction)
    ρ = Density (g/cc)
    b = Langmuir shape factor (psi^{-1})
   Pi = Initial reservoir pressure (psia)

Most of these parameters are either acquired during drilling, logging, and coring or are estimated from analogs. The equation can also yield remaining gas by substituting current reservoir pressure for initial reservoir pressure since the other parameters are reservoir characteristics that don't change substantially over the life of a well. If you plot gas in place as a percentage of the OGIP vs. reservoir pressure as a percentage of initial reservoir pressure, you get a curve like that in Figure 14-3.

For this well, region 1 ends somewhere around 60 percent of initial reservoir pressure remaining. At this point you have produced 7 percent of OGIP. This period is characterized by relative insensitivity to flowing

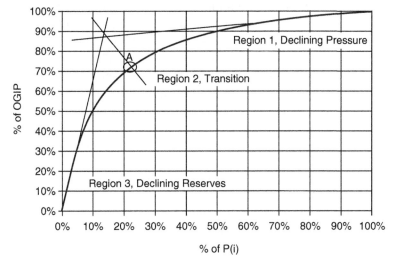

Figure 14-3: San Juan Basin Fairway Isotherm (developed from author's data and OGIP equation)

bottomhole pressure. Often in region 1, wells will not respond to wellsite compression or to deliquification techniques.

The Transition Region is a very large portion of a well's life and it is difficult to characterize. For the well in Figure 14-3, the transition period starts with 93 percent of OGIP remaining and ends with about 36 percent of OGIP remaining. During this time the reservoir pressure has dropped from 60 percent of initial pressure to 7 percent of initial— for this well that corresponds to a pressure drop from 1,100 psia to 130 psia. Region 2 is a difficult period to characterize because at the start of it the well will perform much like it did during region 1 and at the end it is acting like a late-life region 3 well. For determining necessary equipment it is useful to inscribe a straight line on the data for both region 1 and region 3—at the point that they cross draw another line from the intersection back to a point normal to the curve (point A). With point A defined you have set the place where the well will tend to stop acting like a region 1 well and start acting like a region 3 well.

Region 3 is characterized by declining reserves with fairly stable reservoir pressure. It is clear that successfully recovering this gas requires very careful management of flowing bottomhole pressure. Seemingly infinitesimal changes in either wellhead pressure or fluid level can cause significant changes in production.

The well in Figure 14-3 is a San Juan Basin "Fairway" well that had an OGIP of 28.2 BCF and an initial reservoir pressure of 1,819 psia. At the point where it leaves region 1 it had produced for only 7 months and had made 180 MMCF. It stopped acting like a region 1 well at 5 years, 5 months after it had produced 8.9 BCF. This point coincided with a peak production rate of 10.5 MMCF/d, and shortly after this peak both wellhead compression and deliquification were required to sustain production rates. It entered region 3 at 11 years, 3 months after producing 16.9 BCF and the average gas rate was still 2 MMCF/d. The last data available at the time of publication, the well had been producing 15 years, 11 months and was at 5 percent of initial reservoir pressure (91 psia), it had produced 19.2 BCF (68% of OGIP), and the production rate was 620 MCF/d.

The well in Figure 14-3 is only unique in the San Juan Fairway because of the amount of high-quality data that was collected over its entire producing life. Other than that it is a fairly average San Juan Fairway well, but is not necessarily representative of CBM projects in other basins. When other basins have moved farther down their isotherms it will be possible to verify the late-life procedures that will be necessary in those fields.

14.3.2 Flow within a CBM Reservoir

A conventional reservoir is assumed to be more or less homogeneous and gas will tend to flow to a wellbore from 360° in a pore-to-pore Darcy flow. CBM has minimal porosity and limited communication from one micro-cleat to the next. The best flow-communication through the cleat system is along "face cleats," which typically run vertically and tend to align themselves from cell to cell along the axis of maximum external stress on the coal. Butt cleats intersect the face cleats at 90° in the direction of least external stresses, they do not tend to align, and they typically have a minimal contribution to gas production.

Commercial flows within a CBM reservoir generally rely on a system of "channels" that have been created within the coal seam either through geological movements or designed stimulations. Wells where the stimulations have been aligned to cross the maximum number of face cleats have consistently outperformed wells that ignore the lay of the coal stresses.

These channels are analogous to pipe flow, and the pressure traverse from the wellbore out to considerable distances can be essentially constant. Since desorbtion is a pressure-swing phenomenon, the extent of these channels has an overriding effect on flow rates. If the system of channels and channel-branches is extensive and robust, then the flow rate will be high. On the other hand, without these channels the well's ability to reflect low wellbore pressures deep into the reservoir will be limited and the production rate and/or total recovery will be low.

There is limited empirical evidence in at least two basins that injecting gas into the coal (e.g., for CO_2 sequestration) will tend to scour channels that then can be used to produce CBM at rates significantly higher than the injection-well produced at prior to the injection. This has been shown in both the San Juan "Type II" wells (i.e., wells in the San Juan Basin with moderate gas production, high water production, and fairly low CO_2) and in the Piceance Basin, but operators see it as too expensive to inject high pressure gas at high flow rates for stimulation.

14.3.3 CBM Contamination

Adsorption sites can hold any molecule that fits. The typical adsorbtion-site diameter in CBM fields is perfectly suited to CO_2 molecules, but methane and nitrogen also fit reasonably well. Heavier hydro-carbons do not fit well on the sites and it is rare to see significant hydrocarbon-concentration heavier than methane adsorbed to the coal (although interbedded sand-lenses are reasonably common and they can hold any gas or liquid that their pore structure can accommodate).

CO_2 is a very common contaminate, and estimating both peak CO_2 and a CO_2 production profile is a difficult but useful exercise. In the Fairway of the San Juan Basin there were a very limited number of pressurized cores taken during initial field development, but all of them showed a bulk CO_2 in the neighborhood of 18 percent. Initial production from these wells mostly had 6 to 8 percent CO_2 and the question was "where is the missing 10 to 12 percent?" There is no definitive answer to that question, but close observation of the CO_2-levels over most of the life of several wells has yielded some clues. The CO_2 increased gradually over time toward 25 to 29 percent, and then started to decrease. This would suggest that the actions of drilling, completing, and producing the gas has uncovered a very large number of unsaturated adsorption sites near the wellbore by fracturing the coal. These new sites would preferentially adsorb CO_2 because it is a better fit, and you would be effectively filtering CO_2 from the combined 82 percent methane, 18 percent CO_2 stream. As pressures come down, the disproportionate CO_2 content in the near-wellbore will cause CO_2 production to increase with time. As operators anticipated very-low-pressure and vacuum operations, there was a real concern that the CO_2 would continue to increase until a large quantity of gas would have to be abandoned in place because it would become uneconomic to treat it.

A careful material balance in the San Juan Fairway has indicated that the "missing" CO_2 in the early days has a total mass that can be determined. Also, if the original portion of CO_2 in the reservoir is known, then the total mass of CO_2 in the reservoir is also known as a fraction of the OGIP. This analysis showed that there is a point where the missing CO_2 has been recovered and the production stream should drop back toward the *in situ* CO_2 level of 18 percent. This peak was predicted at 25 to 29 percent and has been seen in a number of wells.

14.3.4 Coal Mechanical Strength

An important point that is common to all CBM plays is that coal is soft. The friability (i.e., the ability of the material to resist shear forces) of coal has been reported as low as 15 psi in the San Juan Basin Fairway; other coals are stronger, but still very weak. This characteristic is responsible for the observed fact that significant quantities of coal solids are produced along with the CBM. It also accounts for the observed tendency of a coal bed to "heal" around an inclusion. For example, you can force a shovel into a coal face, but a short time later it will be very difficult or even impossible to remove the shovel. This self-healing characteristic has been observed in coal mines since the 1800s.

In the San Juan Basin Fairway, the technique of cavitation has been used very successfully to allow high-velocity gas-flow to sculpt large downhole cavities by causing the coal to fail and then transporting the failed coal to surface. Success of cavitations outside the San Juan Basin Fairway have been very limited, probably because the friability of the coals in other basins allows the coal to resist failure much longer than it can resist with the Fairway coal.

Every sort of hydraulic fracturing has been tried on CBM wells, and most of them have had both successes and failures. It is likely that many of the failures have relied on small liquid volumes and large propant volumes and the coal was able to heal itself around the propants.

14.4 CBM PRODUCTION

During the time that a well's reservoir pressure is above Point A on Figure 14.3, it is fairly forgiving. Flowing bottomhole pressure needs to be below the pressure corresponding to Point A in Figure 14-3 (400 psia for the well in Figure 14-3), but efforts to drop it significantly below the Point A pressure will have minimal impact on production rate. For the well in Figure 14-3, early-life flowing-tubing pressure was around 160 psia so the well could tolerate about 500 ft of water above the formation and still produce approximately its maximum rate. During the region 1 period, deliquification and/or compression may be required to get the flowing bottomhole pressure below Point A, but often it is installed unnecessarily because the operator says "that is the way we do it."

After Point A, managing flowing bottomhole pressure is critical to maximizing ultimate recovery. Since most wells reach Point A with more than 70 percent of OGIP remaining, it is worthwhile to build initial wellsite facilities with the anticipation of low-pressure operations. Separation equipment should have "blowcases" to allow low-pressure liquid to be boosted easily to pressures required to enter tanks or pipelines.

Gas pipe should have appropriate manifolds to allow inexpensive installation of wellsite compression. The wellsite should be set up to facilitate whatever sort of deliquification strategy you have adopted. For example, running a single line from the wellhead to the separator will require that any pumped liquids be commingled with produced gas at the wellhead for the run to the separator—this sort of multiphase flow is very energy inefficient.

Single-digit flowing bottomhole pressures are achievable in CBM wells, but pressures that low require you to understand and minimize every tiny pressure drop up the wellbore and across the location.

Prior to any well-specific decisions you should have a couple of detailed plans—you need to know how you are going to get water off the formation when the reservoir pressure is under 100 psig, and you need to know what your gathering system pressures are going to be (and how they are going to be maintained). These strategies should be clearly documented and available to everyone making drilling, completion, deliquification, or facilities decisions.

14.4.1 Deliquification Plan

In a field like Horseshoe Canyon in Alberta, Canada, you can be certain that a well will never need help getting liquid to surface; then it is reasonable to design wells with small casing. If you think there is a reasonable chance the well will need deliquification equipment sometime in its life, then you will need more real estate downhole. Doing this analysis prior to spudding the first well goes a long way toward a wellbore design that will work for the life of the well. For example, one of the major operators of the San Juan Basin Fairway completed a large number of wells with cased-and-frac'd completions using small casing, and the wells significantly underperformed relative to offsets. The company revisited their plans and decided to sidetrack many of the wells and redrill many others with larger casing and cavity completions. Production increased significantly. Late in the life of the wells the sidetracked holes presented a difficult deliquification problem.

The deliquification plan should have three parts:

- Initial deliquification
- Mid-life deliquification, and
- Late-life deliquification.

Initial Deliquification

Start-up water can be a very large volume. The discovery well in the San Juan Basin Fairway free-flowed 1,600 bbl/day of water for almost six months before the operator gave up and plugged the well. The replacement well flowed 500 bbl/day for several months before settling down to 5 bbl/MMCF for many years. Many wells don't have the gas rate required to move that kind of water volume and require help. Some operators have had good success running a rod pump or an ESP to pump the well for an initial period until the initial flush-production of water is finished.

The initial deliquification plan needs to consider the equipment will be used if a well needs help getting started, the motive power for that equipment, and what will be done with possibly excessive water volumes. The equipment that will be used presents a minimal risk that technology will pass it by since this period begins immediately. Choice of some sort of plunger pump, jet pump, PCP, or ESP should really be based on expertise within the company and local support—all of them have a good potential to work. This equipment choice should be an important part of the well's casing and wellhead design.

Gas flow during initial deliquification can be intermittent and fairly low. Relying on well production to supply fuel to an engine-driven rod pump might be considerably less effective than you would hope. Pulling fuel gas from a nearby pipeline or nearby producing well have both worked very well. Bringing electric power to a wellsite often has the benefit of providing control options (see later) and cathodic protection that would be more difficult without electricity. Regardless of the source of power/fuel, the plan needs to include an explicit description of that power source and should have contingency plans if the primary power/fuel source is blocked by external forces.

Mid-Life Deliquification

As the well approaches Point A on the Langmuir Isotherm, the well will usually still flow into fairly high line pressures, but the flow rate will often have dropped to the point where some water management is required. One important part of mid-life water management is wellbore configuration. An operator should ask "why is this here?" and "can that job be done with a smaller pressure drop?" about every component of the downhole and wellhead equipment. If there is tubing in the well, you need to know why. Often operators say it is in the well for water management, but then put a check valve on the tubing flowline. A check valve requires some small amount of differential pressure to open, so if you are flowing gas up both the tubing and the tubing-casing annulus then it is unlikely that any gas will flow up the tubing against the resistance of the check valve. For example, a 3,000 ft deep well with 2-3/8 inch tubing inside 7-inch casing is flowing 1 MMCF/d into 100 psig wellhead pressure through both the tubing and the annulus with no water production. If all the gas is flowing up the annulus, then the pressure loss due to friction should be on the order of 0.1 psi—about one-quarter the pressure required to lift the check valve, and nothing is going to flow up the tubing. On the other hand, if you tried to flow the same 1 MMCF/d just up the 2-3/8 in tubing, the friction drop would be on the order of 575 psig. One possible strategy would be to run 3-1/2 in tubing, which would allow 1 MMCF/d to flow with a bottomhole pressure around 350 psig.

The key to success in the middle of the well's life is understanding the consequences of any selected option or direction. Spending time with a nodal analysis program at the design stage will help you understand how you can get a specific target pressure at the coal face when the production rate is at your target values.

Late-life Deliquification

To recover reserves from a low pressure reservoir requires very low flowing bottomhole pressure. A consequence of very low pressures is that a lot of water will evaporate and move as water vapor. In many wells evaporation will be adequate by itself to remove all the water inflow. Late in the life of all CBM wells, evaporation will represent a significant water volume. If a pump is set up for a given water rate, and half that rate moves by evaporation, then the pump will begin to experience difficulties such as gas locking or cavitation. Again, the issues can be successfully managed as long as they are fully understood and anticipated.

Typically, the late-life plan will specify significantly smaller fluid-handling capacity and gentler equipment. For example, it is reasonable to use a standard oil-field beam unit to drive a rod pump in the initial deliquification period, but late-life deliquification seems to do better with pneumatic or hydraulic surface equipment that can provide slower strokes and can take longer to reverse rod direction. It is not clear why the gentler pump action works better, but it is reasonably clear that it does.

14.4.2 Gathering Plan

A CBM producer who manages and controls the gathering system that the wells produce into will always do better than the same producer flowing to a third party. There are two reasons for this: first, the economic analysis for system modifications will use gas sales prices instead of gathering fees; second, the incentives for steady pressures in the system are obvious and tangible to the well operator and they aren't quite so immediate to a third party. The first reason can be very significant. A project that will add 3 MMCF/d of gas selling for $6/MCF has a lot more attraction than a project that will add $1,050/day at $0.35/MCF—the payouts are much shorter and their ranking on net present value (NPV) will be much higher.

The second point is really an alignment of the well operator's goals with the gathering system. In an ideal world, the field techs that operate the wells would also be responsible for pigging and have some involvement in compressor operations. This alignment helps the tech identify when line pressure at a particular well is creeping up due to liquid in the gas line and allows the operator to organize pigging the line that is starting to have a problem. Pigging lines is hard, dirty work that no one likes. Third-party gatherers will generally run pigs on a rigid schedule or never run pigs.

These points are valid only when the gathering system is operated by the same people (or at least with the same supervisor) as the wells. In situations where a large company has a separate division that operates gathering systems, the benefits are completely lost and performance is typically worse than a true third party.

If the CBM operator also operates the gas-gathering system, then prior to field development he or she should develop a staged gathering-system plan. If the wells produce into a third-party gas-gathering system then the producer needs to develop a compression strategy.

Initial System Layout

As a field is developed, the only thing that is certain is that the production forecasts will be wrong. Occasionally, the entire field production profile can be estimated, but any particular well is subject to having significantly more or less production rate than forecast. The wells can't produce until they have a route to market. You can't know what a well will make until it has passed its initial deliquification period.

The only reasonable approach to initial system layout in a CBM field is to assume that every well will have "average" production rate for both water and gas and design the piping to accommodate those rates. It is certain that most wells will not flow at this rate, but that can't be helped. What is also certain is that late in the field's life, significant quantities of liquid will flow up the wellbore as water vapor and much of this vapor will subsequently condense in the gathering system. Every line needs a technique to remove condensation. Simple "pigging valves" are effective on lines that are 6-inch and smaller. Larger lines require more elaborate pigging facilities. In any line, removing the water will improve the efficiency of the flow and will reduce the horsepower that must be deployed to overcome parasitic pressure drops. In steel lines, removing the standing water will prevent the formation of corrosion cells and can significantly increase the life of the system.

Water Strategy

Virtually all CBM wells will produce some amount of water during their entire lifecycle. Some CBM water is quite suitable for surface discharge into rivers and streams. This is an environmental and regulatory consideration, but if the water is suitable for surface discharge, then it is better not to aggregate wellsite water but to discharge it as close to the wellsite as practical. Many small introductions of foreign water to a stream will have a much smaller impact on the stream's biology than pumping an aggregated volume at one point.

Most CBM water is not suitable for surface discharge and must be disposed of. Disposal options are outside the scope of this document, and they must be developed in consultation with environmental, legal, and engineering experts. For any disposal option the water must be transported to the disposal facility. The trade-off that must be considered within a gathering plan is "do I spend capital dollars to aggregate wellsite water or do I spend expense dollars to haul it?" The answer to this question is never simple. One approach that seems to minimize the difficulties is to install "transfer points" and pipe the water from the wellhead to these points. This technique allows efficient use of water hauling while reducing capital. With transfer points, you can install enough tankage to allow less frequent visits by water haulers with larger trucks. If water production becomes excessive (which is an economic consideration) then it may be reasonable to install pumps and run a water line from the transfer points to the disposal facilities.

Government regulators are beginning to "strongly encourage" the use of water-gathering systems instead of water hauling. Water trucks have a significant negative impact on roads, create very real risks to the public, and are very fuel-inefficient. Although water-gathering lines typically leave a more-or-less permanent mark on the landscape, the mark has a lower total impact than ongoing water-truck traffic.

Pressure Targets with Time

Dewatered early-life wells are reasonably easy to produce. They may not need any deliquification help and reservoir pressures are high enough to flow into moderately high line pressure. There is a point in the life of every CBM well that it changes from easy to very difficult. A pressure analysis over the life of the well can show that in the early days, the well can reach mainline pressure with a fairly small gathering system and a central delivery point designed for 10 compression ratios. As the well approaches Point A on the Langmuir Isotherm you may need another stage of compression to get to 40 compression ratios. Late in the well's life you could easily need 1,000 ratios or five stages of compression to get from required wellhead pressure to mainline pressure.

Since pipe loses efficiency as pressures decline, it is generally suboptimum to try to achieve very low wellhead pressures from distant central compressor stations. One approach that has been very effective in several operations is to build an initial gathering system for all the wells being average with the piping funneling toward a single compressor station. Produce the field for a year or so and develop a set of debottlenecking projects to try to equalize the wellhead pressures. As those debottlenecking projects are designed, pick sites for straddle or booster compression stations. As wellhead pressures begin approaching Point A in Figure 14-3, start designing the straddle sites. After the straddle sites are in service for a year or so, begin implementing your late-life strategy. This will be a combination of wellhead compression and (possibly) some sort of mechanical deliquification.

14.4.3 Wellbore

CBM wellbores look much like the wellbore required to produce any "dry" gas reserves. For any operation that anticipates operating at low pressure, the operator should look at each component of the downhole equipment and ask, "Why is this here, and is this the best equipment/size to do that job?" That includes everything from X-nipples to the tubing. For example, if your primary deliquification method is going to be evaporation, then any tubing at all will increase your velocities up the wellbore and unnecessarily add to the pressure drop due to friction.

Horizontal and highly deviated wells are becoming more common in CBM every year. Success with unlined horizontal laterals has been limited due to the frequency of lateral collapse caused by the weak mechanical strength of the coal. It can be demonstrated that the hoop strength of a coal bore decreases with increasing bore diameter so unlined horizontal laterals should be as small a bore-hole as production velocities will allow.

Removing liquid from low-pressure horizontal wells is a serious problem that has not been adequately solved. Some fields have had good success with orienting the lateral up dip and normal to the face cleats. This allows pumps or foamers to be set in the horizontal portion of the wellbore without imposing a large hydrostatic force on the formation.

14.4.4 Flow Lines

It can be said with a great deal of confidence that a single-phase flow line will be more efficient than a multiphase line. Consequently, a separate flow line for each potential flow stream is generally a very good idea. A well that is planned for mechanical pumping should have a flow line from the tubing and a separate line from the casing. They may both end at a wellsite separator or they may go different places, but they shouldn't be joined except in a piece of equipment that is capable of separating them permanently.

14.4.5 Separation

CBM operations generally call for a two-phase separator to remove liquid from the gas, but typically do not need to further separate the liquid into oil and water. For a small added cost, it is a good idea to provide two inlet nozzles to the separator to allow the tubing flow line to enter separately from the casing flow line. This minimizes the mixing of the two streams prior to the separator and will generally result in better liquid removal.

In anticipation of very low separator pressures late in the well's life, it is also a good idea to anticipate forcing the water out of the separator. If your late-life compression strategy includes wellhead compression then a separator with an integral blowcase is a good choice. The blowcase will accumulate liquids and periodically the compressor-discharge pressure is used to blow the liquid out of the blowcase. In the absence of wellhead compression a pump can be used in conjunction with a dump valve to pump out a chamber. The pump-chamber can physically be a blowcase, so the initial wellsite separator can easily be the same vessel for either strategy.

14.4.6 Compression

Transportation lines typically have normal operating pressure on the order of 1,000 psig or 70 bar(g). Late-life CBM wells require flowing wellhead pressures under 10 psig. Translating these two required pressures from one to the other requires compression. A combination of machines to provide over 70 compression ratios and sometimes significantly more ratios (i.e., from 14 in Hg to 1,000 psig is 144 compression ratios) indicates a range of technologies.

For example, a two-stage reciprocating compressor with a suction pressure of 100 psig and a discharge pressure of 1,000 psig is a very efficient piece of equipment. Efficiency drops considerably if you drop the suction pressure to 40 psig (increase compression ratios to 18.5 at sea level) and requires a three-stage compressor. Multistage reciprocating compressors work best with a very narrow suction-pressure range, so the variability caused by water sloshing in the gathering system or liquid-level changes in wellbores will create problems within the compressor.

An effective strategy is staged compression. With this strategy you start your field production with central compressor station suction at a fixed value. For example, you can set the station up for 100 psig suction. This station can be expected to run with this suction pressure for the entire life of the field. At some point the wells will need lower pressures and this can often be provided with single-stage reciprocating compressors located at strategic "straddle compressor sites" that were described in the gathering strategy. The straddle sites can be designed for an inlet pressure around 40 psig and a discharge pressure consistent with 100 psig inlet to the central stations. Later in the life, very low wellhead pressures can be provided with flooded-screw compressors that handle varying suction pressure very well and work effectively with line pressures consistent with the straddle site design suction pressure.

14.4.7 Deliquification

Deliquification of low-pressure CBM wells uses the techniques described in the rest of this book. The key to success is understanding the minimum Net Positive Suction Head Required (NPDH-r) for the pump you want to use. For example, a hydraulic jet pump requires approximately 300 psig pump intake pressure to prevent cavitation. Consequently, whereas an early-life jet pump can be very effective in CBM, after point A on Figure 14-3 jet pumps do not have a place in CBM operations.

One deliquification technique that is unique to a very-low pressure operation is evaporation. Refer to Figure 20-3 in the *GPSA Field Data Book* to see how much liquid water will evaporate at low pressures and moderately high temperatures. This will often be higher than the liquid inflow rate and it is possible to rely on evaporation to satisfy all the well's deliquification requirements. Phase-change scale issues that are discussed in Chapter 6 are an important consideration, but these issues vary from field to field.

14.5 REFERENCES

1. Simpson, D. A., Lea, J.F., and Cox, J.C. "Coal Bed Methane Production, SPE 80900," presented at SPE Production and Operations Symposium, March 2003.

2. Bradley, H. B. *Petroleum Engineering Handbook, Third Edition*, Society of Petroleum Engineers, 1992.

3. Stephens, M. M. *Natural Gas Engineering, Second Edition*, Mineral Industries Extension Services, School of Mineral Industries, The Pennsylvania State College, 1948.

4. Simpson, D. A. and Kutas, M. "Producing Coalbed Methane at High Rates and Low Pressures, SPE 84509," presented at SPE Annual Technical Conference and Exposition, October 2003.

5. Limerick, S. H. "Coalbed Methane in the United States: A GIS Study," Energy Information Administration, http://www.searchanddiscovery.net/documents/2004/limerick/images/limerick.pdf.

PRODUCTION AUTOMATION

Cleon Dunham, Oilfield Automation Consulting and Greg Stephenson, eProduction Solutions

Cleon Dunham, BSAE, Cornell U.,1964, joined Shell Oil Company where he worked in Facilities, Reservoir, Production, and Computer Control Engineering. He focused on automation of production operations and artificial lift, and his last five years before retirement in 2000 were spent in Shell International E&P in The Netherlands helping coordinate Shell's worldwide production automation and artificial lift. After retirement he founded Oilfield Automation Consulting (OAC, www.oilfieldautomation.com) and the Artificial Lift Research and Development Council (ALRDC, www. alrdc.com). OAC is a consulting company focused on automation. ALRDC is a nonprofit organization focused on artificial lift R&D, and industry information sharing. A primary activity is helping to organize international conferences and workshops on artificial lift methods.

Greg Stephenson is the Product Line Manager for Artificial Lift at eProduction Solutions, Inc. He directs and administers the overall product strategy for the optimization of artificial lift systems including marketing, business development, engineering development, and commercialization of new hardware and software products. Greg, BSPE Texas Tech U., has over ten years experience in the areas of production engineering, design and optimization of artificial lift systems, completions engineering, training, and product line management.

15.1 INTRODUCTION

Many gas fields have large numbers of wells. Many of these are in remote locations. Many experience liquid loading and require deliquification to obtain desired gas production rates and ultimate recoveries. The combination of these conditions can make manual surveillance of gas wells difficult, control challenging, and optimization almost impossible. For these and other reasons, many companies are turning to production automation systems to improve the management of their gas well operations.

Surveillance. Production automation systems are used to monitor gas well production. This includes measuring gas production rates, gathering related information such as pressures, temperatures, and such, and monitoring the performance of artificial lift equipment. These measurements and monitoring are used to determine gas production volumes and provide the surveillance and problem detection needed for problems to be addressed.

Control. Automation systems are used for control. This is particularly pertinent when artificial lift systems are used for gas well deliquification. Control systems are important—some would say essential—to use plunger lift systems, pumping systems, gas-lift systems, chemical injection systems, and so on. It can be virtually impossible to manually perform the necessary control in the way needed and with the timeliness required.

Optimization. And, systems are used for optimization. The goal of gas well optimization is to maximize both current production rates and ultimate gas recoveries while minimizing capital, operating, and maintenance costs. The question is: how can the minimum amount of energy and manual effort be expended to produce the maximum amount of gas, on a sustained basis?

15.1.1 Gas Well Deliquification

Later in their life, gas wells begin to load with liquid and need methods of artificial lift and other methods to remove water and other liquids so gas can flow in the presence of the loading. Beam lift needs pump off control. PCPs and ESPs need to maintain an optimal fluid level. Gas-lift requires gas injection controls and optimization. Plunger lift cycles must be monitored and optimized. If surfactants are used, application can be automated. In short, operation of gas wells, as they become liquid loaded, requires many or most of the same capabilities required to automate, monitor, and optimize oil wells.

15.1.2 Gas Well Dewatering

Production automation systems are being used by many companies; but in many cases the companies acknowledge that they are not gaining the benefits they expect. Some systems are not sufficiently reliable; some are underutilized; some are too difficult to understand. In some cases, personnel are not properly trained to use or support them.

Another important factor is that companies are becoming more sophisticated in their selection of artificial lift systems to deliquify their wells. Some wells are better served by plungers; some by chemical systems; some by wellhead compression; some by pumping systems; some by gas-lift; and so on. Each of these systems requires different surveillance, control, and optimization methods. Production automation systems are being called upon for surveillance, control, and optimization of a range of artificial lift systems. This must be in a way that is understandable and usable. There cannot be separate systems for each type of artificial lift; they must be integrated into one approach.

This chapter will cover automation equipment, general applications that are available in most production automation systems, special applications that are designed specifically for each type of artificial lift, some issues that must be considered when planning a production automation system, and finally some case histories.

Note that production automation systems are sometimes referred to as SCADA (Supervisory Control and Data Acquisition) systems. In reality true production automation systems contain much more than SCADA capabilities. They also contain information analysis, logic to diagnose problems, and production optimization capabilities.

This chapter is long, but it only scratches the surface of the overall topic of production automaton for gas operations. If someone is interested in actually pursuing an automaton project or enhancing an existing automation system, they should contact an automation Service Company or Consultant.

15.2 BRIEF HISTORY

Production automation systems, in one form or another, have been used since the 1950s. There was much development of systems for surveillance, control, and optimization of selected forms of artificial lift in the 1970s to the 1990s. However, significant advancements in automation of gas well operations have occurred only in the past few years. A brief history of this development is presented in this section.

15.2.1 Well-Site Intelligence

Since the mid-1980s, dramatic developments have occurred in the area of well-site intelligence. The development of microprocessor technology has made it practical to place devices at the well-site with capabilities similar to those of personal computers. This has opened up opportunities for real-time monitoring, control, and optimization of artificial lift systems. Numerous application-specific field devices have been developed that allow for autonomous control and optimization of artificial lift systems. For instance, in the area of plunger lift, well-site intelligence has evolved from simple time cycle-based control to sophisticated condition-based control logic to self-tuning algorithms that minimize the need for direct intervention from personnel. This has improved the viability of this lift method in remote operations. Similarly, pumping systems can now be equipped with motor controllers or variable frequency drives that can start, stop, speed up, or slow down the pump based on the real-time condition of the well. These devices can monitor the state of the well using both surface and down-hole measurements, such as pressure, temperature, load, current draw, and other parameters. With such data, the controllers are able to detect and diagnose abnormal operating conditions and take corrective action, thus protecting equipment from damage and possible failure.

In the early days of automation, host systems were able to provide fairly simple functionality such as trending of data and basic control capabilities. Most systems were custom-built for the end user at significant expense. Over time, systems were developed with ever-increasing levels of sophistication. Such systems enabled operators to detect, diagnose, and address the problems by changing operational parameters or even redesigning artificial lift equipment to better suit field conditions. In addition, software was developed that allowed operators to manage by exception. This meant that instead of reviewing every well on a daily basis, operators could focus their efforts on those wells where there was a known problem or opportunity for improvement, greatly improving operational efficiency. With the development of PC technology, these concepts were incorporated in off-the-shelf commercial software that could be run on Windows™-based PCs and be deployed with minimal, if any, need for customization. As a result, many operators who could not previously justify the expense of implementing an automation project were able to gain access to this technology and put it to use in their fields. Today, host systems continue to evolve. Where systems have traditionally focused on helping producers improve operational efficiency, new systems are being developed to help engineers maximize the value of the asset. Such systems are intended to link the reservoir to the wells to the gathering system to the facility to the sales point. By utilizing real-time enabled engineering tools, engineers can use such systems to uncover hidden performance trends and better manage the asset.

15.2.2 Communications

Early automation systems typically used hard-wired or telephone-based connections between RTUs located at the wellhead and host systems at a central location. In many cases this came at considerable expense and posed serious logistical challenges. As systems developed, operators migrated to other communication technologies such as microwave, spread-spectrum, and licensed radios. This solved many of the logistical challenges faced in the field, but also required considerable capital investment. Over time, other communications options have opened up, including cellular (CDPD/CDMA), satellite, and fiber optic. Each of these has proven to be a key enabler. Cellular has been particularly useful in automating remote fields with little infrastructure, due to its ease of installation, low cost, and broad coverage. Unfortunately, not all locations in the world have cellular coverage. For such areas, satellite communications have proven to be a useful tool, but also bear a considerable price. In assets where operators wish to transmit extremely large quantities of data and desire maximum band-width, fiber optic is proving to be another useful, yet expensive tool. There are pros and cons to each of the communications options; with so many options available, however, it is now possible to find a fit-for-purpose communications solution for almost any application in the world.

15.2.3 System Architecture

Early automation systems typically consisted of a number of simple RTUs deployed at well sites and connected via hardwire to a central host system on a mini computer. These were custom-built installations that required considerable up-front development and support. Over time, technologies have evolved that have replaced the various components of these systems and offer a variety of options to operators. In many cases, operators deploy off-the-shelf RTUs, which communicate via licensed or spread-spectrum radio to a host system on a Windows™ server. Users then interface with this system using client software or via company intranet using an Internet browser. In many corporate environments, operators deploy large-scale SCADA systems that utilize RTUs and PLCs that communicate directly with a distributed control system (DCS) and archive data in historians for future retrieval. Such systems may even communicate with other enterprise data sources to gain access to well testing, accounting, or other data. In many gas deliquification projects—where well sites are remote, minimal infrastructure is in place, and it is important to minimize cost—increasingly operators are choosing to go another route. In such applications, web-hosting services are proving to be a popular option. Web-hosting allows operators simply to connect a cellular radio or satellite transmitter to the well-site RTU and transmit data to a third-party service. Operators then are able to view their data in an Internet browser with a preconfigured interface. The entire automation infrastructure is managed by the third-party web-hosting service and operators pay a low monthly fee. Although this is not a practical solution in every part of the world or in every corporate environment, it is proving to be a popular choice for operators of gas deliquification projects, particularly in North America.

15.3 AUTOMATION EQUIPMENT

Automation equipment consists of the hardware and software used to implement production automation systems. There are many components and many suppliers of these components. The purpose of this section is not to evaluate or judge the various brands or suppliers, but to provide information about the equipment that is available and some insights into what works well.

15.3.1 Instrumentation

The core components of any production automation system are the instruments used to measure gas production variables of pressure, temperature, flow rate, and so on. In addition, special instruments are needed to measure variables required for some artificial lift systems. For example, the load on the polished rod and the position of the beam are required measurements for sucker rod pumping systems. These special instruments are discussed in the appropriate sections.

Several technologies are available including analog current instruments, direct current voltage instruments, digital instruments, instruments designed to work with Foundation Fieldbus, and others. In general, all transmitters that are obtained from reputable companies are rugged and reliable, and all are reasonably priced.

Pressure Transmitter

Figure 15-1 shows a typical pressure transmitter that measures gauge pressure and transmits a 4–20 mA (milli-amp), 1 to 5 Vdc (volt direct current), or digital output signal. It is used to measure tubing (production pressure), casing pressure, line pressure, separator pressure, and so on.

The signal is transmitted to a remote terminal unit (RTU) or a programmable logic controller (PLC) (see Section 15.3.4) where it is converted into engineering units of psi, kPa, °F, °C, MCF/Day, M^3/Day, and so on.

Differential Pressure Transmitter

Figure 15-2 shows a typical differential pressure transmitter. It is used to measure the pressure drop across an orifice meter or similar device. For gas well production, this often is used to measure the gas flow rate.

Figure 15-1: Pressure Transmitter

Figure 15-2: Differential Pressure Transmitter

Temperature Transmitter

Figure 15-3 shows a typical temperature transmitter. It is used to measure the temperature of the produced gas. For gas well production, the temperature of the gas must be known to accurately calculate the flow rate.

Multivariable Transmitter for P, DP, and T

Figure 15-4 shows a typical multivariable transmitter that measures pressure, differential pressure, and temperature with one device. This device may reduce overall cost since only one device must be installed to perform the functions of three separate measurements.

Transmitters can use analog signals, voltage signals, digital signals, Foundation Fieldbus, or other methods (see Table 15-1); analog current and digital are the most common.

Figure 15-3: Temperature Transmitter

Figure 15-4: Multi-Variable Transmitter for P, DP, and T

Table 15-1
Types of Signal Outputs

Type of Signal Output	Description
Analog Voltage	The output voltage is a simple (usually linear) function of the measurement.
Analog Current	Often called a transmitter. A current (4–20 mA or any other analog current output) is imposed on the output circuit proportional to the measurement. Feedback is used to provide the appropriate current regardless of line noise, impedance, etc. This output is useful when sending signals long distances.
RS232/RS485	The output of the transmitter sends out a serial communications signal.
Parallel	A standard digital output protocol (parallel) such as a printer port, Centronics port, IEEE 488, etc.
HART® Protocol	HART® (Highway Addressable Remote Transmitter) is a method of transmitting data via Frequency Shift Keying on top of the 4–20 mA process signal to allow remote configuration and diagnostic checking. HART® is a registered trademark of the HART Communication Foundation.
PROFIBUS	PROFIBUS is an open Fieldbus standard for use in manufacturing and building automation, as well as process control.
DeviceNet	Utilizing CAN protocol, DeviceNet is a network designed to connect industrial devices such as limit switches, photoelectric cells, valve manifolds, motor starters, drives, and operator displays to PLCs and PCs.
Foundation Fieldbus	Fieldbus or Foundation Fieldbus is a generic term used to describe a common communications protocol for control systems and/or field instruments.
Ethernet	A very common method of networking computers in a LAN. Ethernet will handle about 10,000,000 bits-per-second and can be used with almost any computer.
Analog Frequency or Modulated Frequency	The output signal is encoded via amplitude modulation (AM), frequency modulation (FM), or some other modulation scheme such as sine wave or pulse train, but the signal is still analog in nature.
Special Digital (TTL)	Any digital output other than standard serial or parallel signals. Simple TTL logic signals are an example.
Switch/Alarm	The "output" is a change in state of a switch or alarm.
Other	Other unlisted, specialized, or proprietary outputs.

Types of Signal Outputs from Transmitters

Analog and voltage instruments require a pair of wires between each instrument and the RTU or PLC. Digital and Fieldbus instruments can be "daisy-chained" with several instruments on one cable. Some systems can use wireless communications between the instrument and the RTU or PLC. Since there are many choices, a trained Instrument Engineer should evaluate the options and recommend the right system for each application.

15.3.2 Electronic Flow Measurement

Often, the custody transfer point in gas deliquification applications is at or near the wellhead. For this reason, there is an additional requirement in these applications to provide custody transfer quality gas measurement. To address this need, the industry has adopted a measurement standard from the American Petroleum Institute, called API MPMS 21.1. This standard sets forth requirements for the measurement of gas flow rates as well as storage and transmission of such data. A range of RTUs exist in industry that are designed, tested, and certified to comply with this standard. These devices are commonly referred to as Electronic Flow Measurement (or EFM) devices.

System Description

An EFM system consists of three major elements, defined as *primary*, *secondary*, and *tertiary* devices, respectively. Primary devices refer to the meter itself. These could be orifice, turbine, venturi, or any other form of gas flow meter. Secondary devices include electromechanical transducers that convert the physical inputs of the meter (i.e., pressure, temperature, differential pressure) into an electrical signal. Tertiary devices refer to the flow computer, which takes the electrical inputs from the secondary devices and uses them to calculate a flow rate.

Algorithms

In addition to differential pressure metering algorithms such as those defined in A.G.A. Report Numbers 3, 5, 7 and 8, EFM devices must also perform algorithms that account for the effects of sampling and calculation frequency during periods of fluctuating flow.

Sampling Frequency

In general, an EFM device must sample data from end devices once every second. However, there are exceptions. If the RTU collects data at a frequency that is greater than once per second, these inputs may be averaged using techniques specified in API MPMS 21.1. Also, if sampling frequency is slower than once per second, these values may be used if it can be demonstrated that the difference in uncertainty between the slower sampling rate and one-second sampling rate is no more than .05 percent.

Data Availability

EFM devices are required to collect and retain a minimum amount of data to ensure that gas flow rate calculations are performed accurately, and to provide an audit trail of system operation and quantity determinations. These devices are generally expected to retain hourly averages of all key values as well as the associated configuration parameters and totalized values for each gauge-off (e.g., 24-hour) period.

Audit and Reporting Requirements

EFM devices provide an audit trail in the form of daily and hourly quantity transaction records, algorithm identification, configuration logs, event logs, corrected quantity transaction records, and test records for the metering equipment. This audit trail provides support for the current and prior quantities reported on the measurement and quantity statements as well as the ability to make reasonable adjustments when gas measurement equipment has stopped working, is deemed to be out of calibration, or in cases where parameters were incorrectly entered into the RTU.

Equipment Installation

All EFM equipment is required to be installed in a manner that is consistent with the practices described in API MPMS 21.1. Affected equipment includes the transducers (or transmitters), gauge lines, RTUs, communications, peripherals, and cabling.

Equipment Calibration/Verification

An EFM system is required to be calibrated such that the system as a whole will provide no more than +/−1% uncertainty over the expected range of temperatures and pressures for the installation. EFM components requiring calibration/verification include static pressure transmitters, differential pressure transmitters, temperature transmitters, pulse generators and counters, online analyzers, and densitometers. These calibrations must be performed once per quarter.

Security

All EFM systems are expected to provide specific safeguards pertaining to access, integrity of logged data, algorithm protection, protection of original data, memory protection, and error checking.

15.3.3 Controls

Automatically Controlled Valves and Accessories

Automatic control valves are used in a variety of gas deliquification applications including plunger lift, gas lift, hydraulic lift, and well testing. These devices generally are classified as either fluid-operated or electrically operated. Generally, fluid-controlled valves are either diaphragm operators or fluid cylinders. In automation applications, these devices are equipped with transducers or related equipment, which allows them to accept the various inputs and protocols described in Table 15-1.

Fluid Controlled Valves

Generally, fluid cylinder operators are used in valves requiring a 90° bend, and diaphragm operators are used in valves that have angle, butterfly, globe, or Saunders-style valve bodies.

A variety of fluids may be used to actuate fluid controlled valves. Generally, natural gas is used in oilfield applications. However, other fluids may be used in cases where a suitable natural gas source is not available. These include compressed air, nitrogen, or hydraulic fluid.

Figure 15-5 shows a typical on-off style globe valve. These are commonly used as dump valves on separators or flowline valves in plunger lift applications.

Figure 15-5: On-Off-Style Globe Valve

Figure 15-6: Throttling-Style Globe Valve

Figure 15-6 shows a typical throttling style globe valve. These are used in applications requiring variable control of through-put such as gas lift injection.

Figure 15-7 shows a typical electro-pneumatic transducer. These are used in conjunction with automatic control valves and use an analog input (generally 4–20 mA) and convert this to a proportional pneumatic pressure output in order to adjust the position of the valve.

Figure 15-8 shows a globe valve equipped with an Electro-Pneumatic transducer for actuation.

Electrically Controlled Valves

Two general forms of electric operators are commonly used in oilfield applications. These are generally classified as electric-solenoid and electric motor operators.

Electric-solenoid operators are used to adjust the longitudinal motion of a valve stem and generally are limited to valves of 2″ diameter and smaller. Electric motor operators are used in a variety of valve types, but generally require additional accessories to be installed such as torque

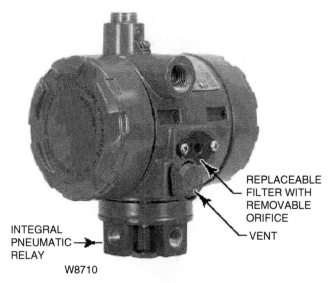

INTEGRAL
PNEUMATIC
RELAY

REPLACEABLE
FILTER WITH
REMOVABLE
ORIFICE

VENT

W8710

Figure 15-7: Electro-Pneumatic Transducer

Figure 15-8: Globe Valve and Transducer

limiters or limit switches to prevent damage to the unit. In addition, electric motor actuators require the use of a rack and pinion assembly to convert the motor's rotary movement to longitudinal displacement.

Production Safety Controls

A variety of devices are used in oilfield applications to ensure that equipment operates under fail-safe conditions. These devices are commonly referred to as production safety controls. Typical production safety controls include high-pressure/low-pressure safety shut-in valves, excess flow valves, pressure relief valves, pressure and temperature switches, and pump-off controls.

Motor Controllers

Motor controllers are devices that regulate the operation of an electric motor. In artificial lift applications, motor controllers generally refer to those devices used in conjunction with switchboards or variable frequency drives to control the operation of the prime mover. Motor controllers often include a manual or automatic means for starting and stopping the motor, selecting forward or reverse rotation, speeding up or slowing down, and controlling other operational parameters. In addition, motor controllers can provide protection for the artificial lift system by regulating or limiting the torque, and protecting against overloads and faults. Many motor controllers contain additional capabilities such as data collection and data logging as well as application-specific control logic.

Figure 15-9 shows a typical motor controller for electric submersible pumping applications. This device receives and displays data from downhole gauges as well as surface electrical parameters. This data can then be used to adjust the operation of the pump according to changing conditions.

Figure 15-9: Artificial Lift Motor Controller

Switchboards

The switchboard is basically a motor control device. The switchboards range in complexity from a simple motor starter/disconnect switch to an extremely sophisticated monitoring/control device.

There are two major construction types: electromechanical and solid state. Electromechanical construction switchboards provide basic over-current and under-current protection to the artificial lift system. Monitoring these features allows for protection of the artificial lift system from damage caused by conditions such as pump-off, gas lock, tubing leaks, and shut-off operations. The solid-state switchboards incorporate a solid state motor controller that allows more elaborate and accurate protection from a much greater list of potential problems. In addition, most solid state controllers incorporate data logging functions.

A valuable switchboard option, particularly in ESP operations, is the recording ammeter. Its function is to record, on a circular strip chart, the input amperage to the prime mover. The ammeter chart record shows whether the unit is performing as designed or whether abnormal operating conditions exist. Abnormal conditions can occur when a well's inflow performance is not matched correctly with pump capability or when electric power is poor quality. Abnormal conditions indicated on the ammeter chart record are primary line voltage fluctuations, low current, high current, and erratic current.

Figure 15-10 is an example of a motor switchboard for an electric submersible pump.

Figure 15-10: Motor Switchboard

Variable Frequency Drives

A variable frequency drive changes the capacity of the artificial lift system by varying the motor speed. By changing the power frequency supplied to the motor and thus motor RPM, the capacity of the pump is changed in a linear relationship. Thus, well production can be optimized by balancing flow performance with pump performance. This applies to both long-range reservoir changes and short-term transients such as those associated with high-GOR wells. This may eliminate the need to change the capacity of a pump to match changing well conditions, or it may mean improved run life by preventing cycling of the system. This capability is also useful in determining the productivity of new wells by allowing evaluation and measurement of pressure and production values over a range of drawdown rates. The change in frequency can be made manually or automatically. The VFD can automatically adjust the operating frequency to maintain a target pressure, flow rate, current, or other set points when operating in a "closed loop" mode.

Figure 15-11 is an example of a variable frequency drive for progressing cavity pump applications. This device receives both electrical and other production data and can use this information to change pump speed to optimize performance or prevent damage to the artificial lift system.

Figure 15-11: Variable Frequency Drive

15.3.4 RTUs and PLCs

RTUs

A Remote Terminal Unit (or RTU) is an electronic device utilizing a microprocessor, which links objects in the physical world with an automation system. This is accomplished by transmitting telemetry data to the system and/or changing the physical state of connected objects based on control messages received from the automation system. RTUs share many common characteristics with PLCs, but in general, tend to be designed to handle a smaller number of points and will often contain application-specific control logic.

One way of looking at an RTU is as a small computer sitting at the wellsite that is ruggedized to handle field conditions and has input and output capabilities for talking to the field equipment. Many RTUs have been customized with application-specific control logic to allow them to perform specific functions in the field. Examples of these include rod pump controllers, plunger lift controllers, data loggers, and a variety of other application-specific devices.

An RTU is comprised of several major components. These include: (1) a communications interface, (2) a microprocessor, (3) nonvolatile memory, (4) environmental sensors, (5) override sensors, and (6) a bus, which is used to communicate with devices or interface boards. This bus is commonly called a field bus or device bus.

A variety of standards (or protocols) are used to communicate with RTUs. These include both generic and proprietary protocols. Perhaps the most widely used generic protocol is MODBUS. Others include ODBC, OPC, and ISO Controller Area Network (ISO 11898). Examples of proprietary protocols include Weatherford's Baker 8800 protocol and Allen-Bradley's data highway.

RTUs can have a number of different types of interface boards. These interface boards can be either digital or analog and can come with inputs only, outputs only, or a combination of the two. These main types of interface boards often are abbreviated AI (Analog Input), AO (Analog Output), DI (Digital Input), or DO (Digital Output). Interface boards are connected to physical objects using wires.

RTUs often have application-specific logic programmed into firmware and/or software. This control logic, sometimes referred to as "wellsite intelligence," allows for autonomous control based on changing conditions, without the need for instructions from a host system. Such control logic is useful for executing functionality that is data-intensive, time-sensitive, or is required for fail-safe operation of equipment. Such functions would not generally be practical to carry out remotely from a host system. Examples of RTU-based control logic include PID loops, which measure and maintain a given flow rate by adjusting a valve's position; gas measurement; pump-off control; data logging; and a variety of others.

Figure 15-12 is an example of a generic RTU. Typical of many RTUs, this device contains a microprocessor, multiple communication interfaces, support for eight AIs, two AOs, eight DIs, and eight DOs. This device is typical for a stand-alone, single well control application. Also pictured are the associated instrumentation and cabling, solar power array, battery back-up, and radio.

Figure 15-12: Remote Terminal Unit

PLCs

A Programmable Logic Controller (PLC) is a digital computer that is designed specifically for the automation of industrial processes. PLCs have the same basic components as RTUs, yet differ in both form and function. Although they share many characteristics with RTUs, PLCs tend to be more scalable, interface with more end devices, and are less likely to contain customized control logic. Typical oilfield applications for PLCs include such tasks as automatic well testing, scanning multiple end devices, and other process control tasks. In a typical oilfield automation system, PLCs may be used to interface with a number of RTUs to collect data or adjust set-points in the controllers (operating the RTUs in a master-slave relationship) and, in turn, communicate that data to a Distributed Control System (DCS).

PLCs originated in the automotive industry in the late 1970s. They were a replacement for the relay logic used to control machinery. Their advantage over the relay logic was that they were programmable and that the program could be changed relatively easily; relay logic is hard-wired, takes up significant space, and is not easily changed. The ladder logic language, which is still popular in PLCs, is the same as the ladder logic drawings used for relay logic wiring and hence is well understood by electricians.

PLCs were originally very large, expensive, and suitable only for large manufacturing plants. They were capable only of binary logic (no analog) and were aimed at machine control. Early PLCs had minimal communications functionality beyond providing a port to plug in the

Figure 15-13: Control Panel with PLC

programming terminal. The operator interface was mostly provided by switches and lights hard-wired to PLC I/O. Over time, PLCs were developed with communications ports to allow them to talk to one another and to provide computer-based operator interfaces, giving rise to the MODBUS protocol. Today, both PLCs and RTUs have evolved to a point where, in many cases, the lines have blurred between the different devices, making the distinctions less meaningful.

Figure 15-13 is an example of a typical PLC-based control panel. Typical of most PLCs, components are rack-mounted and are scalable, in that they allow for the addition of blocks of I/O or processors.

15.3.5 Host Systems

The "host" computers in production automation systems provide many important functions. These are covered in Sections 15.4 and 15.5.

General Automation Systems

There are companies that make "host" systems for the general automation market. These systems, as they come from the factory, provide most of the general applications discussed in Section 15.4, but don't contain the unique applications described in Section 15.5. It may be possible to add some of the unique applications at a cost, but usually the supplier of the "host" system is not able or interested in doing this; it will be necessary to use an independent software supplier. The general applications will typically need to be configured to meet the specific requirements of each location, but they don't need to be developed and tested from scratch.

Equipment Specific Systems

Other companies produce "host" systems specifically designed for the oil and gas production industry. These systems will typically contain most of the general applications and some (most won't contain all) of the unique applications. In some cases, the unique applications are designed to work with specific RTU/PLC logic and capabilities and/or artificial lift systems that are provided by the same company; and they may or may not support RTU/PLC logic or artificial lift systems produced by other companies.

Home Grown Systems

A third category of "host" systems are produced by the operating companies themselves. Very few operating companies develop their own systems. But when they do, they tend to focus on the specific types of production equipment, artificial lift systems, and production automation equipment that are important to them.

Generic Oil and Gas Systems

There are a few companies with the primary business of developing "host" production automation systems. To make their systems attractive to a wide range of customers, they try to support as many different types of production equipment, production automation components, and artificial lift systems as possible. It is difficult for one company to be expert and provide good capabilities for all forms of production and artificial lift systems, but some do a reasonable good job of this.

It may be difficult for an operating company to know what type of system to choose. Often it may seem attractive to use a "host" system provided for the general automation market. However, this may not be wise in that it may not be possible to obtain the unique applications that may be of significant value in gas well production. Or, it may seem attractive to buy a host system that supports specific artificial lift equipment. This should be scrutinized carefully if there is a likelihood of using other types of artificial lift, or even artificial lift systems from other suppliers. Since most operating companies aren't going to develop their own system, the best approach may be to work with a company that focuses on building generic host systems for oil and gas production.

15.3.6 Communications

Communications are required at several levels in production automation systems:

- Between the instruments and controllers and the RTU or PLC
- Between the RTU/PLC and the host automation system
- Between the host automation system and the general user community
- Between the host automation system and other computer systems
- Between the other computer systems and the user community

Instrument to RTU

Communication between the instruments and controllers and the RTU or PLC is normally over a twisted pair cable. This cable may be placed in conduit. Often, many pairs of wires are installed in a large cable that connects from the RTU to several instruments. Normally, as in the case of analog current or voltage signals, a single pair of wires goes directly from the instrument to the RTU or PLC. In some cases, as in the case of digital or Fieldbus transmitters, a single wire may be connected from the RTU or PLC to many instruments or controllers. In a few cases, wireless communications are used.

This communication is one of the weakest links in the production automation system. Wires may be cut, damaged, or shorted. Fortunately, it is easy to tell if there is a communication outage. For example with analog current transmitters, a value of 4 mA represents a zero (0) value of the signal being measured. If the analog current signal goes to 0 mA, this signifies a communication outage. Similar indications exist for voltage, digital, and Fieldbus signals.

RTU to Host

There are many alternatives for communicating between the RTUs or PLCs and the host automation system. These include hardwire, radio, microwave, spread spectrum, satellite, and others.

When low-speed communications (up to 9,600 bits per second) are used, it is common for the Host system to poll the RTUs for information. A typical polling frequency might be once every 15 or 20 minutes. Normally, the host would ask the RTU for its status. If all is OK, the response is short. If there are problems, the RTU may respond with a larger set of information that describes the situation.

With higher-speed communications, it is possible to poll much more frequently. Also, the possibility exists for the RTUs to "report by exception." In this case the RTUs don't need to wait until they are polled. If they detect a problem, they can initiate the communication and send the information to the host.

Table 15-2 lists the physical methods that may be used to communicate between RTUs and PLCs in the field and a host production automation system.

Table 15-3 lists the physical standards that are employed to connect the communication systems between the RTUs and PLCs and the host computers.

Table 15-2
Methods of Communication between RTUs and Host

Method of Communication	Brief Description	Pros	Cons
Hardwire	Physically wired connection from RTU/PLC to host computer.	Very fast. Always open to communication.	Wire may deteriorate or be damaged over time. Can be expensive.
Telephone	Telephone line hardwired between RTU/PLC and host computer.	Can handle communication over very long distances.	Monthly bill. Occasional lack of service. Slow speed.
Radio	FCC regulated UHF or VHF licensed frequencies.	30+ miles line of sight communication. 19,200 Kilobytes per second.	Must have line of sight. Need to rent or own tall towers to mount master and repeater stations.
Microwave	High bandwidth wireless communication.	Up to 100 megabytes per second throughput. Ethernet addressability.	High initial cost to install. Must have AC power at all sites.
Fiber Optic	Communication between RTU/PLC and host computer over fiber optic cable.	Very high band width.	Expensive. May be difficult to maintain.
Spread Spectrum	Nonlicensed wireless communication. Can be frequency hopping, direct sequence, or 802.11.	Up to 1 megabyte per second throughput. No fees or licenses. Uses multiple repeaters. Fits well with solar power.	30 mile range. Line of sight. Shared frequency.
Satellite	Wireless communication from remote site to satellite to ground station.	Very high bandwidth. Communicates to or from anywhere on earth.	Very expensive. Latency, time lag.

Table 15-3
Communication Standards

Communication Standards	Brief Description	Pros	Cons
Hybrid	Combining multiple communications devices to build one system.	Allows multiple RTUs communicating by radio to talk to 1 satellite or phone line.	Operators need to be proficient with multiple instruments.
Cell Phones	Wireless data over cell phone network.	No infrastructure needed. Easy installation. Accessible from anywhere.	Recurring monthly phone bills. Not available in some areas. No ability to repair or troubleshoot.
RS-232	Method of enabling serial communications.	Universal RTU communication protocol. Up to 20 Kilobits per second throughput.	50 feet maximum between RTU/PLC and radio. Only one device per cable run. Requires trenching or conduit.
RS-485	Method of serial communications that allows multiple drops.	Allows multiple devices on one cable run. Up to 10 megabits per second throughput.	4,000 feet maximum. Limited to 32 devices. Requires trenching or conduit.
Ethernet	Network standard using coaxial, twisted pair cable, or spread spectrum radio.	Up to 1 gigabyte per second throughput. Multiple conversations at one time.	Not all RTUs/PLCs support Ethernet. 300 feet limit on wired connections.
Combination	Some radios can accept 232 and act as a terminal server or protocol translator to provide the computer with Ethernet (TCPIP) data.		

The languages in Table 15-4 are employed so the RTUs and PLCs and the host computers can communicate with and understand each other.

The methods in Table 15-5 are used to assure that a message that is transmitted from the RTU or PLC is received correctly by the host computer, and vice versa.

Host to Users

As indicated in Section 15.3.5, most host computer systems use desktop personal computers. The primary method of communication for users is direct on the PC using Microsoft Windows or similar tools. Some users are connected to the host computer via a network connection. Some use PC-to-PC links. Some use intranet or internet access.

Table 15-4
Communication Protocols

Communication Languages	Brief Description	Pros	Cons
Modbus	PLC language provides 256 addressable locations. It is very close to being the universal language between RTUs and PLCs and host computers.	Addressability. Common language for multiple vendors and equipment.	Limited addresses, slower baud rates, typically 9600 or 19200.
Modbus RTU	This is a Modbus protocol designed for use with RTUs and PLCs.		
TCPIP	Ethernet language. Allows for multiple conversations at once by giving every packet its own IP address.	Provides error checking, guaranteed data delivery, and polling of multiple units at once.	More overhead. Higher power consumption. No idle mode. Requires a lot of bandwidth.

Table 15-5
Methods of Data Security

Communication Standards	Brief Description	Pros	Cons
BCH	A BCH (Bose, Ray-Chaudhuiri, Hocquenghem) code is used for error detection. The side (RTU/PLC or host computer) that is transmitting data calculates a BCH code based on the data being sent and appends it to the transmission. The receiving side recalculates the code. If the two codes match, the data has been correctly received.	Provides a very high degree of data transmission security. In theory, BCH codes can be decoded to correct communication errors.	Most RTUs and PLCs only use error detection and retransmission. They don't use error correction.
128 bit AES (Encryption)	American Encryption Standard, commonly accepted by multiple industries.	Provides security on outbound messages. Prevents attack from incoming messages.	Requires bandwidth and computing power. Can be hacked. May need to be used in conjunction with other security.
RADUIS (Central Authentication)	Allows system to have conversations with only devices known to be authenticated by system administrator. All others are blocked.	Prevents "rogue" users from entering system. Has a "time out" feature that removes devices from system if they stop transmitting.	Any device that temporarily goes off-line must be manually reentered into the authentication list before it can resume conversation.
MAC Address Filtering	Allows each port to be secure. Only allows conversations with a list of known MAC addresses.	Prevents attack at the port level. Unknown MAC addresses cannot access a device or port on the device.	Complexity in management. New equipment will not be accepted without hands-on intervention.
VLAN Tagging	Virtual LAN allows multiple LANs inside one network. Keeps management data separate and segregated from SCADA data. Each packet has a tag that identifies its LAN and routes information to the proper network.	Allows for multiple secure conversations within the same network at the same time. Users can only access their VLAN.	Inconvenient if you need access to more than one of the VLANs. Data management is more complex.

Host to Computer Systems

Many production automation systems expand their capabilities by connecting to other computer systems for storage of large volumes of information in various types of database systems, access to analysis, design, and simulation software, access to the World Wide Web, and so on. Often these "extra" systems are not provided by the supplier of the production automation system so some form of agreement must be negotiated.

Computer Systems to Users

There are many ways for users to communicate with the computer systems that are part of the extended production automation system. In some cases the information from the computer system is transmitted back to the automation host system so the user can access it there. In other cases, the information may be available via an intranet or Internet connection.

15.3.7 Database

Overview

Automation systems generate vast quantities of data. For this data to be of value to end users, it needs to be handled so that it can be easily stored, retrieved, and displayed at some time. For this reason, databases play an integral role in automation systems and exist in some form or another in virtually every component of the system. A database can be defined as a structured set of records that is stored in a computer so that a program can consult it to answer queries.

Database Models and Schema

For a given database, there is a structural description of the type of facts held in that database. This is known as a schema. A schema describes the objects in a database and the relationship among them. There are different ways of organizing schemas, called database models. The most common database model is the *relational model*. Relational databases arrange all information in tables of rows and columns, where relationships are represented by values common to more than one table. Other database models include the *hierarchical model* and *network model*, which represent relationships more explicitly.

Storage

Databases typically are stored in memory or on hard disk in one of many formats. Data often is stored by category (i.e., data by month, data by well), creating preconfigured views known as materialized views. In some cases, data may be *normalized* to reduce storage requirements and improve extensibility. In other cases, data may be *denormalized* to reduce join complexity and reduce execution time for queries.

Indexing

All databases take advantage of indexing to increase their speed and efficiency. Indexing is a means of sorting information. The most common form of index is a sorted list of contents in a particular table column with pointers to the row associated with the value. Indexes allow a set of rows matching certain criteria to be located quickly.

Real-time Databases

Unlike conventional databases, a real-time database is designed specifically to meet the demands of a system where information is constantly changing. Whereas traditional databases are adequate for handling *persistent* data that generally is unaffected by time, a real-time database must be able to "keep up" with constantly changing conditions. Real-time databases are traditional databases that use an extension to give the additional power to yield reliable responses. They use timing constraints that represent a certain range of values for which the data are valid. This range is called temporal validity. A conventional database cannot work under these circumstances because the inconsistencies between the real world objects and the data that represents them are too severe for simple modifications. An effective system needs to be able to handle time-sensitive queries, return only temporally valid data, and support priority scheduling. To enter the data in the records, often a sensor or an input device monitors the state of the physical system and updates the database with new information to reflect the physical system more accurately.

Figure 15-14 illustrates the difference between how real-time data is processed with a conventional database versus a real-time database. Conventional database protocols, which generally schedule transactions

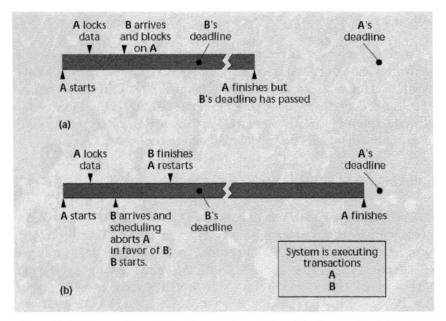

Figure 15-14: Processing of Two Transactions Using (a) Conventional Database Protocols and (b) Time-Cognizant Protocols (after Stankovic *et al.*)

on a first-come, first-serve basis, will let transaction A lock the data and complete, allowing A to meet its deadline. B, on the other hand, will miss its deadline because A's lock on the data prevents B from starting early enough. In contrast, a real-time database with time-cognizant protocols would preempt transaction A and transfer data control to B because B's deadline is earlier. Transaction A would regain control after B completes, and both transactions would meet their deadlines.

FIFO

FIFO is an acronym for first-in, first-out. This describes behavior similar to a queue in which people leave the queue in the order they arrived. In database terminology, this refers to a system in which there is a maximum amount of data that can be stored, and once that limit is reached, the oldest data is overwritten as new data comes into the system. This structure is common in most components of real-time systems. RTUs, PLCs and host systems all contain databases built around this principle.

The implication is these systems cannot archive data for an indefinite period of time. Depending on the size of the database and the quantity of data collected, such databases may be able to store as much as a year's worth of data or as little as a few hours'. For example, consider a system where the database can hold up to 10 million records. If that system were to collect data from a field with 100 wells, each instrumented with 10 analog sensors at one-minute intervals; the database would only be able to store one week's worth of data (100 wells × 10 AIs × 24 hrs/d × 60 min/hr = 1,440,000 records per day). This is an important consideration when designing an automation system to support today's highly instrumented wells. Because of the large quantities of data collected by wells equipped with electronic down-hole gauges and other instrumentation, an FIFO database may not be adequate to handle a system's long-term storage needs. For such an application, it may be necessary to augment the automation system with another type of database designed specifically for this purpose.

Historians

A data historian is a special class of real-time database that is designed to efficiently store and retrieve large sets of real-time data. These are commonly used as a repository for long-term data in automation systems. In addition to the normal characteristics of real-time databases, historians provide internal compression schemes to handle the extensive data storage requirements of an automation system. Further, historians provide tools that allow users to retrieve the data extremely quickly and even perform mathematical operations on the data. Historians are also built with extensibility in mind and can easily be integrated with other real-time data sources, host systems, or enterprise data sources. For this reason, historians often serve as the workhorse of an automation system. Several historians are in use throughout the industry today. The most commonly used historians are PI from OSISoft and Honeywell's Uniformance PHD.

15.3.8 Other

Some production automation systems extend beyond the traditional automation equipment and software of RTUs, PLCs, host computer systems, and databases. There are companies that are not in the automation business, but provide software systems for modeling reservoirs, well inflow, well outflow, nodal analysis, artificial lift systems, well test systems, among others. These software applications may be used for design, analysis, troubleshooting, and optimization of reservoir recovery, well inflow, and artificial lift system behavior. They may be more effective if they are provided with "live" data from an automation system. And the value of the automation system is enhanced if the results of system models and analysis can be fed back to it. For example some alarms are based on a comparison of measured results vs. results predicted by system models. Therefore, some companies are integrating production automation with these other systems.

This integration is possible with the availability of standard software interface systems that allow automation systems and these other systems to communicate with one another, share information, share results, and so on without having to be written by the same company. The interface acts as a translator. The automation system can communicate using its language and data, and the other systems can communicate using their approaches—the interface translates between the two systems.

One of the more common interfaces is the COM object interface system in Microsoft Windows. Many companies can communicate with each others' systems using this standard. COM lets data be exchanged using COM-supported software such as Microsoft Excel, Word, and PowerPoint.

Another common interface standard is the POSC (Petrochemical Open Standards Consortium) interchange format. Several companies including BP, Chevron, ExxonMobil, Shell, Statoil, Halliburton, Invensys, OSISoft, Petroleum Experts, Schlumberger, Sense Intellifield, TietoEnator, and Weatherford are working on the PRODML project to develop a POSC Work Group Agreement. PRODML is a shared solution for upstream oil and gas companies to optimize their production.

The "good news" is that any application (database, simulation software, design program, surveillance program, optimization program, etc.) that is COM or POSC compatible can be interfaced to a production automation system. This is good news for the operating company; they can have access to the capabilities. It is also good news for the automaton companies and the other service providers; they can offer their products without needing to develop interfaces with every different other system.

15.4 GENERAL APPLICATIONS

Automation systems are used in many industries for many purposes. Suppliers of these systems provide a wide range of general capabilities that are appropriate for use across a wide range of applications. This is good news for gas well operators. For the most part, these general applications come ready for use "off the shelf." They must be installed and configured for each specific field and set of wells, but they do not need to be developed and tested; that has already been done.

If a field has only flowing gas wells, these general applications, once properly configured for monitoring and reporting gas well production, may provide 90 percent of the production automation requirements. Of course, this is rarely the case. Many fields require one or more forms of artificial lift. Some of the required special applications for artificial lift and unique gas well needs are covered in Section 15.5. That is, these general applications are necessary, but they are not sufficient if special gas production requirements exist, or if artificial lift systems are used.

The purpose of this section is to discuss some of these general applications and how they can be used for management of gas production wells and systems.

15.4.1 User Interface

Production automation systems start with a user interface. Figure 15-15 shows an example of what a user interface might look like for an automation system for gas well deliquification. This is based on the graphical user interface standard developed by Microsoft Windows.

The purpose of the user interface is to allow the operator to easily navigate and select the specific reports, plots, or other needed functions

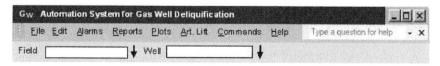

Figure 15-15: Example User Interface for Automation System for Gas Well Deliquification

or capabilities for specific wells or groups of wells. All production auto-mation systems typically use an approach that is similar to the standard developed by Microsoft Windows. Each system will look slightly differ-ent, have different specific pull-down options, have different methods for selecting specific field(s) and well(s), and so on. All systems should have a Help capability.

15.4.2 Scanning

A typical gas field may have tens or hundreds of wells. In many cases, an RTU or PLC is installed at each well. A primary function of the host production automation system is to scan each RTU/PLC on a periodic basis to collect pertinent real-time information for alarming, reporting, plotting, analysis, and so forth. Real time means that the information is live and indicates the condition of the well right now. The primary goal is to collect pertinent information automatically and have it available for processing and display, so the operator can focus on data analysis, not on data acquisition.

Typically, the host system communicates with (scans) each RTU/PLC on a preset frequency. For example, if radio communications are used, it may scan each device once each 15 to 20 minutes. With other com-munication systems, it may scan more frequently. Typically, a limited amount of information is uploaded from the RTU/PLC to the host on each scan. This might consist of a few data words to indicate the status of the well (e.g., are there any outstanding alarms?), the current gas flow rate and pressure, and so on.

If there is a problem (alarm), the host may then automatically upload more information to help define the alarm condition, or this function may be left to the operator.

In addition to periodic routine scans, the host system may also perform special scans at certain times of day, or in conjunction with certain events or conditions. For example, the system may upload a full set of data just before morning report time. Or, it may upload specified data in conjunction with a well test.

15.4.3 Alarming

A typical gas field may have hundreds or thousands of instruments to measure pressure, temperature, flow rate, and such. In principle, alarms can be defined for every instrument. An alarm is an indication that there is or may be a problem. In theory, the operators need to be made aware of all alarms so they can initiate the necessary action to address the alarm condition.

Although this is the theory, it may be counterproductive to actually implement all possible alarms. For example, most production automation systems allow configuration of, among others,

- High and low alarms—an alarm exists if the process variable (e.g., pressure) is above the high alarm limit or below the low alarm limit.
- High, high and low, low alarms—an alarm exists if the variable is above or below the high, high or low, low limits.
- Rate of change alarms—an alarm exists if the value of the variable changes too fast or too slow.

If eight or ten alarm conditions are defined for each variable, the system may generate hundreds or thousands of alarms per day. No operator can properly deal with this many alarms. So, in reality, many of them are ignored. This is not good, especially if there was one or more *real* alarms in the mix.

Rather than use the "standard" types of alarms, as indicated earlier, it is preferable to design the alarm system to make it pertinent for gas well operations. There may be specific alarms that are pertinent for gas wells and there are specific alarms that are pertinent for each type of artificial lift. The first type is discussed in this section. The second type is discussed in Section 15.5.

Before designing actual alarms, it is useful to consider three classes of alarms:

- Class I—simple alarms such as high and low alarms
- Class II—combination alarms where combinations of variables are used to indicate specific alarm conditions that are pertinent for gas wells
- Class III—performance alarms where the values of measured variables are compared with values that are estimated or derived from models of well or system performance

Class I Alarms

Some Class I alarms are pertinent for gas well operations. For example, a zero signal from an instrument may indicate that the instrument or the wiring to the instrument has failed. Figure 15-16 shows a typical alarm display taken from a well test facility. The following information is shown on this panel:

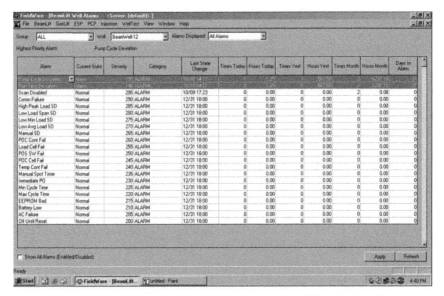

Figure 15-16: Current Alarm Display

- Alarm name
- Current state
- Severity—this number is assigned so alarms can be sorted by their severity
- State change time—when the point last changed state from normal to alarm
- Times in alarm—number of times this point has been in alarm today
- Hours in alarm—total time this point has been in alarm today
- Similar information is shown for yesterday, so far this month, and the previous month

Class II Alarms

Class II alarms are designed specifically to indicate problems with gas wells. They are based on a combination of field measurements. These alarms must be designed and configured for gas well operations but once they are, they can be used in many gas well production automation systems. Typically, these alarms are much more informative than simply determining that the pressure is too high or too low. Two examples are:

- Gas line blocked or frozen. This may occur if the wellhead pressure is above normal and the production rate is below normal. If there is a wellhead or separator temperature measurement, this may also be used as part of the combination.

- Gas line leaking or broken. This may occur if the well head pressure is normal, the flow rate is normal or higher than normal, and the line or separator pressure is below normal.

Class III Alarms

Class III alarms are generated when the value of a measured variable differs from a value that is calculated or estimated by a model of the well or system. These are commonly used for artificial lift and are discussed in Section 15.5. An example for general gas well application is a well flowing below critical velocity. This may occur if the calculated gas flow velocity, based on measured flow rate and pressure, is less than the calculated critical flow velocity.

15.4.4 Reporting

Production automation systems can produce several types of reports. The most common are:

- Current reports: Reports of current information on individual wells or groups of wells.
- Daily reports: Reports that summarize the wells' production and performance for a day.
- Historical reports: Reports that summarize wells' production and performance for the past week, month, or longer.
- Special reports: Reports that contain special information such as well tests, etc.
- Reports for unique applications: Reports that are unique for special applications, as discussed in Section 15.5.

Current Reports

Current reports can show information as of the last scan of the RTUs, or a special scan can be forced so the report contains true current information. The reports can contain measured values such as production rate, pressure, and such. They can contain calculated information such as gas-oil ratio, liquid-gas ratio, critical flow rate, and more. Information on the reports can be sorted by categories such as well name or highest to lowest production rate. Columns can be totaled, for example, to show the total gas production rate for a group of wells. Columns can be averaged, for example, to show the average production rate. In addition to rates and pressures, reports can show alarm and status information, downtime information, actions performed by the operators, among others.

Daily Reports

Most systems produce a set of reports at the end of each "production day." The production day may end at midnight or at some time early in the morning. These reports typically show the current production rate, the total production for the day just ended, the production for the previous day, and the cumulative production so far in the month. In addition to production, daily reports may show downtime for the most recent day and downtime so far in the month. As with current reports, these reports can be sorted, totaled, averaged, and so on.

Historical Reports

Most systems can produce historical reports. Usually, this is a report of items such as the production rates, pressures, critical velocities, and downtime for a well. The report may contain average values for the past few months and daily values for each day in the current month. There may also be historical reports of well tests.

Special Reports

Well test reports are in the special category because they don't contain daily information. A well test is reported when it occurs, and usually this is periodic; for example, once per week or once per month. Well test reports may show the current well test and a previous test for comparison.

Unique Application Reports

Reports are produced for most forms of artificial lift. These are discussed in pertinent places in Section 15.5.

Reports can be displayed on the automation screen, they can be printed, or they can be accessed by any system that is in communication with the automation system. Reports can be requested manually, scheduled automatically at some time, or produced in association with some event. For example, many locations schedule daily reports for automatic printing early in the morning, before people arrive, so they are available for the "morning meeting."

Reports can list all of the wells or conditions in an area. Or, they can list exception conditions. For example, a report may list only those wells where the production is too low, or those wells where the gas flow velocity is below the critical velocity. Exception reports are popular since they allow the operators to focus on problem wells and not have to sort through information on wells that are all right.

15.4.5 Trending and Plotting

In general, any measured or calculated variable that can be reported can be plotted with a trend plot. A trend plot is a plot of one or more variables vs. time. Most production automation systems provide a general trending capability; variables can be plotted by configuring the trend plots. In addition, some other types of plots are possible, for example, so-called xy plots, where one variable is plotted vs. another. And, many of the unique applications use various trends and other types of plots. These are discussed in Section 15.5.

Various adjustments are possible on trend plots. The time scale can be adjusted to show duration from minutes to months or years. The y-axis can be adjusted to show one, two, or many variables on the same plot. The axis can be adjusted from 0.0 to a maximum value, or it can be telescoped so the range of data fills the plot.

Most trend plots have a zoom feature so the operator can zoom in on a smaller time window and/or vertical set of data. Most trend plots support a color and/or a line style coding system so different variables can be color coded and can use different line styles if the plots are printed on a black and white printer.

Some trend plots are static; that is, they display data that has already been collected or calculated. Other plots are dynamic; that is, the data on the plot is updated automatically when a new set of data is scanned from the RTU. Some trend plots show only the data points; some show the data points connected by lines; some show only the lines. Some plots show only the measured or calculated data; some also show trends in the data; some show only historical data; some show both historical and projected future values.

Trend Plot of Two Variables

Trend plots can be very useful in spotting changes in variables over time and in evaluating data. Figure 15-17 is an example trend plot that

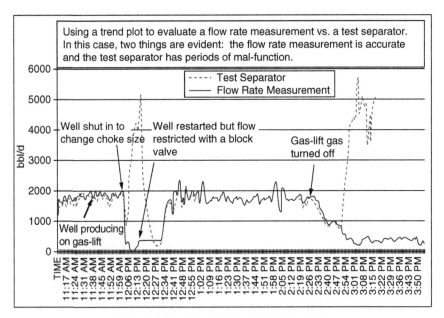

Figure 15-17: Trend Plot of Two Variables

was used to evaluate a flow rate measurement system versus test separator readings. The accuracy of the measurement was confirmed and some problems were detected with the test separator measurements.

15.4.6 Displays

"A picture is worth a thousand words." This saying is attributed Fred R. Barnard in the December 8, 1921 advertisement in the trade journal *Printers' Ink*. Production automation systems make use of this by featuring schematic displays of systems, facilities, and wells. The displays typically contain information about the operation of the item(s) being displayed such as pressures, temperatures, flow rates, alarm and status information, and so on.

Displays may be of several types: unique, generic, static, dynamic, and interactive.

Unique

Unique displays portray a given specific system or set of equipment. An example may be a specific production facility. The display may show a schematic of the facility with pertinent pressures, temperatures, flow rates, alarm and status information, and such.

Generic

Examples of generic displays are for a gas well or an artificial lift system. Figure 15-18 is an example of a generic schematic display of a gas-lift well. The display shows a typical gas-lift well with pertinent pressures, temperatures, flow rates, and other information.

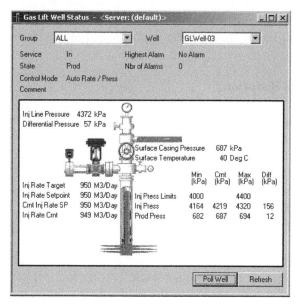

Figure 15-18: Example of a Schematic Display of a Gas-Lift Well

Static

A static display shows information that was collected on the last scan of the RTU(s) that provide information that is shown on the display.

Dynamic

The information on a dynamic display is updated each time new information is obtained from the pertinent RTU(s). Also, some dynamic plots have "live" graphics that show the equipment moving, liquid flowing, and so on.

Interactive

An interactive display is one where the operator can enter parameters or commands on the display. The appropriate action is taken such as downloading the parameters to the appropriate RTU(s), issuing the desired commands, and performing the desired calculations, among others.

Most production automation systems provide a tool kit for constructing the displays and populating them with the tag numbers for the information to be shown on the display. For unique displays, specific tag numbers are used. For generic displays, generic tag numbers are used so that the displayed information depends on the specific well selected for display.

15.4.7 Data Historians

Some production automation systems collect huge amounts of data. If a system serves 1000 wells; collects pressure, temperature, and flow rate once per minute; and calculates critical velocity once per minute, this is 5,760,000 pieces of data per day. Most systems collect, calculate, and store much more than four data items per well per scan.

Most automation systems are designed to focus on real-time information. They are concerned primarily with current rather than historical operations. If they store historical information, it is usually summary data, like hourly or daily average values. However, it is often valuable to store and be able to access detailed, minute-by-minute historical information, sometimes for months or years.

Data historians use special data compression techniques to store huge volumes of data and make this readily available. Real-time data is transmitted from the automation system to the historian. The historians typically provide special methods or techniques for access to the data. For example, one company provides a tool called a *processbook*. Operators can create interactive graphical displays that extract information from the historian on an as-needed basis. The information can be saved and shared with other people who have access to the system via an intranet or Internet connection.

Specific examples of uses of historians are beyond the scope of this book. However, if there is interest, most automation companies can provide access to an historian system.

15.5 UNIQUE APPLICATIONS FOR GAS WELL DELIQUIFICATION

As indicated in Section 15.4, production automation systems usually come with several general applications that can be applied for gas well monitoring, control, and optimization. However, these general applications never address specific requirements for artificial lift or other unique gas well capabilities. Therefore, special or unique applications are required.

Again there is "good news" for gas well operators. Production automation systems for several types of artificial lift have already been developed for use in oil production. Systems for sucker rod pumping, progressing cavity pumping, ESP pumping, and gas-lift exist and have been developed, tested, and used extensively. In many cases, these can be applied for gas wells with only minor modifications.

However, these forms of artificial lift address only a fraction of gas well production. Other major production methods include plunger lift, chemical injection, and wellhead compression. For these and others, new capabilities intended for use on gas wells are required. Some of these have been developed and some are in various stages of research and development.

The purpose of this section is to discuss these unique applications, how they can be used for monitoring, control, and optimization of gas well operations, and some of their benefits and challenges. Some of these systems are very comprehensive. For example, to fully describe an automation system for plunger lift, sucker rod pumping, ESP pumping, gas-lift, and so on, a full-length book would be required for each. So the purpose here is to "hit the high spots." If more information is desired on automation for a specific form of artificial lift, it must be obtained from the authors or an appropriate service company or operating company that uses the system.

15.5.1 Plunger Lift

Plunger lift is a low rate artificial lift method, common in gas well deliquification applications but also in some oil applications. The method requires no outside energy source; it uses the well's natural energy to lift fluids (and the plunger) to the surface. The systems can be installed without a rig, provide easy maintenance, can be deployed at extremely low cost, are tolerant of deviation, and can produce a well to near deple-tion. Some limitations of plunger lift are that the system requires specific gas/liquid ratios to function; components are sensitive to solids; and the system can be labor-intensive, requiring surveillance to work properly. Historically, the need to regulate flow as well as surveillance require-ments and the labor intensive nature of plunger lift have made it one of the most heavily automated lift methods in the industry.

Measurements

To effectively monitor, control, and optimize a plunger lift installa-tion, the automation system must collect a number of surface parame-ters. These include the tubing head pressure, casing head pressure, flowline pressure, differential pressure across an orifice union, gas flow rate (calculated from differential pressure and flowline pressure), and an indication of plunger arrival.

Figure 15-19 is a system illustration depicting the instruments and controls in a plunger lift system. These include: (1) a tubing pressure transducer, (2) a casing pressure transducer, (3) a differential pressure transducer, (4) an orifice union assembly, (5) a plunger lift controller,

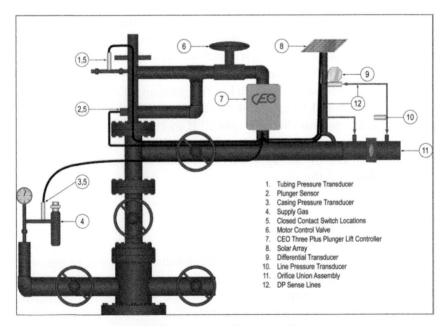

1. Tubing Pressure Transducer
2. Plunger Sensor
3. Casing Pressure Transducer
4. Supply Gas
5. Closed Contact Switch Locations
6. Motor Control Valve
7. CEO Three Plus Plunger Lift Controller
8. Solar Array
9. Differential Transducer
10. Line Pressure Transducer
11. Orifice Union Assembly
12. DP Sense Lines

Figure 15-19: Plunger Lift System Illustration

(6) an automatic control valve, (7) a flowline pressure transducer, and (8) closed contact switches.

Control

Most plunger lift systems operate on time-based control algorithms utilizing control of one or two surface valves to control the movement of the plunger. When the well is closed, bottom-hole pressure builds up. When the well is open, this pressure forces the plunger to the surface carrying fluid on top of the plunger. The plunger lift controller monitors and records the time between plunger cycles and makes adjustments to the time-based control algorithm.

Unique Hardware

Plunger lift uses specialized RTUs called plunger lift controllers (see Figure 15-20). These devices monitor and control the well based on

Figure 15-20: Typical Plunger Lift Controller

internal control logic; in some cases, they communicate with a host system to transmit data and accept commands. In addition, because plunger lift controllers often are used in gas deliquification applications, they can either communicate with an API 21.1-compliant Electronic Flow Measurement (EFM) device or have EFM capabilities integral to the controller itself. In controllers equipped with telemetry, this flow measurement data can then be transmitted to a host system.

Historically, plunger lift has been viewed as a labor-intensive form of artificial lift due to the significant requirement for surveillance and adjustments to the system to optimize production. For this reason, many operators tended to avoid this artificial lift method. The development and evolution of plunger lift controller technologies has reduced the amount of labor required for normal operation and enabled this lift method to become more widely used throughout the industry.

Unique Software

Plunger lift controllers contain special software that provides a computerized means for opening and closing the control valve based on programmed responses or sets of parameters. A variety of options are available for these devices including: (1) actuation based on pressure and flow, (2) time cycle control, (3) self-adjusting models, and (4) telemetry support.

Plunger lift controllers have evolved from simple time cycle controllers to time cycle controllers with plunger arrival recognition to auto adjusting time cycle controllers. As a result, controllers have become self managing plug-and-play devices; and plunger lift systems no longer require the constant supervision they once did.

Various forms of control logic are available in plunger lift controllers. The following are some commonly used operating parameters for plunger lift well control using a flow and pressure-operated control system.

On Pressure Limit Control

Controller initiates *on* cycle when the following conditions are met.

1. Tubing pressure ≥ on pressure limit — Looks for tubing pressure to exceed a set point
2. Casing pressure ≥ on pressure limit — Looks for casing pressure to exceed a set point
3. Tubing – line ≥ on pressure limit — Looks for tubing pressure to build to a set point above line pressure
4. Casing – line ≥ on pressure limit — Looks for casing pressure to build to a set point above line pressure
5. Foss and Gaul Calculations — Looks for casing pressure to reach a calculated value
6. Load Factor — Looks for the casing-tubing/casing-line pressure ratio to be a factor of less than 40%

Off Pressure Limit Control

Controller initiates *off* cycle when the following conditions are met.

1. Plunger has arrived — Used on oil wells
2. Casing pressure ≤ off pressure limit — Looks for casing pressure to fall below a set pressure
3. HW ≤ off pressure limit — Looks for flow rate to fall below a set differential in inches of water
4. Flow Rate — Sometimes a calculated Turner Rate
5. Casing-tubing ≥ off pressure limit — Looks for differential between casing and tubing pressure to increase
6. Casing To Tubing Sway — Looks for the casing and tubing pressure to start moving apart from each other
7. Casing to Line ≥ off pressure limit — Looks for differential between casing and line pressure to increase

Specialized Alarms

Plunger lift systems provide a variety of alarms to notify users of adverse operating conditions or potential opportunities to enhance performance. Some of the more interesting alarms are:

- High plunger velocity. This could indicate a number of issues. More commonly, a high velocity indication could be the result of either: (1) a failure of the plunger to reach bottom (i.e., due to a wellbore obstruction, such as hydrates) or (2) a plunger that is coming up dry (i.e., inadequate inflow or inadequate shut-in time).
- Number of runs per day. An excessive number of cycles per day could be an indication that the plunger is wearing out, resulting in reduced efficiency.

- High tubing head pressure. In certain cases, an extended shut-in period could result in an excessively high tubing head pressure. If an on-cycle were initiated under these conditions, the resulting high pressure slug could potentially damage the separator or cause other upsets to the surface facilities.

Surveillance

Plunger lift surveillance is performed by continuously checking for special alarm conditions and reviewing trend plots. Figure 15-21 shows a typical surveillance panel from a plunger lift automation system. It contains status information, trend plots, and operating parameters for a plunger lift installation.

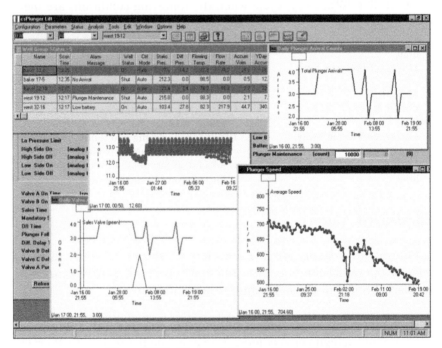

Figure 15-21: Plunger Lift Surveillance Panel

Analysis

Once data is captured in the plunger lift automation system, this data is used to perform analysis to optimize the performance of the system.

One of the key off-line analyses is a determination of whether to switch from continuous plungers to conventional plungers. In addition, an evaluation of gas velocities and pressures can assist in determining whether the well is operating below the critical velocity for liquid loading. In addition to these offline analyses, some plunger lift controllers perform continuous real-time analysis of pressures and velocities to make continuous adjustments to system parameters and optimize production.

Design

Plunger lift is unique in that it lacks the same rigorous design requirements that are common to other lift methods. Application engineering techniques generally consist of: (1) evaluation of flowing well conditions to determine if the well is a candidate for plunger lift, (2) evaluation of the most suitable type of plunger to use, and (3) selection of equipment that is most appropriate for the conditions.

Optimization

Through surveillance of key operating parameters, it is possible to determine if opportunities exist to improve system performance. For example, by monitoring casing pressure, one might identify an opportunity to reduce casing pressure and, in turn, increase formation drawdown. Another key parameter is gas flow rate. In many cases, there is an upper limit to the rates that can be accurately measured by the gas meters at the sales point. If the gas flow rate exceeds this amount, the operator may not be compensated for all the gas that is transferred to the pipeline. So, instantaneous flow rates should not exceed this threshold. Finally, through evaluation of key parameters, users can adjust the shut-in time so sufficient pressure builds to ensure plunger arrival while ensuring that the well is not shut-in for an excessive duration.

15.5.2 Sucker Rod Pumping

Many thousands of oil wells are produced by sucker rod pumping. This is also a common method of artificial lift for gas well deliquification. Automation of sucker rod pumping systems for oil wells started in the 1970s; it is a very well advanced technology. Virtually the same sucker rod automation systems are used for gas wells. Figure 15-22 shows a

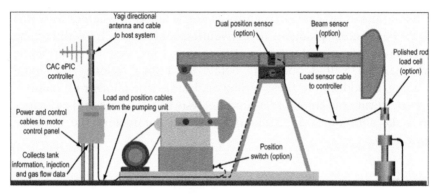

Figure 15-22: Schematic of Sucker Rod Pumping System Equipped for Automation

schematic of a typical sucker rod pumped well equipped for automation.

Measurements

The primary measurements are the load on the polished rod and the position of the polished rod. The load usually is measured with a polished rod load cell mounted on the top of the polished rod or a strain gauge mounted on the beam. The polished rod load cell is preferred as it is more accurate. The position normally is measured with an angular position transducer, an accelerometer or mercury switch mounted on the beam, or a position switch mounted on the A-frame. The accelerometer or beam mounted switch is preferred; it is more accurate.

Some systems use a dual measurement system that measures the load and position with one device. This is lower cost, but not as accurate as a polished rod load cell and separate position measurement.

A number of secondary or optional measurements may be used: a vibration switch to detect unit vibration; a stuffing box leak detector to detect liquid accumulation in the stuffing box; a tubing or flowline pressure transducer; a casing pressure transducer; and other measurements if there are local gas flow meters, and so on.

Control

The primary means of automatic control for sucker rod pumping wells is pump-off control. The system detects when the pump is no longer filed with liquid on the downstroke, which means that the pump didn't completely fill on the upstroke. This is illustrated in Figure 15-23, where the load on the polished rod is carried by the rod part way into the downstroke, and not transferred to the tubing. When the traveling valve strikes the fluid level in the pump barrel part way into the downstroke, some pounding may occur; this is referred to as *fluid pound*. Compare the larger card plot (full card) with the thinner card plot (pumped off) case. The upper set of plots is measured at the surface; this is the *surface card*. The lower set of plots is calculated at the depth of the downhole pump; this is the *pump card*. The difference in length between the surface and downhole is due to rod stretch.

In most cases, the pump-off detection is based on the surface card. However, some systems calculate the downhole card on every pump stroke and make this determination on the downhole or pump card.

This method of control has disadvantages. It depends on some degree of pump-off occurring on every pump cycle. When pump-off occurs, the downhole pump and rods are stressed. Also, if a well is producing gas, which of course is the case for gas wells, the fluid in the pump barrel may be mostly gas and the well may be in a continual state of pump-off. If there is very little liquid compared with the amount of gas, the pump may become gas locked and fail to pump liquid at all.

An alternative method of control is with a variable speed drive. In this case, the speed of the pump is adjusted to keep the liquid level just

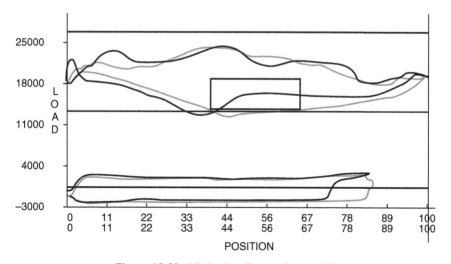

Figure 15-23: Method to Detect Pump-Off

above the pump intake so pump-off doesn't occur. In a gas well with a small amount of liquid production, even this method may be difficult to use.

Unique Hardware

Sucker rod automation uses special RTUs known as rod pump controllers or pump-off controllers (see Figure 15-24). These units are unique in that they contain only the hardware needed for the required measurements and special logic to monitor and control sucker rod pumping wells. Typically, these units are installed in special NEMA 4 (environmentally sound) cabinets, have battery back-up, and are outfitted for radio communications with the host computer system.

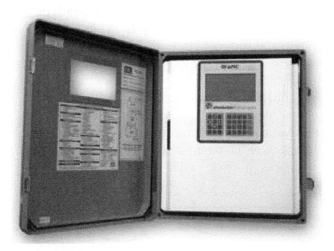

Figure 15-24: Typical Rod Pump Controller

Unique Software

Rod pump controllers contain unique software for managing pump operations. To determine if a well is pumped off, they must collect and process at least 20 pairs of load and position data points per second. Some units process much more data than this. The units check for a number of specific sucker rod pumping alarms (see the next subsection) and check for shut-down conditions. For example, if the rod load is too high, this may indicate a stuck pump.

They check for pump-off on every pump stroke by comparing the load at some preset position with a predefined pump-off load limit. If the load is above the limit, implying that the load is still being carried by the traveling valve and rods, the pump is stopped. It will remain off for a predetermined pump-off idle time and then automatically restart.

Since most rod pump controllers communicate with the host automation system by radio, pump cards are buffered in the unit's memory so they can be transmitted to the host at relatively slow radio transmission speed.

Specialized Alarms

Sucker rod pumping automation systems perform specialized alarming. A few of the more interesting ones are:

- Run time too short. This implies that the well is not producing as much as expected. There may be an inflow problem.
- Run time too long. This implies that the well is producing more than expected, or that the pump is leaking.
- Position signal problem. This implies that the measured position is not sufficiently close to the predicted position.
- Calibration problem. This implies that the design program is not calibrated with the measured data.

Surveillance

The primary purpose of surveillance is to detect problems so they can be addressed. Some sucker rod pumping automation systems have a very interesting method for problem detection.

Over the years, a large library of problem cases has been compiled. Each surface or downhole pump card is compared with the cases in the library and the best fits are reported to the operator. Thus, for example, the morning report may contain a notation for a well where the pump is sticking, or the pump is leaking, or the pump has gas interference. An example of this comparison is shown in Figure 15-25, where the actual surface card (wavy) is compared with a library card (smooth). The list on the left shows possible library cards and how well each fits with the surface card.

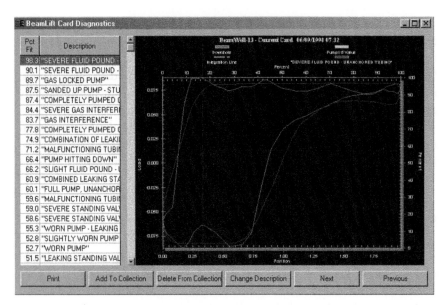

Figure 15-25: Comparison of Surface Card with Library

Analysis

Sucker rod pumping automation systems focus on analysis of the surface pumping unit, the rod string, and the downhole pump. They determine such things are the torque on the gear box, the stress on each taper of the rod string, the downhole pump fillage, and so on. Each calculated value is compared with a target or limit value so problems can be detected. For example, the peak torque is compared with the torque rating of the gear box.

Another result of the analysis is the degree that the pumping unit is in or out of balance. If a unit is out of balance, this can contribute to overload of the gear box. It can also cause inefficiency and excess power usage.

Design

There are excellent sucker rod pumping design programs. Some production automation systems can calibrate the design program to the analysis program by matching the calculated downhole card with the one produced by the design program. When the program is calibrated it can produce very accurate designs for determining the effect of pump size, pump speed, stroke length, and more on such operating parameters as beam loads, torque, rod loads, and so on.

Some automation systems have the design program as part of the system. In other cases, the design program is part of a separate system and information must be shared between the automation system and the system where the design program is run.

Another aspect of design, when sucker rod pumping is used for gas well deliquification, is how best to deal with the gas. In practice, the best way is to set the pump intake below the perforations. The gas tends to rise up the annulus and the liquid flows down into the pump intake. This can be augmented by a gas separator. This is normally a pipe below the pump intake that further assists with gas/liquid separation.

In addition, there are special pump designs that are better equipped to handle gas. An example is shown in Figure 15-26. A small vent hole is placed in the pump barrel. The pump has twin plungers. When the pump plunger travels on the downstroke it pushes some of the gas through the vent hole into the annulus. This can prevent the pump from becoming gas locked.

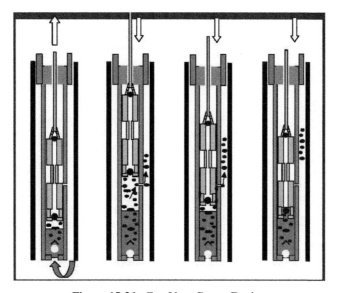

Figure 15-26: Gas Vent Pump Design

Optimization

Optimization of a sucker rod pumping system addresses the following questions:

- What is the optimum pump size?
- What is the optimum stroke length?
- What is the optimum pump speed?
- What is the optimum motor size?
- What is the optimum pumping unit size?

Some automation systems are well equipped to help answer these questions. The first step is to calibrate the design program. Then, the design program can be run for many cases of different pump sizes, stroke lengths, pump speeds, motor sizes, and unit sizes. The optimum solution is usually the one that produces the desired amount of liquid with the lowest capital and operating costs.

15.5.3 PCP Pumping

Progressing cavity pumping (PCP) is gaining in use for deliquification of gas wells. Figure 15-27 shows a schematic of a rod-driven PCP installation. Some of the data items on the schematic aren't activated.

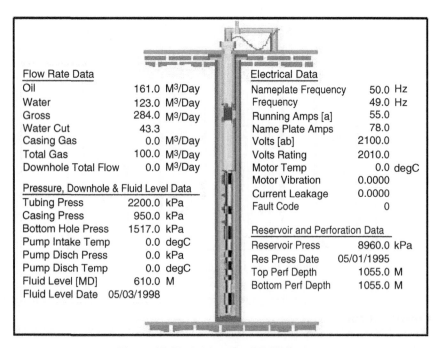

Flow Rate Data		
Oil	161.0	M³/Day
Water	123.0	M³/Day
Gross	284.0	M³/Day
Water Cut	43.3	
Casing Gas	0.0	M³/Day
Total Gas	100.0	M³/Day
Downhole Total Flow	0.0	M³/Day

Pressure, Downhole & Fluid Level Data		
Tubing Press	2200.0	kPa
Casing Press	950.0	kPa
Bottom Hole Press	1517.0	kPa
Pump Intake Temp	0.0	degC
Pump Disch Press	0.0	kPa
Pump Disch Temp	0.0	degC
Fluid Level [MD]	610.0	M
Fluid Level Date	05/03/1998	

Electrical Data		
Nameplate Frequency	50.0	Hz
Frequency	49.0	Hz
Running Amps [a]	55.0	
Name Plate Amps	78.0	
Volts [ab]	2100.0	
Volts Rating	2010.0	
Motor Temp	0.0	degC
Motor Vibration	0.0000	
Current Leakage	0.0000	
Fault Code	0	

Reservoir and Perforation Data		
Reservoir Press	8960.0	kPa
Res Press Date	05/01/1995	
Top Perf Depth	1055.0	M
Bottom Perf Depth	1055.0	M

Figure 15-27: Schematic of PCP System

PCPs' primary advantages are:

- They can do a better job of handing gas than most other types of pumping systems. Because the flow through the pump is continuous, there is less likelihood of gas locking.
- They can do a better job of handing solids than most other types of pumping systems. There are no valves and seats that can be eroded by solids.

Their primary disadvantages are:

- They have temperature limitations. The elastomers used to construct the pump stators have a limited temperature range, although this range is continually being extended.
- They have difficulty with highly aromatic liquids, again due to limitations imposed by the elastomers. This can be a problem if a gas well produces condensate.
- They have limitations in the amount of head it can produce, thus the depth of application may be limited.

PCPs are implemented in two different ways:

- The traditional and most used way is to drive the PCP with a sucker rod from the surface. The rods are rotated at approximately 600 RPM to drive the PCP pump.
- The other method that is rapidly gaining popularity is use of a downhole electrical submersible pump motor. A gear reducer is used to convert the 3600 RPM ESP motor output to 600 RPM for the pump.

These differences have an impact on measurements, control, and other issues.

Measurements

For surface driven PCPs, primary measurements are tubing pressure, tubing temperature, casing pressure, flow rate (if possible), torque on the sucker rod, and RPM of the sucker rod.

For downhole electrical submersible pump motor driven PCPs, measurements include surface pressures and flow rate, and in most cases downhole instruments are used to measure pump intake pressure, pump intake temperature, and other variables.

Control

With PCPs, it is critical that the well not be allowed to pump off, as this will increase the temperature in the pump and may lead to elastomer swelling. With surface drives, some units use fixed speed drives, but the pump must be operated to prevent pump-off. An enhanced method is control of the rod RPM with a variable speed drive (VSD). With the VSD, the pump discharge can be limited to keep the fluid level above the pump. The challenge is to know where the fluid level is. This normally is addressed by taking periodic fluid level measurements, although there are methods to determine the fluid level on a continuous basis, especially if downhole pressure is measured.

With a downhole drive, the pump is controlled by using a variable speed drive to control the RPM of the ESP motor. With downhole electrical motors, it is common to include a downhole measurement system, so the pump intake pressure is known and the fluid level can be determined from this.

Temperature is an important limiting variable. Most PCP control systems can stop the pump or change its speed if the temperature is too high. Torque is also a limiting value. If the torque on a rod driven system is too high, this may indicate a stuck pump.

Unique Hardware

For rod-driven PCPs, a drive head is required to rotate the sucker rods. See Figures 15-28 and 15-29 for examples of drive heads. In addition, a wellhead RTU is used for data acquisition and control; and a variable speed drive may be used to control the rotational speed of the drive rods.

For downhole driven PCPs, a downhole electrical motor is connected to the PCP via a gear reducer and seal section. This is shown in Figure 15-30. In this case, a wellhead RTU is required for data acquisition and control; and a variable speed drive may be used to control the speed of the PCP pump. Also, it is normal in this case to include a downhole measurement system to measure pump intake pressure, pump intake temperature, pump discharge pressure, and other variables.

Figure 15-28: Surface PCP Drive Systems

Figure 15-29: Surface PCP Drive Unit

Figure 15-30: Electrical Submersible PCP

Unique Software

Wellhead RTUs for PCPs contain unique software to control the pump and protect the pump with special shut-down conditions; they also gather and process data for pump alarms, surveillance, and optimization. These specialized RTUs are available from various service companies, so it would be counterproductive to build one from scratch.

There is also special host software for PCP analysis and optimization.

Specialized Alarms

A feature of PCP automation is the detection of special alarm conditions:

- A hole in the tubing or a worn pump may be detected when torque drops, RPM increases (if not controlled), and production declines. The pump intake pressure will increase.
- Pump-off may be detected when torque increases gradually, then levels out, and then sharply increases, eventually leading to a low-speed shut-down. During this time RPM remains unchanged and production declines.
- A rod failure may be detected when torque drops, then levels out, and then sharply increases, eventually leading to low-speed shut-down. Differential between pump intake pressure and pump discharge pressure declines.
- A plugged flowline or WAX buildup may be detected when torque increases but production drops.
- A gas flow restriction or gas pressure build-up in the annulus may be detected when the gas pressure increases and pump intake pressure stays the same as the pump speed adjusts.
- Periodically Pump Intake Pressure (PIP) and Pump Discharge Pressure (PDP) are collected with the pump stalled. A hole in tubing or fluid slippage can be detected when an increase in PIP and a decrease in PDP are observed. Casing gas pressure may also be measured during the stall test.
- A history of break-out torque at every start-up is recorded to analyze any variations over time; this helps in detecting pump slippage and failure.

PCP surveillance is performed by continuously checking for special alarm conditions and reviewing trend plots. Figure 15-31 shows a typical surveillance panel from a PCP automation system. It contains status information, trend plots, and operating parameters for a PCP installation.

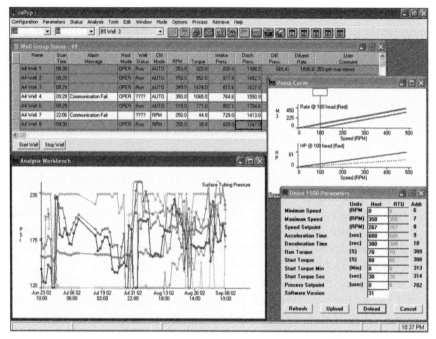

Figure 15-31: PCP Surveillance Panel

Analysis

Some PCP automation systems incorporate a set of pump performance curves from the manufacturers and compare current pump operation with theoretical operation. This permits the operator to spot pumps that are worn or are not operating up to specification for some reason.

Figure 15-32 shows a typical PCP analysis panel.

Figure 15-33 shows the inputs to this analysis capability; Figure 15-34 shows the outputs.

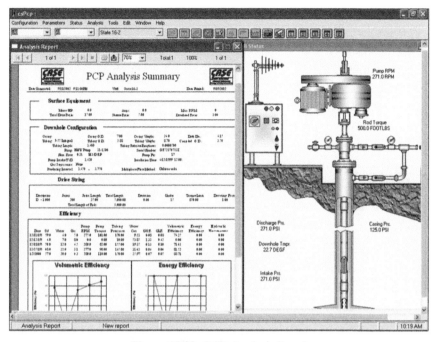

Figure 15-32: PCP Analysis Panel

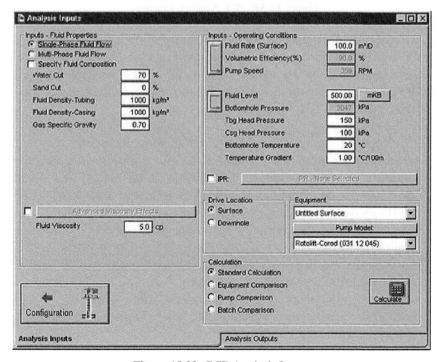

Figure 15-33: PCP Analysis Inputs

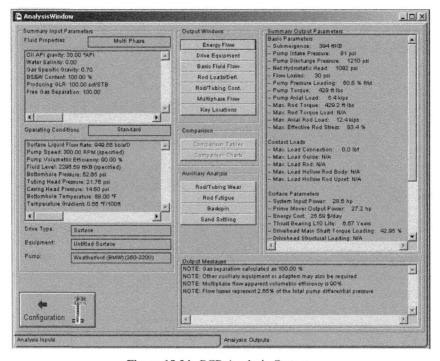

Figure 15-34: PCP Analysis Outputs

Another form of analysis may be performed with the PCP-RIFTS (PCP Reliability Information and Failure Tracking System). This is a system sponsored by an industry consortium for collecting of PCP failure data and providing tools for analyzing the causes of failures. Objectives of PCP-RIFTS are to:

1. Develop "standards" related to PCP utilization, including:
   - Standard practices and guidelines to collect, classify, and analyze run-life and failure data, such as the system developed by C-FER for Electric Submersible Pumps (ESPs) in the ESP-RIFTS JIP (http://www.esprifts.com) (see Section 15.5.4).
   - Standard ways to evaluate elastomers with respect to the main failure mechanisms.
   - Standard ways to name and classify elastomers that would help in their selection for specific applications.
   - Standard pump inspection and failure reporting practices.
2. Further investigate ways to reduce the failure frequency of the most severe mechanisms. This might include:

- Investigation, collection, and review of past field trials.
- Conducting controlled laboratory and field tests on some of the relatively new PCP technologies, such as pump-off control systems and rotor coating materials.

PCP-RIFTS is not part of a production automation system, but some automation systems can interface with RIFTS and provide information to it. Feedback from RIFTS to automation systems is yet to be developed.

Design and Optimization

Unlike sucker rod automation systems, most PCP automation systems don't contain PCP design and optimization software. However, there is at least one (and there may be more) excellent PCP pump design and optimization program available in the industry. This program also contains a PCP analysis tool. Some work may be required to link PCP automation systems to this analysis and design program. When this is done, it should be possible to calibrate the PCP design program to current PCP operating information and then use this for accurate PCP designs and optimization.

15.5.4 ESP Pumping

Electrical submersible pumping (ESP) is the method of choice for many oil production applications where it is necessary to produce a large volume of liquid. It is also used quite extensively for deliquification of gas wells where large volumes of water must be produced. Figure 15-35 shows a schematic of an automation system for an ESP system. This particular schematic is of a relatively complex system designed for wells that produce gas and some sand.

The primary advantages of ESP systems are that they:

- Can produce large volumes of liquid
- Can produce from great depths

The primary disadvantages are:

- Difficulty handling large volumes of gas.
- Difficulty handling solids.
- Difficulty with high temperatures—the motor must be cooled.

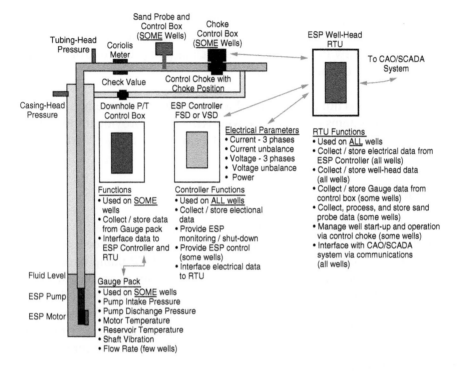

Figure 15-35: Schematic of ESP Automation System

ESP systems use two types of drive systems—fixed speed drive (FSD) and variable speed drive (VSD).

FSD systems are used on the majority of ESP systems. However, most operators try not to turn ESPs off and on. Too many restarts may damage the system. So, special precautions must be used on wells with FSDs that may cycle due to gas production.

Many newer applications use VSDs to provide added control flexibility, especially if the well produces gas, as of course gas wells do. The choice of control system has an impact on the required instrumentation, hardware, and software.

Measurements

Three types of measurements are input into the automation system on most ESP installations:

- Surface instruments are used to measure tubing-head pressure, casing pressure, temperature, and some others such as flow rate (on some wells), and sand production rate (on some wells).

- Electrical parameters are measured by the FSD or VSD controller. Typically many parameters are available from the controllers; the most commonly recorded are current (sometime all three phases), voltage (sometimes all three phases), current unbalance, voltage unbalance, and power consumption.
- Downhole instruments are used on almost all ESP wells that have been installed in the past few years. These are used to measure pump intake pressure, pump intake temperature, motor winding temperature, pump discharge pressure, motor shaft vibration, and downhole flow rate (on some wells).

Control

Three primary types of control are used on ESP systems:

Start, Stop, and Safety Shutdown

The FSD or VSD controller provides start, stop, and safety shutdown capabilities. ESPs can be started manually or automatically, locally or remotely. They can be stopped manually or automatically, locally or remotely. And, ESPs can be shut down if the controller detects an unsafe operation such as low current load, high current load, low or high voltage, unbalance, and so on. A majority of ESPs are run with only this form of control. If, for example, the well produces a significant amount of gas, the system may employ a form of pump-off control by shutting down on low current load, waiting for a preset idle time, and then starting again.

ESP automation systems are designed to interface with the FSD or VSD controllers, download parameters to them (e.g., for low current load shutdown limit), transmit commands to them (e.g., for remote start or stop), obtain electrical measurements from them (e.g., current, voltage, etc.), and monitor the status of the ESP system (e.g., numbers of starts and stops).

Control of Wells with FSDs

If the production rate from the well must be limited, this can be done by controlling the pumping rate or the fluid level. It may be necessary to do this to limit sand production or gas interference. The pumping rate can be controlled by controlling the back pressure on the wellhead with a choke or backpressure control valve. This is inefficient but effective. The biggest concern is that the production rate must be kept high enough to provide adequate ESP motor cooling. The fluid level can be controlled by adjusting the back pressure on the annulus or by recycling some liquid back into the well's annulus. Again, both of these methods are inefficient. However, they are used relatively frequently because so many wells use FSD control.

Control of Wells with VSDs

In recent times (certainly in the 2000s), most wells that require production rate control use variable speed drives. Here, the ESP speed and output can be controlled, over a reasonable range, by adjusting the frequency of the electrical current that controls the speed of the ESP. The most common control variable is the fluid level. It can be determined with reasonable accuracy by knowing the surface casing pressure, the measured downhole pump intake pressure, and the composition of the gas and liquid in the annulus. The pump discharge can be controlled to accurately maintain the fluid level at the desired elevation above the pump intake. See Figure 15-36 for an example of how the fluid level is determined based on knowledge of the surface casing pressure and the measured downhole (pump intake) pressure. VSDs also are used on wells with gas production. The ESP can be slowed down and speeded up to work a potential gas lock through the pump.

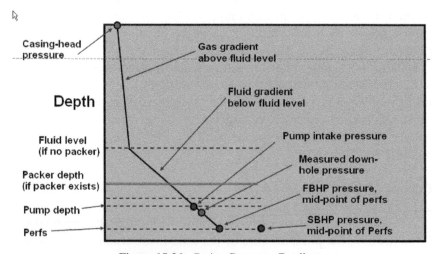

Figure 15-36: Casing Pressure Gradient

Control of Wells on Start-Up

In some cases, especially with sandy or gassy wells, it is important to start them slowly enough to avoid excessive reservoir pressure drawdown and excessively high inflow rates, but fast enough to avoid motor cooling problems.

Figure 15-37 shows start-up guidelines for such cases. The production rate must be kept high enough to avoid downthrust and to provide adequate motor cooling. And, it must be kept low enough to prevent upthrust and high reservoir pressure drawdown that can lead to sand influx or excessive gas production.

Some ESP automation systems provide control for this special start-up mode.

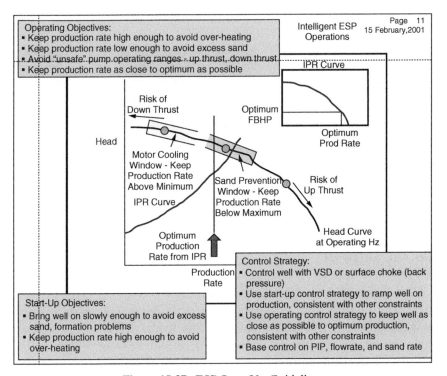

Figure 15-37: ESP Start-Up Guidelines

Unique Hardware

Specialized automation hardware is needed for an ESP system. The "good news" is that some companies make special RTUs for several methods of artificial lift. So although the specific hardware and software are unique for each type of artificial lift, the RTU's physical components, communications interface, and so on are the same or similar for several different types of artificial lift.

In most cases, as shown in Figure 15-38, an ESP RTU is interfaced to an ESP controller and other components. In some cases, companies are working on enhanced ESP controllers that also provide RTU functionality. ESP controllers normally are provided by the supplier of the ESP system, not by the RTU supplier. If use of an enhanced controller to provide both ESP control and RTU functions is desired, check with the ESP supplier. However, make certain that the ESP controller can really provide all the needed RTU functions.

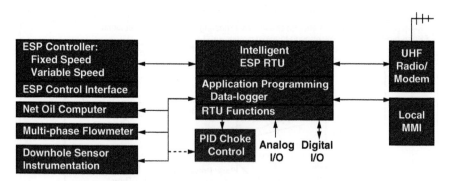

Figure 15-38: Required RTU and Associated Components

Unique Software

Clearly the RTU for ESP automation must contain special software logic. Fortunately this has been very well developed and proven. It would make no sense to develop this from scratch. This special logic interfaces with the ESP controller (either FSD or VSD), reads the surface instrumentation, interfaces with the downhole instrumentation, provides some forms of control when needed (e.g., control of a surface back-pressure choke), and interfaces with the operator via a local interface and communication with the host automation system.

As indicated in the previous subsection, some companies are working on ESP controllers that can provide the functionality of both the controller and the RTU. This has the potential to reduce both cost and equipment footprint at the well. However, the precaution mentioned earlier must be used to be certain the required controller and RTU functionality are provided.

Specialized Alarms

A typical ESP automation system may produce dozens of alarms. Some are relatively typical; for example, high current, low current, high pressure, low pressure, and so on. However, several are special for the ESP operation:

- Low production rate. Risk of overheating motor due to inadequate cooling.
- Very low production rate. Risk of downthrust.
- High production rate. Risk of excessive drawdown that may increase sand or gas production.
- Very high production rate. Risk of upthrust.
- High motor temperature. Another indication of potential motor overheating.
- High motor shaft vibration. An indication of motor–protector–pump alignment problems. This may lead to premature bearing wear or failure.

Surveillance

ESP surveillance is performed by continuously checking for special alarm conditions and reviewing trend plots. Figure 15-39 shows a typical surveillance panel from an ESP automation system. It contains status information, trend plots, and operating parameters for an ESP installation.

In addition to this type of specialized panel, normal alarm reports, status reports, trend plots, and such are used for ESP surveillance.

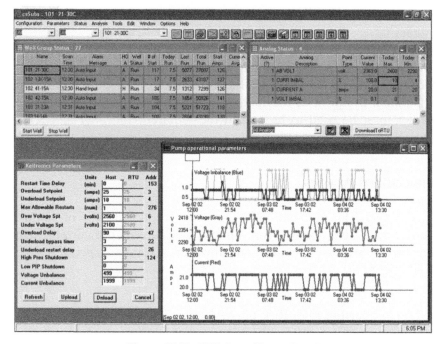

Figure 15-39: ESP Surveillance Panel

Analysis

ESP analysis is performed to answer several questions:

- How well is the ESP well performing?
- How well is the ESP pump system performing?
- How well are the ESP system and the well working together as a system?

Figures 15-40 and 15-41 are designed to help answer the first question. Figure 15-40 shows the pressures and pressure gradients in the well, below and above the pump. It shows the pressure (head) increase created by the pump. It shows the current fluid level. The liquid beneath the fluid level exerts a pressure (back pressure) on the formation, which inhibits inflow from the reservoir to the wellbore.

Figure 15-41 shows the inflow performance relationship (IPR) for the well, based on the Vogel IPR method. It shows the current flowing bottomhole pressure (FBHP), what this would be if the pump were operating perfectly according to the manufacturer's specifications (this is discussed further in the next paragraph), and what this would be if the

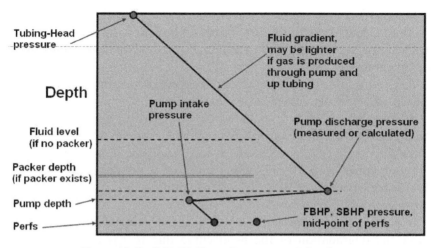

Figure 15-40: ESP Wellbore Pressures and Gradients

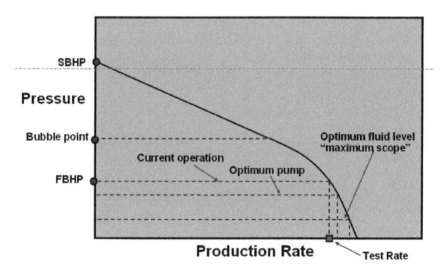

Figure 15-41: ESP Well Inflow Performance Relationship

well were operating with the optimum fluid level. In this example, the well has a relatively high bubble point pressure, so the IPR curve is curved significantly downward below the bubble point. Therefore, the potential increase in production rate by operating at a lower bottomhole pressure is less than it would be if the well had a very low bubble point pressure and the IPR curve were closer to a straight line.

Figure 15-42 is designed to answer the second question about pump performance and the third question about how well the well and pump are working together. It shows the pump's theoretical head curve based on manufacturer's data.

The reduced curve is the actual head curve determined from the measured or calculated head, the measured production rate from a well test, or wellhead flow rate. It shows the downthrust point, the upthrust point, and the optimum operating point. It shows the well's IPR or PI (productivity index) curve; the curve appears different on this plot since the plot shows head vs. production rate rather than pressure vs. rate. The point where the IPR curve and the pump head curve cross is the operating point for the well.

In some ESP automation systems, this plot is expanded further to show the degradation of the head curve due to gas interference or pumping of heavy oil.

Another form of analysis may be performed with the ESP-RIFTS (ESP Reliability Information and Failure Tracking System). This is a system sponsored by an industry consortium for collecting ESP failure data and providing tools for analyzing the causes of failures. As of April 2007, the consortium had 13 member companies. The ESP-RIFTS database contained information on over 26,000 ESP installations in 325 oil fields around the world.

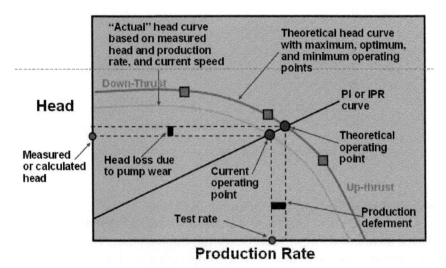

Figure 15-42: ESP Head Curve

Objectives of ESP-RIFTS are to:

- Facilitate sharing of ESP run life and failure information among operating companies.
- Ensure quality and consistency of data in the system.
- Incorporate useful analysis tools.

Benefits of the system are:

- Benchmark the performance of a company's ESPs against that of other operators.
- Improved decision making, based on actual reliability data.
- Enhanced capability to predict run life in new applications.
- Ability to predict workover frequency in existing applications.

ESP-RIFTS is not part of a production automation system, but some automation systems can interface with RIFTS and provide information to it. Feedback from RIFTS to automation systems is yet to be developed.

Design and Optimization

Unlike sucker rod automation systems, most ESP automation systems don't contain ESP design and optimization software, although at least one automation system does allow use of "what if" questions. For example, what will be the effect on production rate if the type of pump, size of pump, or number of pump stages is changed?

However, there are at least two (and there may be more) excellent ESP pump design and optimization programs available in the industry. These programs also contain an ESP analysis tool. Some work may be required to link ESP automation systems to these analysis and design programs. When this is done, it should be possible to calibrate the ESP design program to current ESP operating information and then use this for accurate ESP designs and optimization.

15.5.5 Hydraulic Pumping

Hydraulic lift is an artificial lift method in which the energy to lift fluids to the surface is provided by a power fluid (generally, recirculated pressurized produced fluids). There are two common forms of hydraulic lift—jet pumps and hydraulic piston pumps. Of these, jet pumps are by far the more common and are applicable over a wide range of rates and setting depths. In certain cases, the pumps can be retrieved by reverse-circulating to surface or via wireline. Following retrieval, the pumps can be repaired at the wellsite. The system is suitable for wells with high deviation, making it a choice for offshore applications. Also, jet pumps can be adapted to existing bottomhole assemblies and sliding sleeves. The systems are excellent for producing viscous crude. Limitations include potentially complex completions, high pressure surface equipment, and potentially problematic fluid measurement. Depending on the accuracy of flow measurement, it can be difficult to determine what portion of the flow stream is produced by the formation and what portion is provided as power fluid.

Even though hydraulic lift completion architecture can be complex and varied, two major classes of completion are most common: (1) open completions and (2) closed completions. In an open completion, power fluids are pumped down a conduit to the downhole pump where they commingle with produced fluids and the combined flow stream is produced to the surface. Generally, in such a completion, the power fluid is pumped down the tubing and produced fluids flow up the tubing/casing annulus. In a closed completion, power fluid is pumped down a conduit to the pump and used to lift the produced fluids prior to being pumped up another closed conduit. Both the produced fluids and power fluids are isolated from the tubing-casing annulus. Because of the need to isolate produced fluids from the produced gas, closed completions are more common in gas well deliquification applications. In a typical deliquification application, a closed completion is run without a packer. This enables the power fluid to be pumped down one tubing string, produced fluids to flow up a second parallel tubing string, and gas to flow up the casing-tubing annulus.

The use of hydraulic pumps in deliquification applications is rare. This is due to the following limitations of the system. First, the need for a closed system makes hydraulic lift impractical in many applications where casing sizes are small. Second, the clumping of solids such as coal fines and clogging of the pump intake makes hydraulic lift impractical for coal bed methane applications and other deliquification applications in which a significant amount of solids is produced.

Hydraulic lift is the least common artificial lift method, with worldwide well-count estimated at less than 5 percent of the overall population. As a result, hydraulic lift is one of the most underserved lift methods in terms of automation technology. There are no known real-time host applications or wellsite controllers designed specifically for use in hydraulic lift applications. Instead, where automation exists, operators use existing automation equipment and human/machine interfaces to provide basic surveillance and control capabilities.

Surveillance

Basic parameters for surveillance of hydraulic lift systems include:

- Flowing wellhead pressure
- Wellhead injection pressure
- Wellhead temperature
- Production choke position
- Surface pump suction pressure
- Surface pump discharge pressure
- Surface pump status

Operators may install permanent downhole gauges and monitor parameters such as pump intake and discharge pressures. However, such installations are rare, due in large part to the complex completion configurations required and the ability to reverse-circulate pumps to surface (pump reliability is of less concern than in other lift methods).

Control

Basic control parameters for hydraulic lift systems include:

- Surface pump status (on/off)
- Surface pump speed
- Production choke position

15.5.6 Chemical Injection

In many gas wells, deliquification can be achieved by injection of surfactants. The basic principle is that foam reduces the surface tension between the liquid and gas phases. Because surface tension is directly proportional to the critical velocity for liquid loading, this will reduce the critical velocity needed in the well. Provided the critical velocity can be reduced below the in-situ gas velocity, this may be sufficient to unload the well and prevent future liquid loading from occurring.

Automation of chemical injection systems is limited. In many cases, the surfactant is injected on a periodic basis by means of a spooling unit, where no data is directly uploaded into the automation system. In other cases, a chemical injection system may be installed permanently, whereby a control line is run from the surface to a downhole chemical injection valve. At the surface, a surfactant drum and small injection pump are installed and plumbed into the wellhead. Such systems operate autonomously and require no external control. No specialized host systems exist for monitoring these systems. Should an operator wish to integrate these devices into their automation systems, basic surveillance and/or control would be provided through existing control systems and human/machine interface.

Surveillance

Basic parameters for monitoring chemical injection systems include:

- Injection pump status (on/off)
- Chemical injection rate
- Gas production rate

Control

Control of these systems is limited to turning on and off the injection pump.

15.5.7 Gas-Lift

Gas-lift is the second most used method of artificial lift, after sucker rod pumping. Figure 15-43 shows a schematic of a typical gas-lift system. This is referred to as a "closed loop" system since the gas is compressed, injected, produced, gathered, recompressed, and recirculated around the system.

The primary advantages of gas-lift are:

- It "likes" gas. The more gas a well produces, the better.
- It can be installed in almost any well deviation. For wireline installation of gas-lift valves, the wellbore deviation must not be greater than about 70° from vertical.

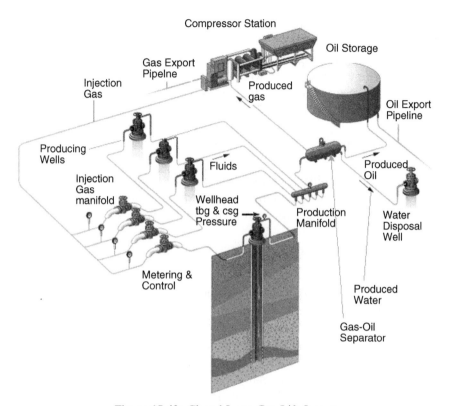

Figure 15-43: Closed Loop Gas-Lift System

- It can be used in any well depth.
- Downhole gas-lift equipment (valves, orifices, etc.) can be installed by wireline; a workover rig is not required.
- It can handle some degree of sand production. There are no parts in the flow stream so sandy fluid can flow up past the gas-lift valves without damaging them.
- In gas wells, gas can be injected to achieve and maintain critical velocity.
- It isn't necessary to cycle wells, as with plunger and pumping systems. Thus, the wells stay on production all the time.

The primary disadvantages or challenges are:

- There must be a source of high pressure gas; this usually is provided by a compression station or another source of high pressure gas.

- A good control system is required to maintain optimum gas-lift operation.
- Diligence is required to keep the gas-lift system design in line with the operation of the well. A common problem is a too-large injection port or orifice that can lead to unstable (heading) operation.

Measurements

For effective gas-lift operation, there are certain required measurements. There are also optional measurements that can enhance the operation. Some typical measurement devices were shown in Section 15.3.

Required measurements are:

- Gas-lift injection rate. This may be measured at the wellhead or at an injection manifold.
- Gas-lift injection pressure. Normally this is the casing-head pressure. It should be measured at the wellhead, downstream of any pressure drop devices such as chokes, control valves, and such.
- Production pressure. Normally this is the tubing-head pressure. It should be measured at the wellhead, upstream of any pressure drop devices such as chokes.

Optional measurements are:

- Gas injection temperature. This may be required to compensate the gas injection rate measurement.
- Production temperature. This is used by some operators to provide surveillance of gas-lift liquid production rates.
- Production rate. In some cases, it is possible to measure (or accurately estimate) the liquid production rate on a continuous basis. This has obvious advantages where it can be provided.
- Downhole variables. In some cases, downhole measurements (and sometimes control) are possible. There are systems that provide downhole pressure, temperature, and gas injection rate measurements. In some of these, the rate of gas injection through the operating gas-lift valve or orifice can be controlled.

Control

There are two fundamental methods of gas-lift: continuous and intermittent. There are multiple variations on each of these.

For continuous gas-lift, the objective is to control the rate of gas injection on a continuous basis and keep it as stable as possible. The three goals are to inject gas as deep as possible, keep it stable, and keep the injection rate optimized, in that order. To keep the injection rate stable, the injection rate at the surface must be in balance with the gas flow capacity of the downhole injection port or orifice.

For intermittent gas-lift, there are two primary control methods: choke control and time-cycle control. With choke control, the rate of gas injection on the surface is controlled by a choke or control valve. The volume (and pressure) of gas gradually increases in the annulus until the pressure is high enough to open the intermittent gas-lift operating valve. When the valve opens, a slug of gas is injected into the production stream (tubing) beneath the accumulated slug of liquid. With time-cycle control, injection is cycled by use of a time-cycle controller on the surface.

The advantage of choke control is that the gas flow rates and pressures remain relatively stable in the gas-lift distribution system. The disadvantage is that the downhole gas-lift injection valve must be changed to change the injection cycle frequency or volume of gas per cycle.

The advantages of time-cycle control are that the injection frequency and the volume of gas per cycle can be controlled on the surface. The disadvantage is that the rates and pressures in the gas-lift distribution system may vary significantly, potentially causing upsets to the system and other wells served by the system.

Unique Hardware

Gas-lift requires use of gas-lift mandrels, valves, and other downhole components. In addition, special gas-lift automation hardware is required.

Gas-lift mandrels are installed in the tubing string. There are two primary types: conventional and side-pocket. Conventional mandrels require that the gas-lift valves be installed in the mandrels on the surface and run with the tubing. Side-pocket mandrels are designed to permit gas-lift valves and other downhole components to be installed with wireline. A schematic of a side-pocket mandrel is shown in Figure 15-44.

There are several types of gas-lift valves. The most common are unloading valves, continuous gas-lift operating valves or orifices, and intermittent gas-lift pilot valves. Unloading valves may be injection pressure operated (the most common) or production pressure operated. Their purpose is to unload liquid the from the well's annulus so gas can be injected at desired operating depth, deep in the well. These valves are intended to be used only for unloading and should remain closed at all other times, so gas can be injected below them in the operating valve or orifice. A schematic of an injection pressure operated unloading gas-lift valve is shown in Figure 15-45.

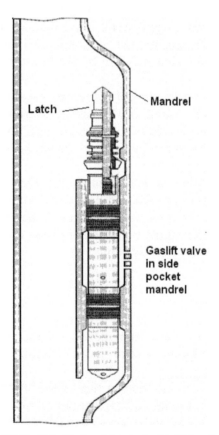

Figure 15-44: Side Pocket Mandrel with Valve Installed

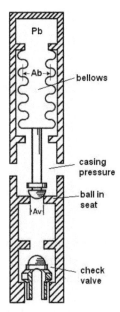

Figure 15-45: Unloading Gas-Lift Valve

Typically, an orifice valve (actually a gas-lift valve with no stem so it can't close) is used at operating depth. Since this valve can't close, it can't throttle the gas-injection rate into the well; it is always fully open. It is important that the size of the orifice be designed to inject the right amount of gas, so there is a balance between the rate injected at the surface and the rate that can flow through the operating valve or orifice. Otherwise, the well will tend to become unstable.

For intermittent gas-lift, a pilot-operated valve normally is used for the operating valve. This valve can instantaneously go "full open" so the slug of gas can rapidly be injected beneath the slug of liquid that has accumulated in the wellbore during the down portion of the intermittent gas-lift cycle.

Typically, facility and wellhead RTUs are used to provide gas-lift measurement and control. When gas-lift injection is measured and controlled at an injection manifold, this normally is done with a facility RTU or DCS at the manifold station. Typically, a small wellhead RTU is used to measure the casing head injection pressure, the production pressure, and potentially the injection temperature, production temperature, production rate, and in some cases, downhole variables.

Unique Software

Specialized software is required for gas-lift automation. This includes software in the RTU(s) to monitor and control, and to provide alarm detection, surveillance, analysis, design, and optimization. This software has been developed by several companies and is available in the industry.

Specialized Alarms

There are several specialized alarms that are important for gas-lift. Some of these include:

- Injection pressure heading. The injection pressure is fluctuating; this will cause inefficiency in the well and may indicate multipoint injection—injection through more than one gas-lift valve all or part of the time.
- Production pressure heading. This almost always occurs with injection pressure heading. It may occur by itself, especially if the tubing size is too large for the current production rate.
- Injection gas freezing. When a pressure drop is taken across a gas control choke or valve, the temperature drops due to the Joule-Thompson effect. If there is water vapor in the gas, it may freeze. The resulting hydrates can block the injection path.
- Gas blowing around. This can occur if gas is being injected through an upper gas-lift valve, or a tubing leak. Gas is being injected but no liquid is being produced.
- Many others. There are many other alarms that are common in gas-lift systems. They are too numerous to list in detail in this section.

Surveillance

Gas-lift surveillance is focused on keeping the wells operating properly and detecting any alarm conditions that need to be addressed. For continuous gas-lift wells this means keeping the gas-lift injection deep, stable, and optimum. Figure 15-46 shows a typical gas-lift surveillance plot with the injection pressure fluctuating or heading. For intermittent gas-lift wells, it means keeping the injection cycles properly timed with the desired volume of gas per cycle. For gas

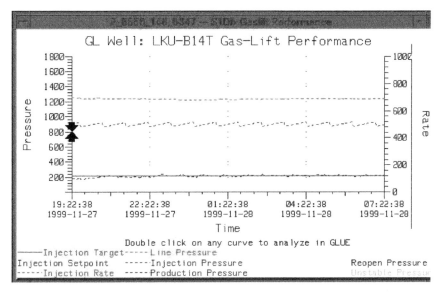

Figure 15-46: Example of Injection Pressure Heading

wells, another objective is to keep the gas injection rate sufficient to maintain critical velocity.

Analysis

The purpose of gas-lift analysis is to determine the root cause(s) of operating problems. A good way to do this is to analyze the performance downhole in the well. Figure 15-47 shows a downhole plot of injection pressure profile(s) and production pressure profile(s) taken at different points on the plot of injection and production pressure vs. time in Figure 15-46.

With this, the operator can determine where gas is being injected from the annulus into the tubing, and if this may be occurring at different depths when a well is unstable or heading, as shown in this example.

Analysis also is used to help determine or diagnose the causes of various gas-lift alarms. For example, there are many "shapes" of injection and production pressure plots vs. time that are indicative of specific types of gas-lift problems.

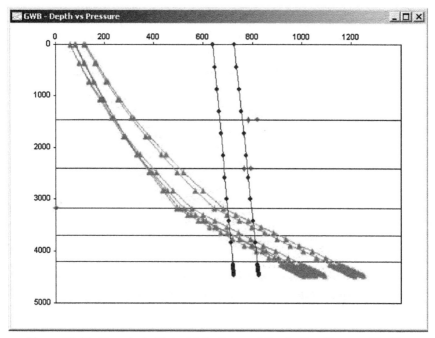

Figure 15-47: Downhole Plot of Injection and Production Pressure Profiles

Design

There are two primary objectives with gas-lift design. The first is to determine the desired spacing of gas-lift mandrels. This is done before the well is completed, or when it is recompleted after a workover. There are several gas-lift design programs available in the industry. There are also several industry courses offered on gas-lift design.

The second objective is to design the unloading gas-lift valves and the operating valve or orifice. Typically, this is done when it is necessary to place a well on gas-lift and it may be redone several times over the life of the well as the well's conditions change. Again, there are several programs available to do this. And there are courses that cover this aspect of design.

An issue is that most gas-lift design programs and courses focus on use of gas-lift for oil wells. Some adjustment, particularly of the selection and design of the operating valve or orifice, may be required for use of gas-lift to deliquify gas wells.

Optimization

In theory, optimization is the process of determining and using the right amount of gas to optimize the economic operation of the well. In practice, for oil wells, it is better thought of as the optimum allocation of gas, given that the amount of gas available for lift will rarely be exactly equal to the sum of the optimum amounts for each well.

For gas wells, two conditions must be considered. If there is a liquid level in the well, the first objective is to remove it, at least down to the maximum depth of gas-lift injection. During this process, classical optimization is not important. The goal here is to unload the well and remove the liquid column. When the liquid column has been removed, the goal becomes to use the optimum amount of gas to maintain critical velocity. Critical velocity is the rate of gas flow (produced gas plus injected gas) that is required to continuously remove liquid from the wellbore and maintain a minimum flowing bottomhole pressure. Critical velocity can be determined by using the Turner or Coleman equations, depending on the well's conditions.

15.5.8 Wellhead Compression

Compression is one of the most effective means of deliquifying gas wells. Benefits of compression are two-fold. First, by reducing the flowing bottomhole pressure, more draw-down is achieved, resulting in a higher production rate. Second, the reduction of in-situ pressure throughout the production string reduces the critical velocity required to remove fluids.

No specialized host systems exist for monitoring wellhead compression systems. Should an operator wish to integrate these devices into their automation systems, basic surveillance and/or control would be provided through existing control systems and human/machine interfaces.

Surveillance

Basic parameters for monitoring wellhead compression systems include:

- Suction pressure
- Discharge pressure

- Suction temperature
- Discharge temperature
- Compressor speed

Control

Control of these systems is limited to turning on and off the compressor and adjusting the compressor speed.

15.5.9 Heaters

Thermal lift is a recent development in the world of gas well deliquification. The method entails the installation of an ESP power cable in the wellbore, either by strapping the cable to the tubing or running inside coiled tubing. The principle is simple: electricity from the power cable generates heat, which in turn raises the temperature of the produced gas above the dew point. The amount of heat generated is a function of the power supplied to the cable, which in turn is dictated by the voltage (and frequency) of the system. Although this method is relatively inefficient, it has proven effective in the field and may be a good choice for certain applications.

No specialized host systems or control systems exist for automating thermal lift systems. Control for such systems can be provided either by adjusting the tap settings in a switchboard or adjusting the operating frequency of a variable frequency drive. Surveillance would be by existing host systems.

Surveillance

Basic parameters for monitoring thermal lift systems include:

- Voltage
- Current
- Frequency
- Gas flow rate
- Flowing wellhead pressure
- Flowing wellhead temperature

Control

Control of these systems is limited to adjusting electrical parameters in the switchboard or variable frequency drive that affects output power. Because current is a function of the operating load (which is negligible), voltage is the key parameter affecting the output power and heat generated by the system. If a switchboard is used, voltage is controlled by adjusting the tap settings in the switchboard. If a variable frequency drive is used, voltage can be controlled by adjusting either the base frequency of the drive or (more likely) the operating frequency. Either of these parameters can be adjusted either manually in the drive or by means of changing set-points from the host system.

15.5.10 Cycling

When wells initially load with liquids, it is sometimes possible to unload them by cycling. Cycling, also known as stop-cocking or intermitting, refers to the process of intermittently cycling the well between flowing and shut-in conditions. When the well is shut in, bottomhole pressure increases and pressurized gas accumulates in the annulus. This increased well pressure pushes all or part of the fluids back into the formation, allowing the well to flow again once the well is opened to production. In essence, cycling is akin to using plunger lift without the plunger. The key consideration in optimizing a cycling well is to maximize the amount of time the well is producing while minimizing the amount of time that the well is flowing below the critical velocity for unloading.

No specialized host systems exist for monitoring wells on cycling production. Should an operator wish to automate such a well, basic surveillance and/or control would be provided through existing control systems and human/machine interfaces.

Surveillance

Basic parameters for monitoring wells on cyclic production include:

- Flowing wellhead pressure
- Casing head pressure
- Gas flow rate
- Production control valve state (open/closed)

Control

Control of these wells is limited to opening and closing an automatic control valve. This can be accomplished by a simple time-cycle controller or through more sophisticated control logic such as that contained in a plunger lift controller.

15.5.11 Production Allocation

Production allocation is the process of determining or estimating the production of each well in a system, based on the actual measured production from the system (e.g., production facility) and the measured or estimated production of each well.

Production automation can assist with this process by obtaining the measurements of actual production from the system and the measurements or estimates of production of each well. The allocation is then a simple mathematical process using the following equation:

$$Qai = Qmi * Sum\ (Qmi)/Qs$$

where:

Qai = Allocated production to well i
Qmi = Measured or estimated production for well i
Qs = Measured production for the production system

This normally works well for gas since often the volume of gas produced by each well is measured with some relative degree of accuracy. It may be a problem when attempting to determine the allocated production of oil, condensate, and water, since these volumes typically are not measured on a continuous basis, and often only with well tests, which may be taken infrequently.

If production rates are "measured" with well tests, it is also necessary to know or determine the "on production" time of each well to calculate the estimated production volume over the time period of the allocation. Again, the production automation system can assist with this if it has some way to determine the actual "on production" time of each well from its surveillance capabilities.

There are some new techniques being considered in the industry to permit the continuous measurement (or at least accurate estimate) of both liquid and gas production rates from each well. When this information can be gathered by a production automation system, it can permit accurate, continuous, or at least daily allocation to each well. Stay tuned for this to become available in the next few years.

15.5.12 Other Unique Applications

Production automation can assist with other gas well applications. One example is automatic adjustment of gas production when this is necessary to meet specific demands for gas delivery. The automation system can not only adjust the production rates to meet the system delivery objectives, it can do this in a way to maximize the efficiency of deliquification. For example, if the overall production rate must be temporarily reduced, it may be possible to do this by pinching back on free flowing wells where liquid loading is not yet an issue, and permitting wells with artificial lift systems for deliquification to continue to work at their optimum level.

15.6 AUTOMATION ISSUES

There are a number of issues that must be understood and addressed in considering, defining, justifying, designing, building, installing, using, and maintaining a production automation system. A common fault of many systems is that focus is placed on purchasing and installing equipment without giving adequate consideration to all the issues. If some are ignored, the system may fail or be under utilized, not due to poor equipment but because of inadequate attention to other details needed for an overall successful system. The purpose of this section is to define and discuss some of the more important issues with the hope that they will be included in the overall production automation project plan.

15.6.1 Typical Benefits

Several types of benefits may be realized from a gas production automation system. For management, the most important are tangible, quantitative, economic benefits that justify the cost of the system and provide a direct economic payout. For others, there are intangible, qualitative benefits that may be as or more valuable but are difficult to quantify in monetary terms. The production automaton system must be designed to provide both types of benefits. The purpose here is to briefly discuss some of the more important tangible and intangible benefits. Actual monetary values can be placed on these only in the context of specific field conditions.

Tangible Benefits

Tangible benefits can be measured in economic terms. Some of the more common are:

- Increased production. Production can be increased by:
  - Early detection of downtime and correction of problems so wells are on production a high percentage of time.
  - Keeping artificial lift systems operating at peak efficiency at all times.
  - Keeping wells flowing (producing) at or above critical velocity or keeping wells pumped off to avoid accumulation of liquid in the wellbore.
  - Keeping wells on production until their true economic limits are reached, thus increasing ultimate recovery from the reservoirs.

- Reduced operating costs. Operating costs can be reduced by:
  - Needing fewer people to monitor and control the wells.
  - Needing less automotive, boat, or helicopter travel to visit the wells.
  - Optimizing the use of energy (e.g., electricity, gas, etc.) to operate artificial lift systems.
  - Optimizing the use of expendables (e.g., chemicals).

- Reduced maintenance costs. Maintenance costs can be reduced by:
  - Keeping artificial lift systems (especially pumping systems) operating within their safe operating envelope.
  - Detecting problems before they become failures.

- Reduced capital costs. In some cases, capital costs can be reduced by:
  - Deploying more expensive artificial lift systems only when needed.
  - Not over-designing artificial lift systems.

- Artificial lift specific benefits. There are some benefits that are unique to each type of artificial lift system. Some of the following have been documented by the industry for oil well production:

- Sucker rod pumping: 7% production increase, 20% energy reduction, 35% maintenance cost reduction.
- Electrical submersible pumping: 3% production increase.
- Gas-lift: 5% production increase, 10% reduction in gas usage, reduced compressor CAPEX.

These benefits are not over and above those listed. They come from such things as reduced downtime, improved efficiency, and the like. However, they confirm that these types of benefits are real.

Intangible Benefits

Intangible benefits are more difficult to quantify, but they may be as or more important. Some of the more common are:

- Safety
  - Operators need to visit wells less frequently. This reduces travel hazards.
  - When there is a problem, operators know about the problem in advance and can be prepared with the right equipment to deal with it.

- Environmental protection
  - Production can be remotely (automatically) stopped if a problem (e.g., leak) is detected.

- Personnel
  - Operators and others gain a better understanding of their wells and equipment.
  - This can be an incentive when seeking to hire good people.

15.6.2 Potential Problem Areas

It is important to be aware of potential problem areas so they can be avoided or addressed. Some of these are best avoided or addressed by using automation experts, either from within the operating company or from a service company or consultant. Some of the potential problem areas are as follows.

Automation System Design

An important consideration in system design is to "keep it simple," but not too simple. The primary objective must be to design a system that can achieve the benefits defined in the preceding section. If the system cannot achieve the required benefits, it will not be fully utilized. If it isn't used, it won't be maintained. If it isn't maintained, it will fail.

The recommended process is as follows:

- Define the gas production operation to be automated.
- Describe the benefits to be achieved.
- Design the system to achieve these benefits.

Instrumentation Selection

As stated earlier, instrumentation is the core of the system. Unless gas production variables can be reliably and accurately measured and controlled, the rest of the system is worthless.

The following process is recommended for instrument selection:

- Define the variables that need to be measured and controlled.
- Select instruments of proven reliability. Reliability is more important than absolute accuracy, and it is more important than initial cost. When purchasing instruments, cheaper is not better.
- Use an experienced instrument engineer to design and implement instrument installation and commissioning.

Automation Hardware and Software Selection

As discussed earlier, there are many suppliers of RTUs, PLCs, and host production automaton computer systems.

In general, RTUs and PLCs should be selected based on the specific application for which they will be used. That is, select an RTU that is superficially designed and programmed to be a rod pump controller for a sucker rod operation. Select an RTU that is specifically designed and programmed to be a plunger lift controller for plunger lift. Select an RTU that is specifically designed and programmed to be a gas-lift controller for gas-lift. This means that if a field uses two or more types of artificial lift, it may have two or more types of RTUs or PLCs. This may cause problems with supply, spare parts, and training of operations and maintenance staff. However, these problems are small compared to the problems that will arise if an attempt is made to "force fit" one type of RTU or PLC to serve various types of operations for which it is not designed. And, if appropriate care is taken in the communications area (see Section 16.6.5), one production automation system can easily support multiple kinds of RTUs and PLCs.

Concerning selection of the host production automation computer system, the recommended approach is to select a system that is designed and programmed to support the types of gas production operations in the field. For example, if a field has flowing wells, plunger lift wells, chemical injection wells, pumping wells, and gas-lift wells, a host system should be selected that is designed and programmed to serve these operations with one common look and feel (e.g., user interface) and one common set of automation software services (e.g., alarming, reporting, plotting, etc.). In addition, the applications for the different types of gas production operations (e.g., artificial lift systems, well test systems, etc.) should communicate with one another when there is a need to share information. This may exclude some host systems that are offered by companies that, for example, focus on one form of artificial lift. But this is far preferable to having to support multiple host systems in the same field.

Environmental Protection

Typically automation equipment for gas production operations must function in extreme environmental conditions such as temperature, humidity, wind, dust, and corrosion. Fortunately, as with instruments and controllers, high quality RTUs, PLCs, and communications equipment are designed to withstand challenging environments. Here again, quality is the key word. This is not a place to accept low bid.

Communications

As discussed earlier, there are many options for communication between the instruments, the RTUs and PLCs, the host computer systems, and other systems. It is not possible to issue general guidelines. It is recommended that a communications system study be performed by a communications expert to select the best combination of communications methods, communications standards, protocols, and data security. Poor communications can lead to lack of system acceptance. This can be avoided by selection of the appropriate communications components.

Project Team

Staffing is discussed in Section 15.6.9. The recommendation here is to select an automation project team that has all the requisite skills, and to equip the team with strong management backing. It is clear that the team must have strong buy-in from operations, but it must also have strong support from management. Without support from both ends, success is doubtful.

Integration into the Organization

An automation system will fail if it is viewed as the property of one engineer, or even one part of the organization. To be successful over the long term, it must be viewed as essential by operations, maintenance, well analysis, production engineering, reservoir engineering, accounting, well services, and management. This is easy to say but challenging to accomplish.

15.6.3 Justification

Each company has its own criteria for justifying and approving projects, so this won't be discussed here. The economic justification is based on the tangible benefits discussed in Section 15.6.1 and the capital, operating, and maintenance costs discussed in Sections 15.6.4 and 15.6.5. Often the intangible benefits are used to enhance the overall justifications of the project. So, a project that is OK based solely on economics may be easily justified if safety, environmental project, and other qualitative benefits are taken into consideration.

The Impact of Time

An important factor in project justification is time. This must be considered in several ways:

- First there is the time value of money. Again, each company has its own way to handle this.
- The increased production benefits will be realized over time.
- The additional production due to achieving greater reservoir recovery will be realized over time. An issue here is acceleration vs. reserve addition. This is discussed later.

- The reduced operating costs will be realized over time.
- Operating and maintenance costs will be incurred over time.
- And over time it is likely that there will be additional capital expenditures. With the rate of obsolescence of equipment, it is likely some or all of the capital components will need to be replaced every few years.

So, to prepare a proper project justification, one must develop a life-cycle plan for realization of benefits and incurring of costs.

Acceleration vs. Increased Recovery

Another aspect of project justification comes in the question of production acceleration vs. reserve addition. If the production automation system helps to produce the same amount of gas, but produce it sooner, this is acceleration. To properly account for this, one must have a before and after production forecast and a forecast of prices and costs over time. None of these are easy to come by.

If the automation system helps to produce more gas by helping to increase ultimate recovery, this may have a significantly higher economic impact. If, for example, the system can help keep liquid out of the wellbore by helping to keep the gas production rate above critical or by helping to keep the well pumped off, there may be an excellent opportunity for increased ultimate recovery from the reservoir.

The Role of Pilot Tests

Some companies are reluctant to expend large sums on an automation project without proving that the projected costs and benefits are real. In former times, pilot projects to test the cost and benefit assumptions were common. Sometimes they were worthwhile:

- Sometimes expected costs were confirmed, although often costs were higher than expected.
- Sometimes expected benefits were confirmed, although often benefits were difficult to measure because many other things were happening in the field at the same time (e.g., more drilling, major workover programs, secondary recovery projects, instillation of new artificial lift systems, etc.).

In general, the biggest benefits from most pilot projects have been the experiences gained by the project staff. Having been through a project once, they have been better equipped to do it again on a larger scale. One of the major shortcomings has been deferral of benefits. It can take a few years to fully define, install, operate, and evaluate a pilot project. While this is happening, the benefits that could be realized in the rest of the field or area are not being obtained.

In modern times, many production automation systems have been installed. Many of them have been successful (see Section 15.7.1). Those that have been less than fully successful (see section 15.7.2) or have been failures (see section 15.7.3) have largely been so due to lack of attention to the issues discussed in this section. If the recommendations and guidelines in this section are followed, it is not necessary to conduct pilot tests of production automation systems for gas production operations.

There is one exception. Although it is not necessary to pilot test an automation system, it may be appropriate to pilot test special new hardware, software, artificial lift systems, and such. Often the supplier of these new systems is very anxious to have the technology tested and proven, so is very willing to participate in field tests, and may even be willing to share in the cost of the test.

15.6.4 CAPEX

A capital expenditure (or CAPEX) is an expenditure that creates future benefits. Companies use CAPEX to acquire or upgrade physical assets such as equipment or property. In accounting, a capital expenditure is added to an asset account or is *capitalized*, increasing the asset's basis—the cost or value of an asset as adjusted for tax purposes.

In performing an economic evaluation for a new automation project, there are two distinct forms of CAPEX to consider: (1) those capital expenditures (or savings) impacting the income stream and (2) those capital expenditures impacting the expense stream.

One of the major objectives of automation is to improve the profitability of an asset by reducing CAPEX over time. An example would be installation of pump-off controllers or other types of wellsite intelligence to prevent wear and tear on downhole equipment, thus increasing run-life and reducing the number of recompletions required over the life of the well. This incremental savings in CAPEX would be an example of a CAPEX that affects the project's income stream.

Such savings in CAPEX are normally realized over an extended period of time.

When deploying an automation project, there are a variety of capital expenditures that impact the expense stream for the project. Generally, these expenses are incurred early in the project, at or prior to the time of deployment. Typical items that affect the project's CAPEX include:

- Instrumentation
- Controls
- Wellsite intelligence (RTUs, PLCs, VFDs, and other controllers)
- Wiring
- Communications infrastructure (radios, towers, fiber optics, modems, etc.)
- Servers
- Desktop computers and/or hardware upgrades
- Software license fees (Host systems, historians, and others)
- Related services that are not treated as OPEX (dependant on corporate accounting practices)

15.6.5 OPEX

An operational expense (or OPEX) is the ongoing cost of running a business, product, or system. Similar to the discussion of CAPEX, there are operational expenses (or savings) that impact the income stream of a project and expenses that impact the expense stream of a project.

In addition to reducing long-term capital expenditures, automation projects also strive to improve an asset's profitability over time by reducing the ongoing OPEX. Such savings in OPEX would impact the project's income stream. Examples of savings include reduction in energy consumption through more efficient operation of artificial lift systems, reduction in man-power requirements, reduced equipment maintenance, and others.

As with CAPEX, there are a variety of operating expenses that are incurred in conjunction with deployment of a new project. These forms of OPEX impact the project's expense stream. Examples include:

- IT (Information Technology) support
- Project management
- Systems integration
- Telecommunications fees

- Web-hosting services
- Software leasing fees

15.6.6 Design

Design of an automation system is a critical step that can have long-term consequences on the asset and the organization. Yet, surprisingly this step is often overlooked, allowing automation systems to be deployed that are too costly, are incompatible with existing infrastructure or overall IT needs, or fail to deliver the benefits that were originally envisioned. For this reason, it is important that careful attention be given to the design of the automation system and that the system is designed with the overall needs of the organization in mind.

There are three major factors that influence the success or failure of a technology project: (1) people, (2) process, and (3) technology. Effective design of automation systems considers the needs of each.

People

For an automation project to be successful, it is important that all the key stakeholders be engaged in the design process at the earliest possible stage. This includes end users such as engineering and operations staff, information technology staff, management, and the various providers of system components. For each of these stake-holders, it is important to determine how the system will be used and what these individuals hope to accomplish with the system. Although this may seem basic, more often than not projects are deployed without truly considering the needs of the people using the systems or the overall goals of the project. This very often leads to automation systems that are a technical success but fail to meet the business needs of the enterprise.

Perhaps the most important group of stakeholders is the actual end users of the technology. Often, new technology can have a disruptive and sometimes threatening impact on individuals in the organization. This can lead to lack of use or misuse of the technology once it is deployed. For end users to adopt new technology, they need to feel they have an emotional stake in the success of the project and that it is in their best interests to see the project succeed. Engaging end users in the functional specification and design of the system yields long-term rewards in the form of a system that is fit-for-purpose and a workforce that is eager to see the project succeed.

For a project to be successful, it must have management buy-in. For this to happen, the project needs to be designed so it supports the organization's business objectives. If, for instance, the organization's goal is to reduce expenses, the system needs to control costs while achieving the technical aims of the project. Also, most successful projects have an executive sponsor. By engaging individuals in the management team in the design of the project, it is far more likely that these key influencers will do what they can to ensure the project's success.

Another key group of stakeholders is the company's information technology team. This group manages the deployment of information technology throughout the organization as well as setting the overall IT strategy for the enterprise. They are uniquely positioned to guide the development and deployment of technology projects. This ensures that existing technology is properly leveraged, reducing the overall cost of the system and minimizing any disruption to IT services while ensuring that the system will continue to work as the underlying infrastructure evolves.

A final group of stake-holders are the actual vendors of the various pieces of automation technology. These organizations typically have significant experience and expertise in the automation domain and first-hand knowledge of what works and what doesn't. By engaging these organizations early in the design process, operating companies can gain insight into what types of solutions are best suited to their particular needs and avoid making costly mistakes.

Process

In any given production operation, certain processes have been established over time to govern how work gets done. Often, these processes are undocumented and may not be recognized by the individuals performing them. Nonetheless, these processes exist and govern all the basic functions of that operation. When new technology is deployed in an asset, it disrupts these processes and challenges people to do their work in new and different ways. If the project is implemented without paying consideration to these underlying processes, it is likely that individuals will go back to doing their work the way they've always done it, causing the technology to be unused. Successful technology projects recognize the need to integrate the technology with company work processes, and include these workflows in the design of the system.

In one such project, a major operator sought to build a real-time system for managing their subsurface maintenance program for a large onshore operation. Prior to deploying the project, this organization spent nearly one year examining the various processes of the organization and mapping out the various roles and responsibilities of those individuals responsible for these activities, particularly as they related to this new system. These processes ultimately were incorporated in the functional requirements specifications for the real-time system. In addition, new roles were created to address organizational gaps, extensive training was provided to users and standard operating procedures were developed that described exactly how key tasks would be performed with the new system and by whom. Although this may seem like a significant amount of work, the result was a technology project that was quickly adopted throughout the organization and easily integrated into their day-to-day business.

Technology

Automation technology is evolving at a startling rate. The result is a myriad of options for virtually every component in any system. This gives operating companies flexibility in designing a system to suit their needs, but also makes the task enormously complex. The challenge is to design a system selecting technology that is fit-for-purpose, rather than choosing technology for technology's sake.

To aid in the evaluation of new technology, the Society of Petroleum Engineers Real-Time Optimization Technical Interest Group (RTO TIG) has proposed a methodology for classifying and assessing various system components. This allows organizations to understand the entire scope of a technology project and identify opportunities for improvement. This methodology divides the automation system into seven major components:

1. Measurement (sensors)
2. Telemetry
3. Data handling and access
4. Analysis
5. Visualization
6. Automatic control
7. Integration and automation

For each of these system components, the technology is assessed and assigned a level for comparison and tracking progress. These levels are defined as follows:

- **Level 1**—ad hoc (manual or disjointed system)
- **Level 2**—multifunctional
- **Level 3**—integrated

Once these major system components are evaluated, a spider diagram can be constructed, such as the one in Figure 15-48.

Each of the system components moves from an initial level (dashed line) to a higher level of maturity, following the implementation of a technology project. The axes represent technology level (0, 1, 2, 3). Movements along any given axis represent some improvement in that technology that may be economic or related to some other non-economic project objective. By using tools such as this one, operators can easily identify which key areas require investment in technology, and budget accordingly. Also, once the project is complete, this provides a benchmark for evaluating the success of the project from a technology standpoint.

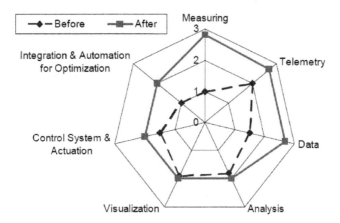

Figure 15-48: Spider Diagram Illustrating Technology Status

15.6.7 Installation

Often, the installation process begins many weeks or months prior to deployment of equipment and software in the field. The various automation vendors painstakingly test the components in their own labs to ensure they will function as designed and work together as a system. Once they are certain that the various systems can function as envisioned, they will work hand-in-hand with the operator's IT department and other service providers to commission the systems and get everything up and running.

Throughout this process, a variety of tasks are performed by numerous individuals. Among these are the following:

- Obtain historical data
- Set-up test servers
- Load historical data into host system
- Site acceptance testing (host system, RTUs, and other end devices)
- Installation of instrumentation
- Installation and configuration of controllers
- Installation and configuration of communication devices
- Commissioning host system
- Customization and configuration of host
- Installation of client software
- User training

15.6.8 Security

Because of the sensitive nature of automation equipment and technology, security is always a concern of operators. This relates both to security of the physical assets and security of proprietary data. To address these concerns, a variety of steps are taken in the course of deploying an automation project.

Field Devices

The hardware that is deployed in the field—particularly in remote locations—can often pose a tempting target for thieves. Items such as solar panels, pressure transducers, RTUs, and copper wiring can all be easily stolen, and often are. To safeguard against theft, operators may take a variety of steps. Devices can be installed inside steel cages, bunkers, or other enclosures where they are out of site and not easily reached. In other cases, operators select hardware that is unlikely to attract the attention of thieves. In some cases, regular patrols are made by security personnel or other company employees to deter theft.

Because of the mission critical nature that is performed by end devices in the field, operators also want to ensure that only authorized individuals are able to adjust settings within those devices. For this reason, virtually all RTUs, PLCs, and other controllers provide security features that require user login, govern who can access what features, and document any changes that were performed to the device settings, who made the changes, and when they were made.

Host Systems

Host systems provide multiple layers of security. The following summarizes the most common security features pertaining to host systems:

- Configuration of the system is based on access privilege governed by the host system itself.
- Users are able to access only those features and perform those functions that are governed by their host system login credentials.
- Any changes made to system settings or data are logged in the system, identifying when the change was made and by whom.
- Systems are deployed within the corporate firewall, often within a corporate intranet. These systems are not accessible over the Internet or from individuals outside the company.
- Prior to logging in to the host system, users must first log in to their corporate network with their normal user credentials.
- In certain cases, the host system can only be accessed via Citrix™ connection or Remote Desktop session, requiring yet another level of user login.
- In many cases, operators employ technologies such as smart cards, tokens with rotating passwords, and even biometrics to further restrict access to the system.

15.6.9 Staffing

At least three teams are recommended for a production automation project. Table 15-6 shows the recommended members of each team and whether or not the members should be in-house staff or may be others, for example, from service companies or consultants.

The three teams are the Steering Team or Steering Committee, the Automation Team, and the Surveillance Team. These teams share some

Table 15-6
Production Automation Teams

Staff Position	In House Staff?	Teams		
		Steering	Automation	Surveillance
Champion	If possible	Facilitator	Advisor	
Management	Definitely	Chair	Chair	
Project engineer	If possible	Yes		
Engineering	If possible	Yes	Yes	Yes
Automation specialists		Yes	Yes	
			Yes	
Technicians	If possible			
Automation support			Yes	
Operations	If possible	Yes	Yes	Yes
Maintenance	If possible	Yes	Yes	Yes
Well analysis	If possible	Yes	Yes	Chair
Well servicing				
Accounting/finance		Yes		
Others/service company			Sometimes	Sometimes

of the same members, they have different responsibilities, and they operate over different time periods.

Steering Committee

The purpose of the Steering Committee is to provide overall priority, justification, direction, and focus for an automation project. It must have representation from a broad spectrum of stake holders, managers or leaders, automation providers, and users. It must be chaired by a member of the operating company's management team to assure strong management support. The Steering Committee may exist over the life of an automation system, but its primary function will be during the early months or years of the project when it is being defined, justified, staffed, and so on.

Companies with successful production automation projects have an automation Champion. This is a person who is strongly committed to the overall automation effort. This person may be in management, engineering, or operations. In most cases, this person will be a member of the operating Company staff, but in some cases, he or she may be from a third party. The representative of management must chair the Steering

Committee, but the Champion will be the facilitator to call meetings, set agendas, "drive" the project schedule, and so on.

Automation Team

The Automation Team is responsible for execution. They define, design, build, test, implement, commission, and maintain the system. Typically the Champion will be an advisor to this team and a Project Engineer will be its chair. This team must exist from the very start of an automation project and must continue, in one form or another, for the entire life of the system, since the jobs of making enhancements, providing maintenance, and training staff are never finished.

This team must have members with special skills from each of the engineering disciplines involved, including applications, instrumentation, communications, hardware, software, and training. Some of these functions may be provided by third-party staff (e.g., experts in instrumentation, communications, hardware, software, and training). However, someone on the project team must oversee each of these, assure that they are performed, and assure that they meet the project objectives.

The team must have members from Operations, Maintenance, and Well Analysis to provide input and feedback as the project is implemented. The system must meet their needs and they must "buy in" to the system.

Surveillance Team

The Surveillance Team (it may have other names) uses the system. They use it every day to monitor, control, and optimize the gas production operations. This team may have many members; many of them may not see themselves as part of a formal team and may not attend team meetings. But, they are the ones who use the system to gain the benefits for the company.

This team, like the others, must have a chair or focal point. In some companies, this person is called a Well Analyst. He or she may also be a Production Engineer, Production Technologist, Well Surveillance Specialist, Automation Specialist, or Lead Operator. The name doesn't matter, but the function is vital. This person must assure that people are continuously assigned and motivated to use the automation system for routine daily monitoring, control, and optimization; that they have the training they need; and they have the support they need from other functions in the company or from third parties for troubleshooting, maintenance, well servicing, system enhancements, training, and so on.

15.6.10 Training

Each person on the three teams defined in the last section must receive training. Three levels of training are defined: aware, knowledgeable, and skilled. Those that must receive each type of training are shown in Table 15-7.

Aware

Awareness training consists of the following:

- Attend "high level" production automation course or seminar.
- Maintain awareness of important production automation issues.
- Have good understanding of:
  - Relative merits of each form of production equipment, artificial lift, production automation economics.
  - Why production automation has been chosen for this field.
  - Interdependencies between production automation system and other production systems in the field.

Table 15-7
Automation Training Requirements

Staff position	Training Level Required			Comments
	Aware	Knowledgeable	Skilled	
Champion			■	Know entire system
Management	■			
Project engineer		■		
Engineering		■		Know entire system
Automation specialists			■	Know components
Technicians			■	
Automation support			■	Know support
Operations		■		
Maintenance			■	Know components
Well analysis			■	Know applications
Well servicing	■			
Accounting/finance	■			
Others/service company	■	■		Depends on job

- Skills and personal characteristics needed by *knowledgeable* and *skilled* production automation staff.
- Value of proper production automation system deployment, including production automation selection, design, installation, operation, optimization, troubleshooting, and surveillance.

Knowledgeable

To be knowledgeable, the following training is needed:

- Attend high-level production automation course or seminar.
- Attend intermediate-level production automation course that provides thorough understanding of production automation selection, design, installation, operation, optimization, troubleshooting, and surveillance.
- Maintain awareness of key production automation technologies and practices:
  - Spend time actually working in one or more facets of production automation.
  - Obtain full set of awareness that is required for the aware level.
  - Have detailed knowledge of both technical and business issues involved in production automation.
  - Have ability to advise people who are directly involved in production automation engineering and/or operations, by assisting them in (1) obtaining needed resources, (2) prioritizing their work, and (3) evaluating the economics of their projects, etc.

Skilled

To be skilled, the following training is needed:

- Attend high-level production automation course or seminar.
- Attend intermediate-level production automation that provides thorough understanding of production automation selection, design, installation, operation, optimization, troubleshooting, and surveillance.
- Attend comprehensive production automation courses that provide thorough and detailed understanding of all of the facets of production automation.
  - These courses should provide significant hands-on training in performing the various aspects of production automation.

- Obtain the full set of awareness and knowledge that are required for the aware and knowledgeable levels of competency.
- Maintain awareness of key production automation technologies and practices by continuing education.
  - Attend company and/or industry production automation workshops and/or seminars and sessions for sharing best practices.
  - Be fully conversant with key recommended practices produced and maintained by various sources in industry.
  - Spend time working under direct tutelage of an expert production automation engineer, well analyst, technician, or operator.
  - Obtain practical, hands-on experience with each aspect of production automation with which the person is involved.
  - Receive feedback on activities performed, in terms of evaluations of actual production automation installations.
  - Develop ability to train new staff in effective production automation engineering and/or operations.

There are a limited number of formal training programs in production automation. However, there are courses available from several service companies on specific subsystems or applications, and there are consultants who offer comprehensive automation system training.

15.6.11 Commercial vs. In-House

In former times, say in the 1960s, 1970s, and early 1980s, there wasn't much choice between commercial vs. in-house for supply of automation systems. There were very few commercial systems. If an operating company wanted a system, it had to develop it, and many operating companies did just that.

Beginning in the middle to late 1980s and 1990s, and certainly in the twenty-first century, the reverse is true. There are several commercial systems and very few operating companies develop their own systems. This is certainly true for automation systems for oil production and is becoming increasingly so for gas.

In current times, many operating companies have no choice. They don't have the staff or internal expertise to design and develop their own systems, so they are dependent on the suppliers. The good news is that operating companies who don't have the staff or expertise can acquire and use good automation systems. The bad news is that often they are constrained to use the systems offered by the suppliers, with little chance to optimize the systems to best meet their requirements.

The best approach, which is happening in a few cases, is a partnership between the operating company and the service company. They work together to fine-tune the system to meet the requirements of the operating company in terms of its specific technical requirements, geographical constraints, staffing needs, and so on. This teamwork approach is recommended whenever a significant gas production automation project is undertaken.

15.7 CASE HISTORIES

There have been some notable production automation success stories, some failures, and some systems that work but have not reached their potential. The purpose of this section is to summarize some of these experiences without providing specific references to companies or locations. The hope is that we can learn from our successes and failures.

15.7.1 Success Stories

Automation has been deployed in upstream oil and gas applications for several decades, with an acceleration in activity over the last 10 to 15 years. Many operators have achieved significant enhancements in operating efficiency. The following are some examples.

Rod Pump Controllers

An operator in the East Linden (Cotton Valley) field in East Texas purchased and installed RPCs, replacing time clocks, on 15 wells. These wells are 10,000 feet deep, producing 42 to 46° API gravity oil, with a 5 to 50 percent water cut. With contract pumpers and using timers set by trial-and-error, this operator saw the upside potential of using RPCs. Before installing RPCs, well problems were identified during daily visits. Timers were used to operate the wells, set based on a single visit to each well during a 24-hour period. The trial-and-error process to set these timers often would result in some wells pounding fluid for long periods of time. Another problem was under-pumping some wells where pumping time was as little as 3 hr/day. Gas break-out during this downtime resulted in the unrecognized need to pump the wells longer to move the gas as well as all available fluid. Factor in any pump wear—also requiring additional run time—and the possibility of under-pumping was very real.

The installation of RPCs eliminated the guesswork and trial-and-error involved in setting time clocks for correct well control. The RPCs also automatically adjusted idle time based on buffered data in the controller from historical cycle times. Observed benefits from the use of RPCs by this operator include reducing rod and tubing failures by 31 percent and electrical cost by 40 percent ($50,000 per year). For wells that were maintaining a high fluid level due to incorrect timer cycles, the RPCs increased fluid production.

Plunger Lift Automation

An operator in the San Juan Basin in northwestern New Mexico implemented a major automation project for a field producing coal bed methane via plunger lift. This project incorporated wellsite control using plunger lift controllers and electronic flow measurement, host software, and telemetry between the host and the wells.

Plunger lift controllers contained self-adjusting algorithms for plunger arrival time, after flow, and shut-in time based on preconfigured parameters to maximize gas production. This resulted in a sustained production uplift of 4 MMSCFD for 40 wells. In 30 tubing flow control installations, an average uplift of 130 MSCFD per well was achieved by automatically controlling the casing valve based on well conditions.

Host System/Workflow Management

A mature asset in the San Joaquin Valley, California was producing over 1000 wells via reciprocating sucker rod pump. Most of these wells were utilizing rod pump controllers that were communicating with a host system. In addition, a variety of tools were used in the asset to manage operational issues such as rig management. This project sought to consolidate various tools that were used throughout the organization into a single web-based platform that would standardize the overall workflow used for surveillance, optimization, and well services management. In addition, the organization sought to create a single "system of record" that would provide an electronic repository for all operational data created during the day-to-day operations of the field. One of the primary goals was to enable the ability to perform score-carding against this data so that the operator could uncover hidden performance trends and enable continuous improvement within the asset.

From a business standpoint, the operator hoped to improve the field's failure rate, minimize downtime and production deferrals, and minimize time to production. Prior to designing this system, the operator conducted a comprehensive examination of the organization's business practices and workflows. Based on this study, a system was specified that incorporated the business practices in place. Rather than adapting the organization to the software, the operator sought to implement a software system that reflected the organization's dynamics.

The tangible evidence of success in this project was the reduction in well failure rate that followed deployment of the software system. The improvement in failure rate was from 0.15 to 0.1 failures per well per year in the field. This corresponds to a cost saving for repairing failures of approximately $0.5 million per year in this field. As a result of this initial success, the system was deployed across the entire business unit. (Scaling this performance improvement up to the whole unit represents an annual saving of $6 million.) Since the software system enables online surveillance as well as automated standardized well service management planning and execution, production deferment on failures should reduce substantially. The time taken to diagnose a problem and schedule the right job to address it is much faster with the new system. An additional means of speeding the repair process is that contractors have access to the system and can see the appropriate scheduled jobs as soon as they are approved for action. An estimate of the annual savings in deferred production for the unit is $3 million.

15.7.2 Failures

People love to talk about success stories, but they rarely discuss failures. This is unfortunate because we learn more from the projects that fail than from the ones that succeed. The following are three cases where the projects' failures yield clear lessons.

Beam Pump Optimization

A real-time optimization project was implemented in an onshore field in Western Venezuela. This system incorporated rod pump controllers, which communicated to a dedicated host system in the field office via radio. Operators were able to use a variety of sophisticated features such as management by exception and downhole pump card diagnostics.

Initially, extensive training was conducted as well as regular support of end users in the field. During the early years of the project, the system was an integral part of the production management process, with numerous documented benefits.

However, over time, personnel transferred out of the asset and much of the local knowledge of the system was lost. As new personnel were brought into the asset, they did not receive training. In addition, the automation provider interfaced primarily with the company IT department, who were principally concerned with the technical performance of the software and servers. As a result, the automation company remained insulated from end users' concerns. As one might predict, this system fell into disuse and, over time, reached a state where well test data was no longer being imported into the system and none of the wells were configured.

The lesson learned is that once a system is deployed, the work is not finished. For an automation system to yield value over the life of the asset, there needs to be an evergreen process for training and support.

PCP Optimization

A real-time PCP optimization system was deployed in a 300-well field in Latin America. This system incorporated electronic downhole gauges, variable frequency drives, automatic well testing using multiphase flow meters, and a sophisticated host system. Although a data historian had been installed in the field, the system was designed such that most of the analog data was fed directly into the host system, bypassing the historian. Data was collected in one-minute intervals from each of the 10 analog instruments per well. As a result, the host system was able to store only approximately two and a half days' worth of data before overwriting its circular buffer. This meant that to preserve this data for future analysis, a technician had to travel to the field location and manually download data to a DVD several times per week. Although it was then possible to find the data from a specific point in time and recover this for analysis (i.e., pressure transient analysis), it was not possible to perform long-term surveillance within the host system. This rendered many of the features of the host system useless and prevented engineers and operators from evaluating any long-term performance trends.

The lesson learned is that it is important to include all key stakeholders in the design of an automation system and ensure that the system architecture supports the needs of the users as well as the project's business objectives.

Gas-Lift Automation

In one project, an operator sought to deploy a real-time optimization project for a large population of wells. The intent was to build an integrated model that included well performance, network performance, and facilities performance while regularly updating the model with real-time data from the automation system and well test database.

After an extensive review of various proposed solutions, the operator implemented a "home-grown" solution in which a team of consultants built a custom human/machine interface that interfaces with the facilities software, well analysis software, and network optimization software. The result was a system that took significantly longer to build and cost significantly more than anticipated, and that required a staff of dedicated full-time consultants to keep the system running. Further, once the system was deployed, many of the intended capabilities were not achieved and use of the system resulted in minimal improvement in fieldwide performance.

The main lesson learned from this case speaks to the issue of "build vs. buy." There are pros and cons to building a system from scratch just as there are for buying one that is available off-the-shelf. However, if a suitable system exists in the market that meets the majority of needs, it is generally better to buy the off-the-shelf solution. This provides at least an 80-percent solution at a fraction of the cost and time to deployment. Also, the burden of supporting this product is born by the vendor rather than by the organization, whereas costs of ongoing improvements to the product are shared by all of its customers.

15.7.3 Systems That Haven't Reached Their Potential

Some automation systems technically are qualified as a success, yet fail to achieve their full potential. Often this is due to lack of use resulting from inattention to soft rather than hard issues. This is often because the individuals expected to use the system are never properly trained in its use, are not engaged in the specification or design process, are fearful that the technology may displace them, or all of these reasons.

In one project, an operator had implemented a successful pilot project in a large onshore asset. This project utilized wellsite intelligence and sophisticated host systems. As a result, they were able to realize substantial reductions in downtime and increases in incremental production. Based on the initial success, they opted to implement the same tools across each of the remaining assets in the business unit. Much to their surprise, none of the other fields achieved measurable improvement in the key performance metrics.

After investigating the use of the system across the business unit, it became clear that the problem was one of managing soft issues. Even though the same technology was deployed across all the assets, the approach to change management was not the same. In the pilot project, a team concept was encouraged where operators and engineers worked as equal stakeholders with common goals that were communicated in daily operational review meetings. All the individuals in the team were fully engaged in the design and implementation of the system and felt that they had a personal stake in its success. However, when the system was implemented in other assets within the business unit, the technology was deployed but the organizational approach was not. The lesson learned was that in technology projects, soft issues can be the difference between success and failure.

15.8 SUMMARY

Considering the value of gas, the costs of staff and services, the negative impact of liquid loading on gas production and ultimate recovery, the requirement to remove liquid loading over the long term, the importance of gas well monitoring, control, and optimization, and the difficulty of performing these manually, the business case for production automation is compelling.

Production automation equipment and applications exist to assist with effective management of gas well operations. This equipment and these applications must be carefully selected, configured, installed, operated, and maintained. This requires attention to many details, including project staffing and training.

The cost of automation systems is low enough that every gas well, and especially every gas well that has an artificial lift system, should be automated.

15.9 REFERENCES

The following references have been used in compiling this chapter. All are acknowledged with gratitude.

Section 15.2

1. Dunham, C. L. and Anderson, S.R. "The Generalized Computer-Assisted (CAO) System: A Comprehensive Computer System for Day-to-Day Oilfield Operations," paper SPE 20366 presented at the 1990 SPE Petroleum Computer Conference, Denver, CO, June 25–28.
2. Dunham, C. L. "Supervisory Control of Beam Pumping Wells," paper SPE 16216 presented at the 1987 SPE Production Operations Symposium, Oklahoma City, OK, March 8–10.

Section 15.3.1

1. Examples of transmitters: http://www.foxboro.com.
2. Information on transducers: http://sensors-transducers.globalspec.com.

Section 15.3.2

1. Requirements for EFM: American Petroleum Institute Manual of Petroleum Measurement Standards, Chapter 21—Flow Measurement Using Electronic Metering Systems, Section 1—Electronic Gas Measurement.

Section 15.3.3

1. Burrell, G. R. *Petroleum Engineering Handbook*, Third Printing, Chapter 16: Automation of Lease Equipment, Society of Petroleum Engineers, 1992.
2. Examples of control valves: http://www.emersonprocess.com/fisher/oilandgas.html.
3. Examples of motor controller, switchboard and VFD: http://www.weatherford.com.

Section 15.3.6

1. Information on communications: http://www.freewave.com/.

Section 15.3.7

1. Buchmann, A. "Real Time Database Systems," *Encyclopedia of Database Technologies and Applications*, Laura C. Rivero, Jorge H. Doorn, and Viviana E. Ferraggine, Eds., Idea Group, 2005.

2. Carpron, H. L. and Johnson, J. A. *Computers: Tools for the Information Age,* 5th ed., Prentice Hall, 1998.

3. Stankovic, J. A., Son, S. H., and Hansson, J. "Misconceptions about Realtime Databases," Cybersquare. University of Virginia. June, 1999.

Section 15.3.8

1. Information on data exchange: http://posc.org/, http://www.energistics.org/posc/PRODML_'07_Work_Group.asp?SnID=6378497.

Section 15.4.1

1. Example of user interface: Microsoft Windows User Interface standard.

Section 15.4.3

1. Example alarm display: Shell Global Solutions, Fieldware, http://www.shell.com/home/Framework?siteId=globalsolutions-en.

Section 15.4.5

1. Information on trend plots: SPE 49463, Real-Time Artificial Lift Optimisation, W.J.G.J. der Kinderen, H. Poulisse, C. L. Dunham, Member SPE, SIEP-RTS, The Netherlands.

Section 15.4.6

1. Example display: Shell Global Solutions, Fieldware, http://www.shell.com/home/Framework?siteId=globalsolutions-en.

Section 15.4.7

1. Example data historian access: http://www.osisoft.com/Products/PI+Process+Book.htm.

Section 15.5.1

1. Information on Plunger Lift: Weatherford Artificial Lift Training Course.

2. Plunger lift images: eProduction Solutions, http://www.ep-solutions.com/.

Section 15.5.2

1. Schematic of sucker rod pumping well, method for detecting pump-off, example of rod pump controller: eProduction Solutions, http://www.ep-solutions.com/.

2. Example of sucker rod surveillance: Shell Global Solutions, Fieldware, http://www.shell.com/home/Framework?siteId=globalsolutions-en.

3. Example of gas vent pump to overcome gas locking: Harbison-Fischer, http://www.hfpumps.com/.

Section 15.5.3

1. Schematic of progressing cavity pumping well: Shell Global Solutions, Fieldware, http://www.shell.com/home/Framework?siteId=globalsolutions-en.

2. Examples of PCP surface drive systems: Baker Hughes Centrilift, http://www.bakerhughes.com/centrilift/PCPS/driveheads.htm.

3. "Thru-Tubing Conveyed Progressive Cavity ESP—Operational Issues, A Short Story," J. Ryan Dunn, *et al.*, 2002 ESP Workshop, Houston, Texas.

4. Special PCP alarms: eProduction Solutions, http://www.ep-solutions.com/.

5. Information on PCP-RIFTS and PCP pump design: C-FER Technologies, http://www.cfertech.com/ and http://www.pc-pump.com/.

6. Information on controls: Burrell, G. R., *Petroleum Engineering Handbook*, Third Printing; Chapter 16: Automation of Lease Equipment, Society of Petroleum Engineers, February, 1992.

Section 15.5.4

1. Schematic of electrical submersible pumping well: eProduction Solutions, http://www.ep-solutions.com/.
2. Casing pressure gradient. Automation Training Program. Oilfield Automation Consulting, www.oilfieldautomaton.com.
3. RTU components: eProduction Solutions, http://www.ep-solutions.com/.
4. ESP surveillance panel: eProduction Solutions, http://www.ep-solutions.com/.
5. ESP surveillance plots: Automation Training Program. Oilfield Automation Consulting, www.oilfieldautomaton.com.
6. Information on ESP-RIFTS: C-FER Technologies, http://www.cfertech.com/.

Section 15.5.5

1. Simpson, D. A., Lea, J. F., and Cox, J. C. "Coal Bed Methane Production," Paper SPE 80900, presented at Production and Operations Symposium, Oklahoma City, OK, March 2003, 23–25 .
2. Information on Hydraulic Lift: Weatherford Hydraulic Lift Course.

Section 15.5.7

1. Schematic of gas-lift system, schematic of gas-lift mandrel, schematic of gas-lift valve, example of injection pressure heading, downhole plot of injection and production pressure profiles. Gas-lift course materials. OGCI Petroskills, http://petroskills.com/.
2. Turner, R. G., Hubbard, M. G., and Dukler, A. E. "Analysis and Prediction of Minimum Flow Rate for the Continuous Removal of Liquids from Gas Wells," *Journal of Petroleum Technology*, Nov. 1969, 1475–1482.
3. Coleman, S. B., Clay, H. B., McCurdy, D. G., and Norris, H. L. III. "A New Look at Predicting Gas-Well Load Up," *Journal of Petroleum Technology*, March 1991, 329–333.

Section 15.5.9

1. Pigott, M. J., Parker, M. H., Vincente, D., Dalrymple, L. V., Cox, D. C., and Coyle, R. A. "Wellbore Heating to Prevent Liquid Loading," SPE 77649, presented at the SPE Annual Technical Conference and Exhibition, San Antonio, TX, September 2, 2002.

Section 15.6.6

1. Mochizuki, *et al.* "Real Time Optimization: Classification and Assessment," SPE 90213, presented at the SPE Annual Technical Conference and Exhibition, Houston, TX, September 28–29 2004.
2. Ormerod, *et al.* "Real-Time Field Surveillance and Well Services Management in a Large Mature Onshore Field: Case Study," SPE 99949, presented at the 2006 SPE Intelligent Energy Conference held in Amsterdam, The Netherlands, April 11–13, 2006.

Section 15.6.9

1. Table of Production Automation Teams: Automation Training Program. Oilfield Automation Consulting, www.oilfieldautomation.com.

Section 15.6.10

1. Table of Production Automaton Training: Automation Training Program. Oilfield Automation Consulting, www.oilfieldautomation.com.

DEVELOPMENT OF CRITICAL VELOCITY EQUATIONS

A.1 INTRODUCTION

This appendix summarizes the development of the Turner [1] equations to calculate the minimum gas velocity to remove liquid droplets from a vertical wellbore.

A.1.1 Physical Model

Consider gas flowing in a vertical wellbore and a liquid droplet transported at a uniform velocity in the gas stream as illustrated in Figure A-1.

The forces acting on the droplet are gravity, pulling the droplet downward, and the upward drag of the gas as it flows around the droplet.

The gravity force is

$$F_G = \frac{g}{g_C}(\rho_L - \rho_G) \times \frac{\pi d^3}{6}$$

and the drag force is given by

$$F_D = \frac{1}{2 g_C} \rho_G C_D A_d (V_G - V_d)^2$$

where

g = gravitational constant = 32.17 ft/s^2
g_C = 32.17 lbm-ft/lbf-s^2
d = droplet diameter

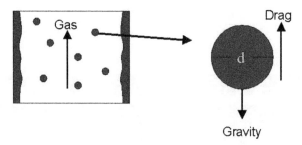

Figure A-1: Liquid Droplet Transported in a Vertical Gas Stream

ρ_L = liquid density
ρ_G = gas density
C_D = drag coefficient
A_d = droplet projected cross-sectional area
V_G = gas velocity
V_d = droplet velocity

The critical gas velocity to remove the liquid droplet from the wellbore is defined as the velocity at which the droplet would be suspended in the gas stream. A lower gas velocity would allow the droplet to fall, resulting in liquid accumulation in the wellbore. A higher gas velocity would carry the droplet upward to the surface and remove the droplet from the wellbore.

Thus, the critical gas velocity V_C is the gas velocity at which $V_d = 0$. In addition, since the droplet velocity is zero, the net force on the droplet also is zero. The defining equation for the critical gas velocity is then

$$F_G = F_D$$

or

$$\frac{g}{g_C}(\rho_L - \rho_G)\frac{\pi d^3}{6} = \frac{1}{2g_C}\rho_G C_D A_d V_C^2$$

Substituting $A_d = \pi d^2/4$ and solving for V_C gives

$$V_C = \sqrt{\frac{4g}{3}\frac{(\rho_L - \rho_G)}{\rho_G}\frac{d}{C_D}} \tag{A-1}$$

This equation assumes a known droplet diameter. In reality, the droplet diameter is dependent upon the gas velocity. For liquid droplets entrained in a gas stream, [2] shows that this dependence can be expressed in terms of the dimensionless Weber number:

$$N_{WE} = \frac{V_C^2 \rho_G d}{\sigma g_C} = 30$$

Solving for the droplet diameter gives

$$d = 30 \frac{\sigma g_C}{\rho_G V_C^2}$$

and substituting into Equation (A-1) gives

$$V_C = \sqrt{\frac{4}{3} \frac{(\rho_L - \rho_G)}{\rho_G} \frac{g}{C_D} 30 \frac{\sigma g_C}{\rho_G V_C^2}}$$

or

$$V_C = \left(\frac{40 g g_C}{C_D}\right)^{1/4} \left(\frac{\rho_L - \rho_G}{\rho_G^2} \sigma\right)^{1/4}$$

Turner assumed a drag coefficient of $C_D = .44$ that is valid for fully turbulent conditions. Substituting the turbulent drag coefficient and values for g and g_C gives

$$V_C = 17.514 \left(\frac{\rho_L - \rho_G}{\rho_G^2} \sigma\right)^{1/4} \ ft/s \qquad (A-2)$$

where

ρ_L = liquid density, lbm/ft^3
ρ_G = gas density, lbm/ft^3
σ = surface tension, lbf/ft

Equation (A-2) can be written for surface tension in dyne/cm units using the conversion lbf/ft = .00006852 dyne/cm to give

$$V_C = 1.593 \left(\frac{\rho_L - \rho_G}{\rho_G^2} \sigma\right)^{1/4} \ ft/s \qquad (A-3)$$

where

ρ_L = liquid density, lbm/ft^3
ρ_G = gas density, lbm/ft^3
σ = surface tension, dyne/cm

A.2 EQUATION SIMPLIFICATION

Equation (A-3) can be simplified by applying typical values for the gas and liquid properties. From the real gas law, the gas density is given by

$$\rho_G = 2.715\gamma_G \frac{P}{(460+T)Z} \, lbm/ft^3 \qquad (A-4)$$

Evaluating Equation (A-4) for typical values of

Gas gravity γ_G 0.6
Temperature T   120   F
Gas deviation factor Z   0.9

gives

$$\rho_G = 2.715 \times .6 \frac{P}{(460+120) \times .9} = .0031P \, lbm/ft^3$$

Typical values for density and surface tension are

Water density	67 lbm/ft³
Condensate density	45 lbm/ft³
Water surface tension	60 dyne/cm
Condensate surface tension	20 dyne/cm

Introducing these typical values and the simplified gas density Equation (A-4) into Equation (A-3) yields

$$V_{C,water} = 1.593 \left(\frac{67 - .0031P}{(.0031P)^2} 60 \right)^{1/4} = 4.434 \frac{(67 - .0031P)^{1/4}}{(.0031P)^{1/2}} \, ft/s$$

$$V_{C,cond} = 1.593 \left(\frac{45 - .0031P}{(.0031P)^2} 20 \right)^{1/4} = 3.369 \frac{(45 - .0031P)^{1/4}}{(.0031P)^{1/2}} \, ft/s$$

A.3 TURNER EQUATIONS

Turner et al. [1] found that for his field data, where wellhead pressures were typically ≥ 1000 psi, a 20 percent upward adjustment to the theoretical values was required to match the field observations. Applying the 20 percent adjustment then yields

$$V_{C,water} = 5.321 \frac{(67 - .0031P)^{1/4}}{(.0031P)^{1/2}} \ ft/s$$

$$V_{C,cond} = 4.043 \frac{(45 - .0031P)^{1/4}}{(.0031P)^{1/2}} \ ft/s$$

However in the original paper by Turner [1], the coefficients were found to be 5.62 for the previous critical water velocity equation and 4.02 for the previous condensate critical velocity, but these values are slightly in error as just developed.

A.4 COLEMAN ET AL. EQUATIONS

Coleman et al. [3] found that Equation (A-3) was an equation that would fit their data. This was without the 20 percent adjustment that Turner made to fit his data at higher average wellhead pressures. So if the "corrected" Turner equations are written without the 20 percent adjustment, then the Coleman equations can be written as follows if the same simplifications and typical data are used as before:

$$V_{C,water} = 4.434 \frac{(67 - .0031P)^{1/4}}{(.0031P)^{1/2}} \ ft/s$$

$$V_{C,cond} = 3.369 \frac{(45 - .0031P)^{1/4}}{(.0031P)^{1/2}} \ ft/s$$

A.5 REFERENCES

1. Turner, R. G, Hubbard, M. G., and Dukler, A. E. "Analysis and Prediction of Minimum Flow Rate for the Continuous Removal of Liquids from Gas Wells," *Journal of Petroleum Technology*, Nov. 1969, 1475–1482.

2. Hinze, J. O. "Fundamentals of the Hydrodynamic Mechanism of Splitting in Dispersion Processes," *AICHE Journal*, September, 1955, 1, No.3, 289.

3. Coleman, S. B, Clay, H. B., McCurdy, D. G., and Norris, H. L. III, "A New Look at Predicting Gas-Well Load Up," *Journal of Petroleum Technology*, March 1991, 329–333.

DEVELOPMENT OF PLUNGER LIFT EQUATIONS

B.1 INTRODUCTION

This appendix summarizes the plunger lift equations developed in [1] for a plunger lifted well as illustrated in Figure B-1.

B.2 MINIMUM CASING PRESSURE

The minimum casing pressure at the moment that the plunger and liquid slug arrive at the surface is given by:

$$P_{C,\min} = (14.7 + P_P + P_{wh} + P_C S_V)(1 + D/K) \tag{B-1}$$

where

P_P = pressure required to lift the plunger, psi
P_C = pressure required to lift one barrel of liquid and overcome friction, psi
Pwh = tubing well head pressure, psig
S_V = liquid slug volume above plunger, bbl
K = factor to account for gas friction below the plunger
D = plunger depth, ft

Approximate values for K and P_C are given in Table B-1.
In Equation (B-1), K is calculated from

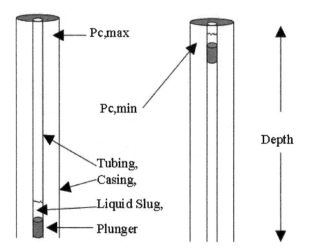

Figure B-1: Schematic of Plunger Lift before Plunger Release and Just as Plunger and Liquid Reach the Surface

Table B-1
Approximate Values for K and P_c from [1]

Tubing Size (inch)	K	P_c
2.375	33,500	165
2.875	45,000	102
3.000	57,600	67

$$K = \frac{Z(T_{avg}+460)(OD_{TBG}/12)(2 \times 32.2 \times 144 \times 3600)}{(144/53.3 \times \gamma_G \times f_{gas} \times V^2)}$$

where

T_{avg} = the average temperature, °F
Z = the average gas deviation factor
γ_G = the produced gas gravity
f_{gas} = friction factor for gas flow in tubing
V = average gas velocity, ft/sec
OD_{tbg} = tubing OD, inches

The factor P_C is calculated from

$$P_C = P_{WEIGHT} + P_{FRICTION}$$

$$P_{WEIGHT} = .433 \times \gamma_L \times L_S$$

$$P_{FRICTION} = \frac{62.4\gamma_L f_L L_S V^2}{(ID_{TBG}/12) \times 2 \times 32.2 \times 144 \times 3600\}}$$

where

γ_L = produced liquid gravity
f_L = friction factor for liquid flow in tubing
L_S = liquid slug length
V = average liquid velocity, ft/sec
ID_{tbg} = tubing ID, inches

The approach for the development of the minimum casing pressure is that when the slug of liquid and the plunger arrive at the surface of the tubing, the casing pressure must support the weight of the liquid and the plunger, the friction in the tubing, the friction between the liquid and the tubing, and the surface tubing pressure.

B.3 MAXIMUM CASING PRESSURE

The maximum casing pressure is given by:

$$P_{C,max} = \frac{A_{ANN} + A_{TBG}}{A_{ANN}} P_{C,min}$$

where

A_{ANN} = cross-sectional area between casing and tubing
A_{TBG} = cross-sectional area of tubing

This approach assumes conservatively that all energy comes from expansion of the gas from the casing into the tubing to surface the plunger. It can be corrected to account for the formation gas that is produced and gas leaking upwardly around the plunger as the plunger is coming up to the surface but is not here or in the original reference. It just assumes that when the gas in the casing expands into the tubing, then the surface casing pressure drops to $P_{c,min}$.

B.4 SUMMARY

This development shows what the casing pressure must be at the top of the casing before the tubing valve is opened. The $P_{c,min}$ is the casing pressure when the liquid and plunger arrive. These equations can be used for an oil well. For a gas well, the plunger is then held at the surface to produce gas for some time until the velocity declines to near a critical velocity or pressures on the tubing or casing are monitored to reach certain cycle values.

B.5 REFERENCE

1. Foss, D. L. and Gaul, R. B. "Plunger Lift Performance Criteria with Operating Experience—Ventura Field," *Drilling and Production Practice*, API (1965), 124–140.

A P P E N D I X   C

GAS FUNDAMENTALS

C.1 INTRODUCTION

This appendix catalogs some commonly used gas fundamental expressions that are useful when operating gas wells.

C.2 PHASE DIAGRAM

A hydrocarbon gas is a mixture of different hydrocarbon molecules in varying composition. The type and amount of each molecular species in the gas determines the mixture properties at a given pressure and temperature.

Critical Temperature T_C is the temperature of a gas above which it cannot be liquified by increasing pressure.

Critical Pressure P_C is the pressure a gas exerts when in equilibrium with the liquid phase at the critical temperature.

Critical Volume V_C is the volume of one pound of gas at the critical temperature and pressure.

Cricondenbar is the highest pressure at which a gas can exist.

Cricondenterm is the highest temperature at which a liquid can exist.

Bubble Point is the pressure, at a given temperature, above which the mixture is 100% liquid.

Dew Point is the pressure, at a given temperature, above which the mixture is 100% gas.

Phase diagram

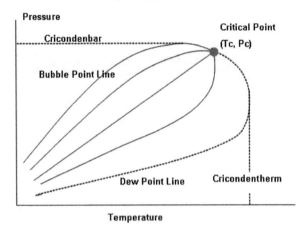

Figure C-1: Typical Gas Well Reservoir Phase Diagram

C.3 GAS APPARENT MOLECULAR WEIGHT AND SPECIFIC GRAVITY

Molecular weight is defined for a specific molecule but not for a mixture of different molecular species. For gas mixtures, the apparent gas molecular weight M is defined to represent the average molecular weight of all the molecules in the gas. Thus, M can be calculated from the mole fraction of each molecular species in the gas as

$$M = \sum_{all\ apecies\ j} y_j M_j$$

where:

y_j = mole fraction of molecule j
M_j = molecular weight of molecule j

Example C-1: Molecular Weight of Air

Dry air consists mainly of N_2 (78%, M = 29.01), O_2 (21%, M = 32.00), Argon (1%, M = 39.94), and minute amounts of other gases.

Estimate the apparent molecular weight of dry air.
Solution:

$$M_{air} = \sum_{all\ apecies\ i} y_i M_i = .78 \times 28.01 + .21 \times 32.00 + .01 \times 39.94 = 28.97$$

The specific gravity γ_G of a gas is the ratio of the gas apparent molecular weight to the apparent molecular weight of air:

$$\gamma_G = \frac{M_G}{M_{air}} = \frac{M_G}{28.97}$$

C.4 GAS LAW

The relationship between pressure, temperature, volume, and density of a real gas is well known and can be described by the gas law equation

$$pV = nZRT$$
$$\rho = \frac{pM}{ZRT}$$

where:

    p = absolute pressure
    V = volume
    T = absolute temperature
    n = number of moles of gas
    R = gas constant
    ρ = density
    M = molecular weight
    Z = gas deviation factor

The gas constant R depends on the units used for the equation as shown in Table C-1.

Table C-1
Gas Constant Values

Units	R
atm-cc/g-mole-°K	82.06
BUT/lb-mole-°R	1.987
psia-ft³/lb-mole-°R	10.73
lbf/ft²/ abs-ft³/lb-mole-°R	1544
atm-ft³/lb-mole-°R	0.73
mm Hg- liters/gm-mole-°K	62.37
in. Hg-ft³/lb-mole-°R	21.85
cal/g-mole-°K	1.987
kPa-m³/kg-mole-°K	8.314
J/kg-mole-°K	8414

Example C-2: Density of Dry Air

Estimate the density of dry air at standard conditions (1 atm, 60F).
Solution:
At standard conditions, air is very nearly an ideal gas with $Z = 1$.
Then

$$\rho = \frac{pM}{ZRT} = \frac{14.65 \times 28.97}{10.73 \times (60 + 460)} = .0761 \, lbm/ft^3$$

C.5 Z FACTOR

An ideal gas would have $Z = 1$. The Z factor, or gas deviation factor, accounts for the deviation of a real gas from ideal gas behavior. The Z factor usually is calculated from correlations based on the gas gravity.

For gas mixtures of chemically similar molecules, the Z factor is correlated with the pseudo-critical temperature T_{pc} and pseudo-critical pressure P_{pc} instead of the actual critical properties.

$$T_{pc} = \sum y_j Tc_j \quad P_{pc} = \sum y_j Pc_j$$

where y_j = mole fraction of gas j.

Note that the pseudo-critical properties are *not* related to the actual critical temperature and pressure of the gas.

For hydrocarbon gases, the pseudo-critical properties are correlated with the gas gravity as

$$T_{pc} = 170.5 + 307.3\gamma_G$$

$$P_{pc} = 709.6 - 58.7\gamma_G$$

For condensate fluids,

$$T_{pc} = 187 + 330\gamma_G - 71.5\gamma_G^2$$

$$P_{pc} = 706 - 51.7\gamma_G - 11.1\gamma_G^2$$

From the Law of Corresponding States, all gases have the same Z factor at the same values of reduced temperature T_r and reduced pressure P_r:

$$T_r = \frac{T}{T_C} \qquad P_r = \frac{p}{P_C}$$

Using this concept, a chart for the Z factor of gas mixtures has been developed by Standing and Katz [1] to give the Z factor for values of T_r and P_r. There are several equations available that have been fitted to this chart to explicitly calculate the Z factor.

One equation from Brill and Beggs [2] and corrected by Standing [3] is

$$z = A + (1 - A)e^{(-B)} + CP_r^D$$

where

$$A = 1.39(T_r - .92)^5 - .36T_r - 0.101$$
$$B = P_r(.62 - .23T_r) + P_r^2[0.066/\{Tr - 0.86\} - 0.037] + .32P_r^6/$$
$$\{EXP[20.723(T_r - 1)]\}$$
$$C = 0.132 - 0.32\log(T_r)$$
$$D = \exp(0.715 - 1.128T_r + 0.42T_r^2)$$

Using the preceding equations, the chart of Figure C-2 was constructed with some sample relationships for Z vs. T_r and P_r.

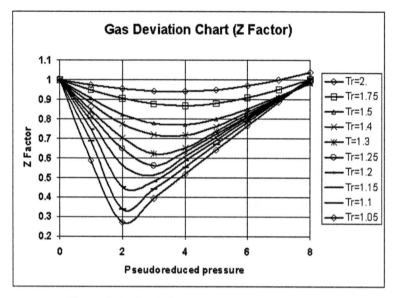

Figure C-2: Gas Z Factor as Function of T_r and P_r

For impurities, corrections can be made to P_{pc} and T_{pc} according to work done by Wichert and Aziz [4]:

$$T'_{pc} = T_{pc} - \varepsilon$$

$$P'_{pc} = \frac{P_{pc}T'_{pc}}{T'_{pc} + \varepsilon(B - B^2)}$$

$$\varepsilon = 120(A^{0.9} - A^{1.6}) + (B^{0.5} + B^4)$$

where

B = mol fraction of H_2S
A = mol fraction $CO_2 + B$

Once these corrected values of T_{pc} and P_{pc} are calculated, then the pseudo-reduced values of T_r and P_r can be used in the preceding equations or in charts using T_r and P_r to find the value of Z.

C.6 GAS FORMATION VOLUME FACTOR

The gas formation volume factor B_g is the ratio of the volume of gas at reservoir conditions to the volume of the same mass of gas at standard conditions. Using the gas law presented earlier, the gas formation volume factor becomes

$$B_g = \frac{0.0283zT}{p} \, ft^3/scf = \frac{5.04zT}{p} \, bbl/Mscf$$

where:

p = psia
T = deg R

C.7 PRESSURE INCREASE IN STATIC COLUMN OF GAS

Consider an incremental vertical distance dh in a static column of gas. Integrating the expression for dp that occurs over distance dh gives an equation for the pressure at depth

$$dP = \rho \frac{g}{g_c} dh$$

$$dP = \frac{2.7 p \gamma_G \dfrac{g}{g_c}}{144(T+460)Z} dh$$

$$\int_{Ptop}^{Pbot} \frac{dP}{P} = \frac{2.7 \gamma_g \dfrac{g}{g_c}}{144(T+460)Z} \int_0^H dh$$

$$\ln(P_{bot}/P_{top}) = \frac{0.01875 \gamma_g \dfrac{g}{g_c} H}{(T+460)Z}$$

$$P_{bot} = P_{top} \times exp\left(\frac{0.01875 \gamma_g \dfrac{g}{g_c} H}{(T+460)Z} \right)$$

The preceding equation for P_{bot} can be used to calculate the pressure increase down an annulus of a gas-lifted or flowing multiphase flow well or to the fluid level in the annulus for a pumping well. It is more accurate if the calculations are broken up into increments and the temperature and Z factor are the averages for each segment of calculation.

C.8 CALCULATE THE PRESSURE DROP IN FLOWING DRY GAS WELL: CULLENDER AND SMITH METHOD [5]

$$\frac{dp}{dl} = \left(\frac{dp}{dl}\right)_{el} + \left(\frac{dp}{dl}\right)_{f} + \left(\frac{dp}{dl}\right)_{acc}$$

OR:

$$\left(\frac{dp}{dl}\right) = \frac{g}{g_c}\rho\cos(\theta) + \frac{f\rho v^2}{2g_c d} + \frac{\rho v dv}{g_c dl}$$

where: θ is the angle from vertical

Ignoring the acceleration term and substituting in the real gas law gives:

$$\left(\frac{dp}{dl}\right) = \frac{pM}{aRT}\left[\cos(\theta) + +\frac{f\rho v^2}{2g_c d}\right]$$

$$v = \frac{q}{A} \quad q = q_{sc}\frac{p_{sc}Tz}{T_{sc}pz_{sc}}$$

substituting gives:

$$\frac{dp}{dl} = \left[\frac{pM\cos(\theta)}{zRT}\right] + \left[\frac{MTzp_{sc}fq_{sc}}{RpT_{sc}^2 2g_c dA^2}\right]$$

or:

$$\frac{pdp}{zRdl} = \frac{M}{R}\left[\left(\frac{p}{zT}\right)^2\cos(\theta)\right] + C$$

where:

$$C = \frac{8p_{sc}^2 q_{sc}^2 f}{T_{sc}^2 g_c \pi^2 d^5}$$

Separating variables gives:

$$\int_{p_{tf}}^{p_{wf}} \frac{\dfrac{p}{Tz}dp}{\left(\dfrac{p}{zT}\right)^2\cos(\theta) + C} = \frac{M}{R}\int_0^{MD} dl$$

Noting $\cos(\theta) = \dfrac{TVD}{MD}$

$$\int_{P_{tf}}^{P_{wf}} \frac{\dfrac{p}{Tz}\,dp}{\dfrac{0.001TVD}{MD}\left(\dfrac{p}{zT}\right)^2 \cos(\theta) + F^2} = 18.75\gamma_g(MD)$$

Where:

$$F^2 = \frac{0.667 f q_{sc}^2}{d}$$

Writing the above equation with grouped terms and breaking the well into only two increments for illustration of length MD/2 gives:

Upper 1/2 of well:

$$18.75\gamma_g(MD) = (p_{mf} - p_{tf})(K_{mf} + K_{tf})$$

Lower 1/2 of well:

$$18.75\gamma_g(MD) = (p_{wf} - p_{mf})(K_{wf} + K_{mf})$$

Where:

p_{wf} = flowing bhp to be solved for
p_{tf} = flowing tubing pressure, input
p_{mf} = flowing pressure mid-way in well

&

$$K = \frac{\dfrac{p}{Tz}}{\dfrac{0.001(TVD)}{MD}\left(\dfrac{p}{Tz}\right)^2 + F^2}$$

The solution can proceed by first calculating Nre, a friction factor f, and p_{mf} by assuming p_{mf} and solving for p_{mf} using the following equation:

$$18.75\gamma_g(MD) = (p_{mf} - p_{tf})(K_{mf} + K_{tf})$$

Since K_{mf} is a function of p_{mf}, the solution is iterative. Once the intermediate pressure is solved for, then P_{wf} can be solved for in the two-segment example. In a real case for accuracy, the solution would be broken into several increments.

C.9 PRESSURE DROP IN A GAS WELL PRODUCING LIQUIDS

One of many correlations for gas wells producing some liquids is the Gray [6] correlation that was developed by H.E. Gray, an employee of Shell Oil Co. API14BM provides insight to Gray's work. It is a vertical flow correlation for gas wells to determine pressure changes with depth and the bottomhole pressure. The method developed by Gray accounts for entrained fluids, temperature gradient, fluid acceleration, and non-hydrocarbon gas components. Well test data are required to make the necessary calculations. As per Gray, for two-phase, pressure drop can be defined from the following equation.

$$dp = \frac{g}{g_c}[\xi\rho_g + (1-\xi)\rho_l]dh + \frac{f_t G^2}{2g_c D_{pmf}}dh - \frac{G^2}{g_c}d\left(\frac{1}{\rho_{mi}}\right)$$

where:

ξ = the in-situ gas volume fraction   h = depth
D = conduit traverse dimension   p = pressure
G = mass velocity g_c = dimensionless constant
ρ = density f_t = irreversible energy loss

Further, as given in API14BM, ξ can be defined as

$$\xi = \frac{1 - Exp\left\{-2.314\left[N_V\left(1 + \frac{205.0}{N_D}\right)\right]^B\right\}}{R+1}$$

$$B = .0814\left[1 - .0554Ln\left(1 + \frac{730R}{R+1}\right)\right]$$

$$N_v = \frac{\rho_m^2 V_{sm}^4}{g\tau(\rho_l - \rho_g)}$$

$$N_D = \frac{g(\rho_l - \rho_g)D^2}{\tau}$$

$$R = \frac{V_{so} + V_{sw}}{V_{sg}}$$

where N_v, N_D, and R are velocity, diameter, and superficial liquid-to-gas ratio parameters, which mainly influence the hold-up for condensate wells. In Gray's method, superficial liquid and gas densities are used and a superficial mixture velocity (V_{sm}) is calculated.

The values of the superficial velocities are determined from:

$$V_s = Q/A$$

The Q values for oil and water are from input of bbls/MMscf for the water and the condensate (oil). The conventional liquid holdup H_l, is found as:

$$H_l = 1 - \xi$$

The final Gray equation is untested by the author for:

1. Flow velocities > 50 ft/sec
2. Tubing sizes > 3 ½ in, tested only for tubing ID 1.049 – 2.992 in
3. Condensate ratios of >150 bbls/MMscf
4. Water ratios > 5 bbls/MMscf

Calculated Result with Dry Gas and Gas with Liquids

The curve in Figure C-3 shows how calculated curves of flowing bhp's appear plotted at the bottom the tubing.

C.10 GAS WELL DELIVERABILITY EXPRESSIONS

C.10.1 Backpressure Equation

Perhaps the most widely used inflow expression for gas wells is the gas backpressure equation [7]:

$$q_G = C_1(\bar{P}_r^2 - P_{wf}^2)^n$$

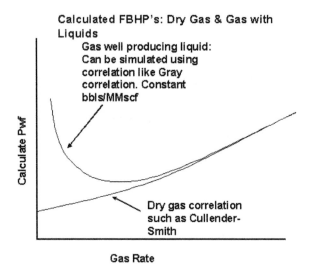

Figure C-3: Tubing Performance with/without Liquids

where:

q_G = gas rate
C = inflow coefficient
n = inflow exponent
P_r = reservoir pressure
P_{wf} = flowing bottomhole pressure

Once values for C and n are determined using test data, the backpressure equation can generate a predicted flow rate for any flowing wellbore pressure, P_{wf}. Since there are two constants, C and n, a minimum of two pairs of pseudo-stabilized data (q_g, P_{wf}) are needed, but usually at least four data pairs (a four-point test) are used to determine C and n to account for possible errors in the data collection.

The equation can be written as:

$$\log(\bar{P}_r^2 - P_{wf}^2) = \log \Delta P^2 = \frac{1}{n}\log q_g - \frac{1}{n}\log C$$

A plot of Δp^2 vs. q_g on log-log paper will result in a straight line having a slope of 1/n and an intercept of q_g = C at ΔP^2 = 1. The value of C can also be calculated using any point from the best line through the data as

$$C_1 = \frac{q_g}{(\bar{P}_r^2 - P_{wf}^2)^n}$$

For high permeability wells where the flow rates and pressures attain steady state for each test within a reasonable period of time (conventional flow-after-flow test), the log-log plot is easily used to generate the needed data. For tighter permeability wells, isochronal [10] or modified isochronal tests and plots can be used where the slope is generated from shorter flow tests and a parallel line is drawn though an extended pressure-rate point for final results.

To assist in calculating the approximate time for pseudo-stabilized flow to occur starting with any flow, the following equation may be used, with $T_{DA} \geq 0.1$. However, for rapidly stabilizing formations, just collect pressure and flow data until they become constant.

$$t_s = \frac{380 \varphi \bar{\mu}_g \bar{C}_t A t_{DA}}{2.637 \times 10^{-4} k_g}$$

Where:

$\bar{C}_t$ = total compressibility, $1/\text{psi} = s_g c_g + s_o c_o + c_f$

with compressibilities and $\bar{\mu}$ evaluated at $\bar{p} = (p_i + p_{wf})/2$.

A = drainage area
φ = porosity

C.10.2 Darcy Equation

Darcy's law for radial flow of a single phase gas is

$$q_g = \frac{7.03 \times 10^{-4} k_g h (\bar{P}_r^2 - P_{wf}^2)}{\bar{\mu}_g \bar{z} \bar{T} [\ln(x) - 0.75 + s_t + D q_g]}$$

Where:

q_g = gas flow rate, Mscf/D
P_r = average reservoir pressure, psia
P_{wf} = bottomhole flowing pressure, psia
x = factor related to drainage area geometry, see Table C-2, below

Table C-2
Factors for Darcy Equation for Different Shapes and Well Positions in a Drainage Area, Where A = Drainage Area and $A^{1/2}/r_e$ is Dimensionless

System	X	System	X
	$\dfrac{r_e}{r_w}$		$\dfrac{0.966\,A^{1/2}}{r_w}$
	$\dfrac{0.571\,A^{1/2}}{r_w}$		$\dfrac{1.44\,A^{1/2}}{r_w}$
	$\dfrac{0.565\,A^{1/2}}{r_w}$		$\dfrac{2.206\,A^{1/2}}{r_w}$
	$\dfrac{0.604\,A^{1/2}}{r_w}$		$\dfrac{1.925\,A^{1/2}}{r_w}$
	$\dfrac{0.61\,A^{1/2}}{r_w}$		$\dfrac{6.59\,A^{1/2}}{r_w}$
	$\dfrac{0.678\,A^{1/2}}{r_w}$		$\dfrac{9.36\,A^{1/2}}{r_w}$
	$\dfrac{0.668\,A^{1/2}}{r_w}$		$\dfrac{1.724\,A^{1/2}}{r_w}$
	$\dfrac{1.368\,A^{1/2}}{r_w}$		$\dfrac{1.794\,A^{1/2}}{r_w}$
	$\dfrac{2.066\,A^{1/2}}{r_w}$		$\dfrac{4.072\,A^{1/2}}{r_w}$
	$\dfrac{0.884\,A^{1/2}}{r_w}$		$\dfrac{9.523\,A^{1/2}}{r_w}$
	$\dfrac{1.485\,A^{1/2}}{r_w}$		$\dfrac{10.135\,A^{1/2}}{r_w}$

r_e = radius of external boundary, ft
r_w = radius of the wellbore, ft
k_g = effective permeability to gas, md
$\bar{z}$ = gas deviation factor determined at the average temperature and average pressure

$\bar{T}$ = average reservoir temperature, °R

$\bar{\mu}_g$ = gas viscosity, cp, determined at the average temperature and average pressure

s_t = total skin

Dq_g = pseudo rate dependent skin due to turbulence or non Darcy flow. This is usually zero for low pressure gas wells that might be liquid loaded.

Neely [8,9] rewrote the single phase flow equation for gas wells as

$$q_G = \frac{C(\bar{P}_r^2 - P_{wf}^2)}{\bar{\mu}\bar{z}}$$

Where:

$\bar{\mu}$ = average gas viscosity that is a function of pressure

$\bar{z}$ = average gas deviation factor that is a function of pressure

C = a constant (not same as C in back pressure equation) and can be determined from a single well test if the shut-in average reservoir pressure is known.

The P_{wf} should be determined from a downhole pressure gauge. The viscosity and Z factor should be determined at the bottomhole temperature and average bottomhole pressure. Then C will not change as rates are varied from the well unless damage sets in such as scale buildup. Using the preceding equation can result in a more accurate inflow expression, showing a correction to a higher AOF compared to the old log-log backpressure equation [11].

C.11 REFERENCES

1. Standing, M. B. and Katz, D. L. "Density of Natural Gases," Trans. AIME, 1942.

2. Brill, J. P. and Beggs, H. D. *Two Phase Flow in Pipes*, The University of Tulsa, Tulsa, OK, 1978.

3. Standing, M. B. "Volumetric and Phase Behavior of Oil Field Hydrocarbon Systems, SPE" of AIME, 8th Printing, 1977.

4. Wichert, E. and Aziz, K. "Calculate Z's for Sour Gasses," Hydrocarbon Proceedings, May, 1972.

5. Cullender, M. H. and Smith, R. V. "Practical Solution of Gas Flow Equations for Wells and Pipelines with Large Temperature Gradients," Trans. AIME 207, 1956.

6. Gray, H. E. "Vertical flow correlation in gas wells," in user manual for API 14B, *Subsurface controlled safety valve sizing computer program, App. B,* June 1974.

7. Rawlins, E. L. and Schellhardt, Am. A. "Back Pressure Data on Natural Gas Wells and Their Applications to Production Practices," Bureau of Mines Monograph 7, 1935.

8. Neely, A. B. "The Effect of Compressor Installation on Gas Well Performance," HAP Report 65-1, Shell Oil Company, January 1965.

9. Greene, W. R. "Analyzing the Performance of Gas Wells," presented the annual SWPSC, Lubbock, TX, April 21, 1978.

10. Fetkovich, M. J. "The Isochronal Testing of Oil Wells," Paper 4529, 48th Annual Fall Meeting of SPE, Las Vegas, NV., 1973.

11. Russell, D. G., Goodrich, J. H., Perry, G. E., and Bruskotter, J. F. "Methods for Predicting Gas Well Performance," *Journal of Petroleum Technology,* January 1965.

Index

Printed and bound by CPI Group (UK) Ltd, Croydon, CR0 4YY

03/10/2024

01040430-0006